5 STEPS TO A 5™

AP Chemistry

2015

5 STEPS TO A 5™

AP Chemistry

2015

John T. Moore, EdD
Richard H. Langley, PhD

New York Chicago San Francisco Athens London Madrid
Mexico City Milan New Delhi Singapore Sydney Toronto

1 2 3 4 5 6 7 8 9 10 DOH/DOH 1 0 9 8 7 6 5 4

ISBN 978-0-07-183851-1
MHID 0-07-183851-1
ISSN 2150-6418

e-ISBN 978-0-07-183853-5
e-MHID 0-07-183853-8

The series editor was Grace Freedson, and the project editor was Del Franz.
Series design by Jane Tenenbaum.

ABOUT THE AUTHORS

JOHN MOORE grew up in the foothills of western North Carolina. He attended the University of North Carolina–Asheville, where he received his bachelor's degree in chemistry. He earned his master's degree in chemistry from Furman University in Greenville, South Carolina. After a stint in the United States Army he decided to try his hand at teaching. In 1971 he joined the faculty of Stephen F. Austin State University in Nacogdoches, Texas, where he still teaches chemistry. In 1985 he started back to school part-time, and in 1991 received his doctorate in education from Texas A&M University. In 2003 his first book, *Chemistry for Dummies,* was published.

RICHARD LANGLEY grew up in southwestern Ohio. He attended Miami University in Oxford, Ohio, where he earned bachelor's degrees in chemistry and mineralogy and a master's degree in chemistry. He next went to the University of Nebraska in Lincoln, where he received his doctorate in chemistry. He took a postdoctoral position at Arizona State University in Tempe, Arizona, then became a visiting assistant professor at the University of Wisconsin–River Falls. He has taught at Stephen F. Austin State University in Nacogdoches, Texas, since 1982.

The authors are coauthors of *Chemistry for the Utterly Confused*, *Biochemistry for Dummies,* and *Organic Chemistry II for Dummies*.

Both authors are graders for the free-response portion of the AP Chemistry exam. In fact, between them, they have almost twenty years of AP grading experience and estimate that together they have graded over 100,000 exams.

CONTENTS

PREFACE

Welcome to the AP Chemistry Five-Step Program. The fact that you are reading this preface suggests that you will be taking the AP exam in chemistry. The AP Chemistry exam is constantly evolving and so this guide has evolved. We have updated the book to match the new AP Chemistry exam. The new exam has an emphasis on sets—a series of questions that refer to the same given information, along with changes in the free-response portion.

The AP Chemistry exam certainly isn't easy, but the rewards are worth it—college credit and the satisfaction of a job well done. You will have to work and study hard to do well, but we will, through this book, help you to master the material and get ready for the exam.

Both of us have many years of experience in teaching introductory general chemistry at the university level. John Moore is the author of *Chemistry for Dummies*, and he and Richard "Doc" Langley have also written *Chemistry for the Utterly Confused*, a guide for college/high school students. Each of us has certain skills and experiences that will be of special help in presenting the material in this book. Richard has also taught high school science and John has years of experience teaching chemistry to both public school teachers and students. Both of us have been graders for the AP Exam chemistry free-response questions for years and have firsthand knowledge of how the exam is graded and scored. We have tried not only to make the material understandable but also to present the problems in the format of the AP Chemistry exam. By faithfully working the problems you will increase your familiarity with the exam format, so that when the time comes to take the exam there will be no surprises.

Use this book in addition to your regular chemistry text. We have outlined three different study programs to prepare you for the exam. If you choose the year-long program, use it as you are taking your AP Chemistry course. It will provide additional problems in the AP format. If you choose one of the other two programs, use it with your chemistry textbook also; but you may need to lean a little more on this review book. Either way, if you put in the time and effort, you will do well.

Now it's time to start. Read the Introduction: The Five-Step Program; Chapter 1, What You Need to Know About the AP Chemistry Exam; and Chapter 2, How to Plan Your Time. Then take the Diagnostic Exam in Chapter 3. Your score will show how well you understand the material right now and point out weak areas that may need a little extra attention. Use the review exams at the end of the chapters to check your comprehension. Also pay attention to the free-response questions. That is where you can really shine, and they are worth almost as much as the multiple-choice part. Use the Rapid Reviews to brush up on the important points in the chapters. Just before taking the exam, review the section on avoiding "stupid mistakes" at the back of this book. Keep this book handy—it is going to be your friend for the next few weeks or months.

Good luck: but remember that luck favors the prepared mind.

ACKNOWLEDGMENTS

The authors would like to thank Grace Freedson, who believed in our abilities and gave us this project. Many thanks also to Del Franz, whose editing polished up the manuscript and helped its readability. Thanks to our colleagues at the AP Chemistry readings for their helpful suggestions.

INTRODUCTION: THE FIVE-STEP PROGRAM

The Basics

Not too long ago, you enrolled in AP Chemistry. A curiosity about chemistry, encouragement from a respected teacher, or the simple fact that it was a requirement may have been your motivation. No matter what the reason, you find yourself flipping through a book that promises to help you culminate this experience with the highest of honors, a 5 in AP Chemistry. Yes, it is possible to achieve this honor without this book. There are many excellent teachers of AP Chemistry out there who teach, coax, and otherwise prepare their students into a 5 every year. However, for the majority of students preparing for the exam, the benefits of buying this book far outweigh its cost.

The key to doing well on the Advanced Placement (AP) Chemistry exam is to outline a method of attack and not to deviate from this method. We will work with you to make sure you take the best path towards the test. You will need to focus on each step, and this book will serve as a tool to guide your steps. But do not forget—no tool is useful if you do not use it.

Organization of the Book

This book conducts you through the five steps necessary to prepare yourself for success on the exam. These steps will provide you with the skills and strategies vital to the exam, and the practice that will lead you towards the perfect 5.

First, we start by introducing the basic five-step plan used in this book. Then in Chapter 1, we will give you some background information about the AP Chemistry exam. Next, in Chapter 2, we present three different approaches to preparing for the exam. In Chapter 3, we give you an opportunity to evaluate your knowledge with a Diagnostic Exam. The results of this exam will allow you to customize your study. In Chapter 4, we offer you a multitude of tips and suggestions about the different types of questions on the AP Chemistry exam. Many times good test-taking practices can help raise your score.

Since the volume of the material to be mastered can be intimidating, Chapters 5 to 19 present a comprehensive review of the material that you will cover in an AP Chemistry course. This is review material, but since not all of this material appears in every AP Chemistry class, it will also help to fill in the gaps in your chemistry knowledge. You can use it in conjunction with your textbook if you are currently taking AP Chemistry, or you can use it as a review of the concepts you covered. At the end of each chapter, you will find both a multiple-choice and free-response exam for you to test yourself. The answers and explanations are included. This will also help you identify any topics that might require additional study.

After these content chapters, there are two complete chemistry practice exams, including multiple-choice and free-response questions. The answers and explanations are included. These exams will allow you to test your skills. The multiple-choice questions will provide you with practice on questions similar to those asked on past AP exams. These are not the exact questions, but ones that will focus you on the key AP Chemistry topics. There are also examples of free-response questions; there are fewer of these, since they take much longer to answer. After you take an exam, you should review each question. Ask yourself, why was this question present? Why do I need to know this? Make sure you check your answers against the explanations. If necessary, use the index to locate a particular topic and reread the review material. We suggest that you take the first exam, identify those areas that need additional study, and review the appropriate material. Then take the second exam and use the results to guide your additional study.

Finally, in the appendixes you will find additional resources to aid your preparation. These include:

- A tip sheet on how to avoid "stupid" mistakes and careless errors
- Common conversions
- How to balance redox equations
- A list of common ions
- A bibliography
- A number of useful websites
- A glossary of terms related to AP Chemistry
- A table of half-reactions for use while answering free-response questions
- A table of equations and abbreviations for use while answering free-response questions
- A periodic table for use when answering any exam questions

The Five-Step Program

Step 1: Set Up Your Study Program

In Step 1, you will read a brief overview of the AP Chemistry exams, including an outline of the topics. You will also follow a process to help determine which of the following preparation programs is right for you:

- Full school year: September through May.
- One semester: January through May.
- Six weeks: Basic training for the exam.

Step 2: Determine Your Test Readiness

Step 2 provides you with a diagnostic exam to assess your current level of understanding. This exam will let you know about your current level of preparedness and on which areas you should focus your study.

- Take the diagnostic exam slowly and analyze each question. Do not worry about how many questions you get right. Hopefully this exam will boost your confidence.
- Review the answers and explanations following the exam, so that you see what you do and do not yet fully understand.

Step 3: Develop Strategies for Success

Step 3 provides strategies that will help you do your best on the exam. These strategies cover both the multiple-choice and free-response sections of the exam. Some of these tips are based upon experience in writing questions, and others have been gleaned from our years of experience reading (grading) the AP Chemistry exams.

- Learn how to read and analyze multiple-choice questions.
- Learn how to answer multiple-choice questions.
- Learn how to plan and write answers to the free-response questions.

Step 4: Review the Knowledge You Need to Score High

Step 4 encompasses the majority of this book. In this step, you will learn or review the material you need to know for the test. Your results on the diagnostic exam will let you know on which material you should concentrate your study. Concentrating on some material does not mean you can ignore the other material. You should review all the material, even what you already know.

There is a lot of material here, enough to summarize a yearlong experience in AP Chemistry and highlight the, well, highlights. Some AP courses will have covered more material than yours, some will have covered less; but the bottom line is that if you thoroughly review this material, you will have studied all that is on the exam, and you will have significantly increased your chances of scoring well. This edition gives new emphasis to some areas of chemistry to bring your review more in line with the revised AP Chemistry exam format. For example, there is more discussion of reactions and the laboratory experience. Each chapter contains a short exam to monitor your understanding of the current chapter.

Step 5: Build Your Test-Taking Confidence

In Step 5, you will complete your preparation by testing yourself on practice exams. This section contains *two* complete chemistry exams, solutions, and sometimes more important, advice on how to avoid the common mistakes. In this edition, the free-response exams have been updated to more accurately reflect the content tested on the AP exams. Be aware that these practice exams are *not* reproduced questions from actual AP Chemistry exams, but they mirror both the material tested by AP and the way in which it is tested.

The Graphics Used in This Book

To emphasize particular skills and strategies, we use several icons throughout this book. An icon in the margin will alert you to pay particular attention to the accompanying text. We use these four icons:

This icon highlights a very important concept or fact that you should not pass over.

This icon calls your attention to a strategy that you may want to try.

This icon indicates a tip that you might find useful.

This icon points to material that is not directly tested on the AP Chemistry exam but may be required by your teacher in high school and certainly by your college teacher. Although you won't find this specific content on the AP exam, knowing it will improve your understanding of chemistry, helping you to better grasp the material that is directly tested on the exam.

Boldfaced words indicate terms that are included in the glossary at the end of this book.

Set Up Your Study Program

CHAPTER 1

What You Need to Know About the AP Chemistry Exam

IN THIS CHAPTER

Summary: Learn what topics are on the test, how the ETS scores the test, and basic test-taking information.

Key Ideas

- Most colleges will award credit for a score of 4 or 5.
- Multiple-choice questions account for half of your final score.
- Points are not deducted for incorrect answers to multiple-choice questions. You should try to eliminate incorrect answer choices and then guess; there is no penalty for guessing.
- Free-response questions account for half of your final score.
- There is a conversion of your composite score on the two test sections to a score on the 1-to-5 scale.

Background of the Advanced Placement Program

The College Board began the Advanced Placement program in 1955 to construct standard achievement exams that would allow highly motivated high school students the opportunity to receive advanced placement as first-year students in colleges and universities in the United States. Today, there are 34 courses and exams with more than 2 million students from every state in the nation and from foreign countries taking the annual exams in May.

The AP programs are for high school students who wish to take college-level courses. In our case, the AP Chemistry course and exam involve high school students in college-level chemistry studies.

Who Writes the AP Chemistry Exam?

A group of college and high school chemistry instructors known as the AP Development Committee creates the AP Chemistry exam. The committee's job is to ensure that the annual AP Chemistry exam reflects what is taught in college-level chemistry classes at high schools.

This committee writes a large number of multiple-choice questions, which are pre-tested and evaluated for clarity, appropriateness, and range of possible answers.

The free-response essay questions that make up Section II go through a similar process of creation, modification, pre-testing, and final refinement, so that the questions cover the necessary areas of material and are at an appropriate level of difficulty and clarity. The committee also makes a great effort to construct a free-response exam that will allow for clear and equitable grading by the AP readers.

It is important to remember that the AP Chemistry exam undergoes a thorough evaluation after the yearly administration of the exam. This way, the College Board can use the results to make course suggestions and to plan future tests.

The AP Grades and Who Receives Them

Once you have taken the exam and it has been scored, your test will be graded with one of five numbers by the College Board:

- A 5 indicates that you are extremely well-qualified.
- A 4 indicates that you are well-qualified.
- A 3 indicates that you are adequately qualified.
- A 2 indicates that you are possibly qualified.
- A 1 indicates that you are not qualified to receive college credit.

A grade report, consisting of a grade of 1 to 5, will be sent to you in July. You will also indicate the college to which you want your AP score sent at the time of the exam. The report that the college receives contains your score for every AP exam you took that year and the grades you received in prior years, except for any that you request withheld. In addition, your scores will be sent to your high school.

Reasons for Taking the AP Chemistry Exam

Why put yourself through a year of intensive study, pressure, stress, and preparation? Only you can answer that question. Following are some of the reasons that students have indicated to us for taking the AP exam:

- Because colleges look favorably on the applications of students who elect to enroll in AP courses
- To receive college credit or advanced standing at their colleges or universities
- To compare themselves with other students across the nation
- For personal satisfaction
- Because they love the subject
- So that their families will be proud of them

There are other reasons, but no matter what they are, the primary reason for your enrolling in the AP Chemistry course and taking the exam in May is to feel good about yourself and the challenges that you have met.

While there may be some idealistic motivators, let's face it: most students take the exam because they are seeking college credit. This means you are closer to graduation before you even start attending classes. Even if you do not score high enough to earn college credit, the fact that you elected to enroll in AP courses tells admission committees that you are a high achiever and serious about your education.

Questions Frequently Asked About the AP Chemistry Exam

What Is Going to Appear on the Exam?

This is an excellent question. The College Board, having consulted with those who teach chemistry, develops a curriculum that covers material that college professors expect to cover in their first-year classes. Based upon this outline of topics, the multiple-choice exams are written such that those topics are covered in proportion to their importance to the expected chemistry understanding of the student. Confused? Suppose that faculty consultants agree that environmental issues are important to the chemistry curriculum, maybe to the tune of 10 percent. If 10 percent of the curriculum in an AP Chemistry course is devoted to environmental issues, you can expect roughly 10 percent of the multiple-choice exam to address environmental issues. Remember this is just a guide and each year the exam differs slightly in the percentages.

How Is the Advanced Placement Chemistry Exam Organized?

Table 1.1 summarizes the format of the AP Chemistry exam.

Table 1.1

SECTION	NUMBER OF QUESTIONS	TIME LIMIT
I. Multiple-Choice Questions	60	90 minutes
II. Free-Response Questions		90 minutes
Multipart Questions	3	20–25 minutes per question
Single-part Questions	4	3–10 minutes per question

The exam is a two-part exam designed to take about three hours. The first section has 60 multiple-choice questions. You will have 90 minutes to complete this section.

The second part of the exam is the free-response section. You will begin this section after you have completed and turned in your multiple-choice scan sheet. There will be a break before you begin the second section. The length of this break will vary from school to school. You will not be able to go back to the multiple-choice questions later.

You will receive a test booklet for the free-response section of the test. You will have 90 minutes to answer seven questions. These questions may cover any of the material in the AP Chemistry course. The free-response section consists of two parts. In both parts, you may use a calculator. There will probably be two lab questions—one an experimental design question and the other question an analysis of data or observations. There will be two questions involving representations of molecules—one involving a conversion between different types of representations, and the other requiring an analysis or creation of an atomic or molecular view explaining a representation. Finally, there will be a quantitative question involving reasoning to solve a problem.

Who Grades My AP Chemistry Exam?

Every June a group of chemistry teachers gathers for a week to assign grades to your hard work. Each of these "Faculty Consultants" spends a day or so in training on a question. Each reader becomes an expert on that question, and because each exam book is anonymous, this process provides a very consistent and unbiased scoring of that question. During a typical day of grading, there is a selection of a random sample of each reader's scores for crosschecking by other experienced "Table Leaders" to ensure that the graders maintain a level of consistency throughout the day and the week. Statistical analysis of each reader's scores on a given question assure that they are not giving scores that are significantly higher or lower than the mean scores given by other readers of that question. All these measures assure consistency and fairness for your benefit.

Will My Exam Remain Anonymous?

Absolutely. Even if your high school teacher happens to read your booklet, there is virtually no way he or she will know it is you. To the reader, each student is a number and to the computer, each student is a bar code.

What About That Permission Box on the Back?

The College Board uses some exams to help train high school teachers so that they can help the next generation of chemistry students to avoid common mistakes. If you check this box, you simply give permission to use your exam in this way. Even if you give permission, no one will ever know it is your exam.

How Is My Multiple-Choice Exam Scored?

You will place your answers to the multiple-choice questions on a scan sheet. The scan sheet is computer graded. The computer counts the number of correct responses. There is no penalty for incorrect answers or for leaving an answer blank.

How Is My Free-Response Exam Scored?

You are required to answer seven free-response questions. The point totals will vary, but there is an adjustment of the points to match the assigned weighting of the question. For example, question #1 may be on a scale of 10 points, while question #2 may be on a scale of 7 points, and question #3 on a scale of 5 points. Since these questions are to count equally, a multiplier will be used to adjust the points to the same overall value.

So How Is My Final Grade Determined and What Does It Mean?

Your total composite score for the exam is found by adding the value from the multiple-choice section to the score from the essay section and rounding that sum to the nearest whole number.

Keep in mind that the total composite scores needed to earn a 5, 4, 3, 2, or 1 change each year. A committee of AP, College Board, and Educational Testing Service (ETS) directors, experts, and statisticians determines these cutoffs. The same exam that is given to the AP Chemistry high school students is given to college students. The various college professors report how the college students fared on the exam. This provides information for the chief faculty consultant on where to draw the lines for a 5, 4, 3, 2, or 1 score. A score of 5 on this AP exam is set to represent the average score received by the college students who scored an A on the exam. A score of 3 or 4 is the equivalent of a college grade B, and so on.

How Do I Register and How Much Does It Cost?

If you are enrolled in AP Chemistry in your high school, your teacher is going to provide all of these details. You do not have to enroll in the AP course to register for and complete the AP exam. When in doubt, the best source of information is the College Board's website: www.collegeboard.com.

Students who demonstrate financial need may receive a refund to help offset the cost of testing. There are also several optional fees that are necessary if you want your scores rushed to you, or if you wish, to receive multiple grade reports.

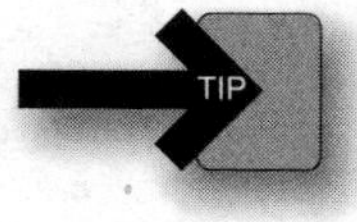

What Should I Do the Night Before the Exam?

Last-minute cramming of massive amounts of material will not help you. It takes time for your brain to organize material. There is some value to a last-minute review of material. This may involve looking over the Rapid Review portions of a few (not all) chapters, or looking through the Glossary. The night before the test should include a light review, and various relaxing activities. **A full night's sleep is one of the best preparations for the test.**

What Should I Bring to the Exam?

Here are some suggestions:

- Several pencils and an eraser that does not leave smudges.
- Black- or blue-colored pens for use on the free-response section.
- A watch so that you can monitor your time. You never know if the exam room will, or will not, have a clock on the wall. Make sure you turn off the beep that goes off on the hour.
- A calculator that you have used during your preparation for the exam. Do not bring a new or unfamilar calculator.
- Your school code.
- Your photo identification and social security number.
- Tissues.
- Your quiet confidence that you are prepared and ready to rock and roll.

What Should I NOT Bring to the Exam?

It's probably a good idea to leave the following items at home:

- A cell phone, beeper, PDA, or walkie-talkie.
- Books, a dictionary, study notes, flash cards, highlighting pens, correction fluid, a ruler, or any other office supplies.
- Portable music of any kind.
- Clothing with any chemistry on it.
- Panic or fear. It's natural to be nervous, but you can comfort yourself that you have used this book and that there is no room for fear on your exam.

You should:

- Allow plenty of time to get to the test site.
- Wear comfortable clothing.
- Eat a light breakfast and/or lunch.
- Remind yourself that you are well prepared and that the test is an enjoyable challenge and a chance to share your knowledge.
- Be proud of yourself!
- Review the tip sheet on avoiding "stupid" mistakes at the back of this book.

Once test day comes, there is nothing further you can do. Do not worry about what you could have done differently. It is out of your hands, and your only job is to answer as many questions correctly as you possibly can. The calmer you are, the better your chances of doing well.

How to Plan Your Time

IN THIS CHAPTER

Summary: The right preparation plan for you depends on your study habits and the amount of time you have before the test.

Key Idea

✪ Choose the study plan that's right for you.

Three Approaches to Preparing for the AP Chemistry Exam

You are the best judge of your study habits. You should make a realistic decision about what will work best for you. Good intentions and wishes will not prepare you for the exam. Decide what works best for you. Do not feel that you must follow one of these schedules exactly; you can fine-tune any one of them to your own needs. Do not make the mistake of forcing yourself to follow someone else's method. Look at the following descriptions, and see which best describes you. This will help you pick a prep mode.

You're a full-year prep student if:

1. You are the kind of person who likes to plan for everything very far in advance.
2. You arrive very early for appointments.
3. You like detailed planning and everything in its place.
4. You feel that you must be thoroughly prepared.
5. You hate surprises.

If you fit this profile, consider **Plan A**.

You're a one-semester prep student if:

1. You are always on time for appointments.
2. You are willing to plan ahead to feel comfortable in stressful situations, but are OK with skipping some details.
3. You feel more comfortable when you know what to expect, but a surprise or two is good.

If you fit this profile, consider **Plan B.**

You're a six-week prep student if:

1. You get to appointments at the last second.
2. You work best under pressure and tight deadlines.
3. You feel very confident with the skills and background you learned in your AP Chemistry class.
4. You decided late in the year to take the exam.
5. You like surprises.

If you fit this profile, consider **Plan C.**

Look now at Table 2.1 and the following calendars for plans A, B, and C. Choose the plan that will best suit your particular learning style and timeline. For best results, choose a plan and stick with it.

Table 2.1 General Outline of Three Different Study Plans

Month	PLAN A (Full School Year)	PLAN B (One Semester)	PLAN C (Six Weeks)
September–October	Introduction to material and Chapter 5	Introduction to material	Introduction to material and Chapters 1-4
November	Chapters 6–7		
December	Chapters 8–9		
January	Chapters 10–11	Chapters 5–7	
February	Chapters 12–13	Chapters 8–10	
March	Chapters 14–16	Chapters 11–14	
April	Chapters 17–19; Practice Exam 1	Chapters 15–19; Practice Exam 1	Skim Chapters 5–14; all Rapid Reviews; Practice Exam 1
May	Review everything; Practice Exam 2	Review everything; Practice Exam 2	Skim Chapters 15–19; Practice Exam 2

Calendar for Each Plan

Plan A: You Have a Full School Year to Prepare

The main reason for you to use this book is as a preparation for the AP Chemistry exam. However, this book can fill other roles. It can broaden your study of chemistry, help your analytical skills, and enhance your scientific-writing abilities. These will aid you in a college course in chemistry. Use this plan to organize your study during the coming school year.

SEPTEMBER–OCTOBER (Check off the activities as you complete them.)

— Determine the student mode (A, B, or C) that applies to you.
— Carefully read Chapters 1–4 of this book. You should highlight material that applies specifically to you.
— Take the Diagnostic Exam.
— Pay close attention to your walk-through of the Diagnostic Exam.
— Look at the AP and other websites.
— Skim the review chapters in Step 4 of this book. (Reviewing the topics covered in this section will be part of your yearlong preparation.)
— Buy a few color highlighters.
— Look through the entire book. You need to get some idea of the layout, and break it in. Highlight important points.
— Have a clear picture of your school's AP Chemistry curriculum.
— Use this book as a supplement to your classroom experience.

NOVEMBER (The first 10 weeks have elapsed.)

— Read and study Chapter 5, Basics.
— Read and study Chapter 6, Reactions and Periodicity.
— Read and study Chapter 7, Stoichiometry.

DECEMBER

— Read and study Chapter 8, Gases.
— Read and study Chapter 9, Thermodynamics.
— Review Chapters 5–7.

JANUARY (20 weeks have elapsed.)

— Read and study Chapter 10, Spectroscopy, Light, and Electrons.
— Read and study Chapter 11, Bonding.
— Review Chapters 5–10.

FEBRUARY

— Read and study Chapter 12, Solids, Liquids, and Intermolecular Forces.
— Read and study Chapter 13, Solutions and Colligative Properties.
— Review Chapters 5–11.
— Evaluate your weaknesses and refer to the appropriate chapters. You may wish to retake part of the Diagnostic Exam.

MARCH (30 weeks have now elapsed.)

— Read and study Chapter 14, Kinetics.
— Read and study Chapter 15, Equilibrium.
— Read and study Chapter 16, Electrochemistry.
— Review Chapters 5–13.

APRIL

— Take Practice Exam 1 in the first week of April.
— Evaluate your strengths and weaknesses. Review the appropriate chapters to correct any weaknesses.
— Read and study Chapter 17, Nuclear Chemistry.
— Read and study Chapter 18, Organic Chemistry.
— Read and study Chapter 19, Experimental Investigations.
— Review Chapters 5–16.

MAY (first 2 weeks) (THIS IS IT!)

— Review Chapters 5–19—all the material!
— Take Practice Exam 2.
— Score your exam.
— Review the tip sheet on avoiding "stupid" mistakes at the back of this book.
— Get a good night's sleep before the exam. Fall asleep knowing you are well prepared.

GOOD LUCK ON THE TEST!

Plan B: You Have One Semester to Prepare

This approach uses the assumption that you have completed at least one semester of AP Chemistry. This calendar begins in mid-year and prepares you for the mid-May exam.

JANUARY–FEBRUARY

— Read Chapters 1–4 in this book.
— Pay careful attention to the Diagnostic Exam.
— Pay close attention to your walk-through of the Diagnostic Exam.
— Read and study Chapter 5, Basics.
— Read and study Chapter 6, Reactions and Periodicity.
— Read and study Chapter 7, Stoichiometry.
— Read and study Chapter 8, Gases.
— Read and study Chapter 9, Thermodynamics.
— Evaluate your strengths and weaknesses.
— Re-study appropriate chapters to correct your weaknesses.

MARCH (10 weeks to go.)

— Read and study Chapter 10, Spectroscopy and Electrons.
— Review Chapters 5–7.
— Read and study Chapter 11, Bonding.
— Read and study Chapter 12, Solids, Liquids, and Intermolecular Forces.
— Review Chapters 8–10.
— Read and study Chapter 13, Solutions and Colligative Properties.
— Read and study Chapter 14, Kinetics.

APRIL

— Take Practice Exam 1 in the first week of April.
— Evaluate your strengths and weaknesses.
— Study appropriate chapters to correct your weaknesses.
— Read and study Chapter 15, Equilibrium.
— Review Chapters 5–10.
— Read and study Chapter 16, Electrochemistry.
— Read and study Chapter 17, Nuclear Chemistry.
— Review Chapters 11–14.
— Read and study Chapter 18, Organic Chemistry.
— Read and study Chapter 19, Experimental Investigations.

MAY (first 2 weeks) (THIS IS IT!)

— Review Chapters 5–19—all the material!
— Take Practice Exam 2.
— Score your exam.
— Review the tip sheet on avoiding "stupid" mistakes at the back of this book.
— Get a good night's sleep before the exam. Fall asleep knowing you are well prepared.

GOOD LUCK ON THE TEST!

Plan C: You Have Six Weeks to Prepare

This approach is for students who have already studied most of the material that may be on the exam. The best use of this book for you is as a specific guide towards the AP Chemistry exam. There are time constraints to this approach, as the exam is only a short time away. This is not the best time to try to learn new material.

APRIL 1–15

— Skim Chapters 1–4.
— Go over Chapter 5.
— Skim Chapters 6–9.
— Carefully go over the Rapid Review sections of Chapters 5–9.
— Complete Practice Exam 1.
— Score the exam and analyze your mistakes.
— Skim and highlight the Glossary.

APRIL 15–MAY 1

— Skim Chapters 10–14.
— Carefully go over the Rapid Review sections of Chapters 10–14.
— Carefully go over the Rapid Review sections of Chapters 5–9 again.
— Continue to skim and highlight the Glossary.

MAY (first 2 weeks) (THIS IS IT!)

— Skim Chapters 15–19.
— Carefully go over the Rapid Reviews for Chapters 15–19.
— Complete Practice Exam 2.
— Score the exam and analyze your mistakes.
— Review the tip sheet on avoiding "stupid" mistakes at the back of this book.
— Get a good night's sleep before the exam. Fall asleep knowing that you are well prepared.

GOOD LUCK ON THE TEST!

Determine Your Test Readiness

CHAPTER 3

Take a Diagnostic Exam

IN THIS CHAPTER

Summary: The diagnostic exam is for your benefit. It will let you know where you need to spend the majority of your study time. Do not make the mistake of studying only those parts you missed; you should always review all topics. It may be to your advantage to take the diagnostic exam again just before you begin your final review for the exam. This exam has only multiple-choice questions. It will give you an idea of where you stand with your chemistry preparation. The questions have been written to approximate the coverage of material that you will see on the AP exam and are similar to the review questions that you will see at the end of each chapter. However, there will be a few questions on content that will not be tested on the AP exam; these questions refer to basic chemistry knowledge that your teacher will expect you to know. Once you are done with the exam, check your work against the given answers, which also indicate where you can find the corresponding material in the book. We also provide you with a way to convert your score to a rough AP score.

Key Ideas

- Practice the kind of multiple-choice questions you will be asked on the real exam.
- Answer questions that approximate the coverage of topics on the real exam.
- Check your work against the given answers.
- Determine your areas of strength and weakness.
- Highlight the pages that you must give special attention to.

Getting Started: The Diagnostic Exam

The following problems refer to different chapters in the book. The important thing is not whether you get the correct answer, but whether you have difficulty with one or more questions from a chapter. If so, then review the material in that chapter. You may use a calculator and periodic table. For each question, circle the letter of your choice.

Chapter 5

1. In most of its compounds, this element exists as a monatomic cation.

(A) F
(B) S
(C) N
(D) Ca

2. Which of the following groups has the species correctly listed in order of decreasing radius?

(A) Cu^{2+}, Cu^{+}, Cu
(B) V, V^{2+}, V^{3+}
(C) F^{-}, Br^{-}, I^{-}
(D) B, Be, Li

3. Which of the following elements has the lowest electronegativity?

(A) F
(B) I
(C) Ba
(D) Al

4. Which of the following represents the correct formula for hexaamminecobalt(III) nitrate?

(A) $[Co(NH_3)_6](NO_3)_3$
(B) $[Co(NH_3)_6](NO_2)_3$
(C) $Am_6Co(NO_3)_3$
(D) $(NH_3)_6Co_3(NO_3)$

5. Which of the following represents the correct formula for hexaamminechromium(III) chloride?

(A) $[Cr(NH_3)_6](ClO_3)_3$
(B) $(NH_3)_6Cr_3Cl$
(C) Am_6CrCl_3
(D) $[Cr(NH_3)_6]Cl_3$

6. The discovery that atoms have small, dense nuclei is credited to which of the following?

(A) Einstein
(B) Dalton
(C) Bohr
(D) Rutherford

Chapter 6

7. ___ $Mn(OH)_2(s)$ + ___ $H_3AsO_4(aq) \rightarrow$
___ $Mn_3(AsO_4)_2(s)$ + ___ $H_2O(l)$

After the above chemical equation is balanced, the lowest whole-number coefficient for water is:

(A) 6
(B) 2
(C) 12
(D) 3

8. Which of the following best represents the net ionic equation for the reaction of calcium hydroxide with an aqueous sodium carbonate solution to form a precipitate?

(A) $Ca^{2+} + Na_2CO_3 \rightarrow CaCO_3 + 2\ Na^{+}$
(B) $2\ Ca(OH) + Na_2CO_3 \rightarrow Ca_2CO_3 + 2\ NaOH$
(C) $Ca(OH)_2 + CO_3^{2-} \rightarrow CaCO_3 + 2\ OH^{-}$
(D) $Ca^{2+} + CO_3^{2-} \rightarrow CaCO_3$

9. A student mixes 50.0 mL of 0.10 M $Fe(NO_3)_2$ solution with 50.0 mL of 0.10 M KOH. A green precipitate forms, and the concentration of the hydroxide ion becomes very small. Which of the following correctly places the concentrations of the remaining ions in order of decreasing concentration?

(A) $[Fe^{2+}] > [NO_3^{-}] > [K^{+}]$
(B) $[Fe^{2+}] > [K^{+}] > [NO_3^{-}]$
(C) $[NO_3^{-}] > [K^{+}] > [Fe^{2+}]$
(D) $[K^{+}] > [Fe^{2+}] > [NO_3^{-}]$

10. Solutions containing this ion give a reddish-brown precipitate upon standing.

(A) Cu^{2+}
(B) CO_3^{2-}
(C) Fe^{3+}
(D) Al^{3+}

Chapter 7

11. $14\ H^+(aq) + 6\ Fe^{2+}(aq) + Cr_2O_7^{2-}(aq) \rightarrow 2\ Cr^{3+}(aq) + 6\ Fe^{3+}(aq) + 7\ H_2O(l)$

The above reaction is used in the titration of an iron solution. What is the concentration of the iron solution if it took 45.20 mL of 0.1000 M $Cr_2O_7^{2-}$ solution to titrate 75.00 mL of an acidified iron solution?

(A) 0.1000 M
(B) 0.4520 M
(C) 0.3616 M
(D) 0.7232 M

12. Manganese, Mn, forms a number of oxides. A particular oxide is 49.5% by mass Mn. What is the simplest formula for this oxide?

(A) MnO
(B) Mn_2O_3
(C) Mn_2O_7
(D) MnO_2

13. $2\ KMnO_4(aq) + 5\ H_2C_2O_4(aq) + 3\ H_2SO_4(aq) \rightarrow K_2SO_4(aq) + 2\ MnSO_4(aq) + 10\ CO_2(g) + 8\ H_2O(l)$

How many moles of $MnSO_4$ are produced when 2.0 mol of $KMnO_4$, 2.5 mol of $H_2C_2O_4$, and 3.0 mol of H_2SO_4 are mixed?

(A) 1.0 mol
(B) 3.5 mol
(C) 2.0 mol
(D) 2.5 mol

14. ____ $KClO_3 \rightarrow$ ____ KCl + ____ O_2

After the above equation is balanced, how many moles of O_2 can be produced from 1.0 mol of $KClO_3$?

(A) 1.5 mol
(B) 3.0 mol
(C) 1.0 mol
(D) 3.0 mol

Chapter 8

15. $Ba(s) + 2\ H_2O(l) \rightarrow Ba(OH)_2(aq) + H_2(g)$

Barium reacts with water according to the above reaction. What volume of hydrogen gas, at standard temperature and pressure, is produced from 0.400 mol of barium?

(A) 8.96 L
(B) 5.60 L
(C) 4.48 L
(D) 3.36 L

16. A sample of chlorine gas is placed in a container at constant pressure. The sample is heated until the absolute temperature is doubled. This will also double which of the following?

(A) potential energy
(B) moles
(C) density
(D) volume

17. A balloon contains 2.0 g of hydrogen gas. A second balloon contains 4.0 g of helium gas. Both balloons are at the same temperature and pressure. Which of the following statements is true?

(A) The number of hydrogen molecules is less than the number of helium atoms in each balloon.
(B) The density of the helium in its balloon is less than the density of the hydrogen in its balloon.
(C) The volume of the hydrogen balloon is less than that of the helium balloon.
(D) The average kinetic energy of the molecules/atoms in each balloon is the same.

18. The volume and pressure of a real gas are NOT the same as those calculated from the ideal gas equation, because the ideal gas equation does NOT take into account:

(A) the volume of the molecules and the attraction between the molecules.
(B) the volume of the molecules and the mass of the molecules.
(C) the attraction between the molecules and the mass of the molecules.
(D) the volume of the molecules and variations in the absolute temperature.

19. Aluminum metal reacts with gaseous HCl to produce aluminum chloride and hydrogen gas. What volume of hydrogen gas, at STP, is produced when 13.5 g of aluminum is mixed with an excess of HCl?

(A) 22.4 L
(B) 33.6 L
(C) 11.2 L
(D) 16.8 L

20. A sample containing the gases carbon dioxide, carbon monoxide, and water vapor was analyzed and found to contain 4.5 mol of carbon dioxide, 4.0 mol of carbon monoxide, and 1.5 mol of water vapor. The mixture had a total pressure of 1.2 atm. What was the partial pressure of the carbon monoxide?

(A) 0.48 atm
(B) 0.18 atm
(C) 5.4 atm
(D) 0.54 atm

21. An ideal gas sample weighing 0.548 g at 100°C and 0.993 atm has a volume of 0.237 L. Determine the molar mass of the gas.

(A) 71.3 g/mol
(B) 143 g/mol
(C) 19.1 g/mol
(D) 0.0140 g/mol

22. If a sample of He effuses at a rate of 30 mol per hour at 45°C, which of the gases below will effuse at approximately one-half the rate under the same conditions?

(A) CH_4
(B) O_3
(C) N_2
(D) H_2

23. The average kinetic energy of nitrogen molecules changes by what factor when the temperature is increased from 30°C to 60°C?

(A) (333 – 303)
(B) 2
(C) 1/2
(D) $\sqrt{303-333}$

Chapter 9

24. Which of the following required energy to produce a gaseous cation from a gaseous atom in the ground state?

(A) free energy
(B) lattice energy
(C) kinetic energy
(D) ionization energy

25. The average _____________ is the same for any ideal gas at a given temperature.

(A) free energy
(B) lattice energy
(C) kinetic energy
(D) ionization energy

26. Which of the following is the maximum energy available for useful work from a spontaneous reaction?

(A) free energy
(B) lattice energy
(C) kinetic energy
(D) ionization energy

27. The energy required to completely separate the ions from a solid is which of the following?

(A) free energy
(B) lattice energy
(C) kinetic energy
(D) ionization energy

28. Oxidation of ClF by F_2 yields ClF_3, an important fluorinating agent formerly used to produce the uranium compounds in nuclear fuels:

$$ClF(g) + F_2(g) \rightarrow 2\ ClF_3(l)$$

Use the following thermochemical equations to calculate $\Delta H°_{rxn}$ for the production of ClF_3:

1. $2\ ClF(g) + O_2(g) \rightarrow Cl_2O(g) + OF_2(g)$ $\quad \Delta H° = 167.5$ kJ
2. $2\ F_2(g) + O_2(g) \rightarrow 2\ OF_2(g)$ $\quad \Delta H° = -43.5$ kJ
3. $2\ ClF_3(l) + 2\ O_2(g) \rightarrow Cl_2O(g) + 3\ OF_2(g)$ $\quad \Delta H° = 394.1$ kJ

(A) +270.2 kJ
(B) −135.1 kJ
(C) 0.0 kJ
(D) −270.2 kJ

29. Choose the reaction expected to have the greatest increase in entropy.

(A) $N_2(g) + O_2(g) \rightarrow 2\ NO(g)$
(B) $CO_2(g) \rightarrow CO_2(s)$
(C) $2\ XeO_3(s) \rightarrow 2\ Xe(g) + 3\ O_2(g)$
(D) $2\ K(s) + F_2(g) \rightarrow 2\ KF(s)$

30. A certain reaction is nonspontaneous under standard conditions but becomes spontaneous at lower temperatures. What conclusions may be drawn under standard conditions?

(A) $\Delta H > 0$, $\Delta S > 0$ and $\Delta G > 0$
(B) $\Delta H < 0$, $\Delta S < 0$ and $\Delta G = 0$
(C) $\Delta H < 0$, $\Delta S > 0$ and $\Delta G > 0$
(D) $\Delta H < 0$, $\Delta S < 0$ and $\Delta G > 0$

Chapter 10

31. Which of the following groups contains only atoms that are paramagnetic in their ground state?

(A) Be, O, and N
(B) Mg, He, and Rb
(C) K, C, and Fe
(D) Br, Sb, and Kr

32. Which of the following could be the electron configuration of a transition metal ion?

(A) $1s^2 1p^6 2s^2 2p^3$
(B) $1s^2 2s^2 2p^6 3s^2 3p^6 4s^2 3d^{10} 4p^5$
(C) $1s^2 2s^2 2p^6 3s^2 3p^6 3d^3$
(D) $1s^2 2s^2 2p^5$

33. Which of the following is the configuration of a noble gas?

(A) $1s^2 1p^6 2s^2 2p^3$
(B) $1s^2 2s^2 2p^6 3s^2 3p^6 4s^2 3d^{10} 4p^6$
(C) $1s^2 2s^2 2p^6 3s^2 3p^6 3d^3$
(D) $1s^2 2s^2 2p^5$

34. Which of the following is the electron configuration of a halogen?

(A) $1s^2 1p^6 2s^2 2p^3$
(B) $1s^2 2s^2 2p^6 3s^2 3p^6 4s^2 3d^{10} 4p^6$
(C) $1s^2 2s^2 2p^6 3s^2 3p^6 3d^3$
(D) $1s^2 2s^2 2p^5$

35. Which of the following is an impossible electron configuration?

(A) $1s^2 1p^6 2s^2 2p^3$
(B) $1s^2 2s^2 2p^6 3s^2 3p^6 4s^2 3d^{10} 4p^6$
(C) $1s^2 2s^2 2p^6 3s^2 3p^6 3d^3$
(D) $1s^2 2s^2 2p^5$

36. This explains why the exact position of an electron is not known.

(A) Pauli exclusion principle
(B) electron shielding
(C) Hund's rule
(D) Heisenberg uncertainty principle

37. This is why nitrogen atoms, in their ground state, are paramagnetic.

(A) Pauli exclusion principle
(B) electron shielding
(C) Hund's rule
(D) Heisenberg uncertainty principle

38. This means that an atomic orbital can hold no more than two electrons.

(A) Pauli exclusion principle
(B) electron shielding
(C) Hund's rule
(D) Heisenberg uncertainty principle

39. Which of the following explains why the 4s orbital fills before the 3d?

(A) Pauli exclusion principle
(B) electron shielding
(C) Hund's rule
(D) Heisenberg uncertainty principle

40. Magnesium reacts with element *X* to form an ionic compound. If the ground-state electron configuration of *X* is $1s^2 2s^2 2p^3$, what is the simplest formula for this compound?

(A) MgX_2
(B) Mg_2X_3
(C) Mg_3X_2
(D) MgX

Chapter 11

41. VSEPR predicts that an IF_5 molecule will have which of the following shapes?

(A) tetrahedral
(B) trigonal bipyramidal
(C) square pyramid
(D) trigonal planar

42. Which of the following does NOT have one or more π bonds?

(A) SO_2
(B) SF_6
(C) O_2
(D) SO_3

43. Which of the following is nonpolar?

(A) IF_5
(B) BrF_3
(C) CF_4
(D) SF_4

44. The only substance listed below that contains ionic, σ, and π bonds is:

(A) Na_3N
(B) NO_2
(C) $NaNO_3$
(D) NH_3

45. Which molecule or ion in the following list has the greatest number of unshared electron pairs around the central atom?

(A) SO_2
(B) CO_3^{2-}
(C) XeF_2
(D) CF_4

46. What types of hybridization of carbon are in the compound acetic acid, CH_3COOH?

(A) sp^3, sp^2, and sp
(B) sp^3 only
(C) sp^3 and sp^2
(D) sp^2 and sp

Chapter 12

47. Which of the following is the best description of the structure of graphite?

(A) composed of atoms held together by delocalized electrons
(B) composed of molecules held together by intermolecular dipole-dipole interactions
(C) composed of positive and negative ions held together by electrostatic attractions
(D) composed of macromolecules held together by strong bonds

48. Which of the following best describes Ca(s)?

(A) composed of atoms held together by delocalized electrons
(B) composed of molecules held together by intermolecular dipole-dipole interactions
(C) composed of positive and negative ions held together by electrostatic attractions
(D) composed of macromolecules held together by strong bonds

49. Which of the following categories best describes $CaCO_3(s)$?

(A) composed of atoms held together by delocalized electrons
(B) composed of molecules held together by intermolecular dipole-dipole interactions
(C) composed of positive and negative ions held together by electrostatic attractions
(D) composed of macromolecules held together by strong bonds

50. Which of the following is applicable to $SO_2(s)$?

(A) composed of atoms held together by delocalized electrons
(B) composed of molecules held together by intermolecular dipole-dipole interactions
(C) composed of positive and negative ions held together by electrostatic attractions
(D) composed of macromolecules held together by strong bonds

51. The critical point on a phase diagram represents:

(A) the highest temperature and pressure where a substance can sublime
(B) the highest temperature and pressure where the substance may exist as discrete solid and gas phases
(C) the temperature and pressure where the substance exists in equilibrium as solid, liquid, and gas phases
(D) the highest temperature and pressure where the substance may exist as discrete liquid and gas phases

52. This explains why copper is ductile.

(A) London dispersion forces
(B) covalent bonding
(C) hydrogen bonding
(D) metallic bonding

53. This is why acetic acid, CH_3COOH, molecules exist as dimers in the gaseous phase.

(A) London dispersion forces
(B) covalent bonding
(C) hydrogen bonding
(D) metallic bonding

54. For the following, pick the answer that most likely represents their relative solubilities in water.

(A) $CH_3CH_2CH_2OH < HOCH_2CH_2OH < CH_3CH_2CH_2CH_3$
(B) $CH_3CH_2CH_2CH_3 < HOCH_2CH_2OH < CH_3CH_2CH_2OH$
(C) $CH_3CH_2CH_2CH_3 < CH_3CH_2CH_2OH < HOCH_2CH_2OH$
(D) $CH_3CH_2CH_2OH < CH_3CH_2CH_2CH_3 < HOCH_2CH_2OH$

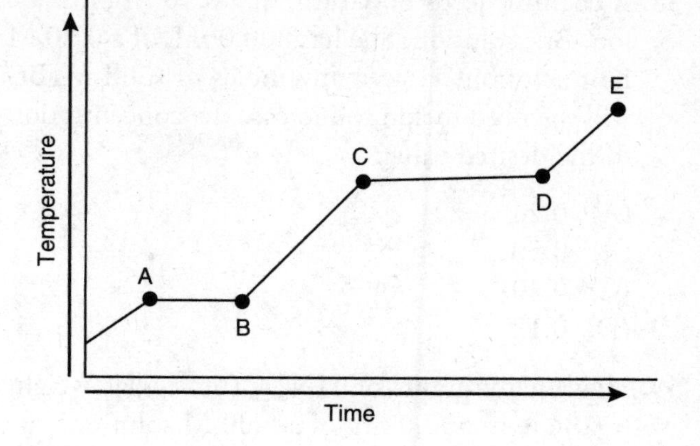

55. The above diagram represents the heating curve for a pure crystalline substance. The solid is the only phase present up to point:

(A) C
(B) B
(C) E
(D) A

56. For all one-component phase diagrams, choose the correct statement from the following list.

(A) The line separating the gas from the liquid phase may have a positive or negative slope.
(B) The line separating the solid from the liquid phase may have a positive or negative slope.
(C) The line separating the solid from the liquid phase has a positive slope.
(D) The temperature at the triple point is the same as at the freezing point.

Chapter 13

57. A solution is prepared by dissolving 0.500 mol of NaCl in 500.0 g of water. Which of the following would be the best procedure to determine the molarity of the solution?

(A) Measure the volume of the solution.
(B) Titrate the solution with standard silver nitrate solution.
(C) Determine the freezing point of the solution.
(D) Determine the osmotic pressure of the solution.

58. A chemist needs 800.0 mL of a 0.50 M bromide ion, Br^-, solution. She has 800.0 mL of a 0.20 M KBr solution. How many moles of solid $MgBr_2$ will she need to add to increase the concentration to the desired value?

(A) 0.24
(B) 0.50
(C) 0.30
(D) 0.12

59. How many grams of HNO_3 (molecular weight 63.0) are in 500.0 mL of a 5.00 M solution?

(A) 31.5 g
(B) 63.0 g
(C) 5.00 g
(D) 158 g

60. If a solution of ethyl ether, $(C_2H_5)_2O$, in ethanol, C_2H_5OH, is treated as an ideal solution, what is the mole fraction of ethyl ether in the vapor over an equimolar solution of these two liquids? The vapor pressure of ethyl ether is 480 mm Hg at 20°C, and the vapor pressure of ethanol is 50 mm Hg at this temperature.

(A) 0.50
(B) 0.76
(C) 0.91
(D) 0.27

Chapter 14

61. Step 1: $2\ NO_2(g) \rightarrow N_2(g) + 2\ O_2(g)$
Step 2: $2\ CO(g) + O_2(g) \rightarrow 2\ CO_2(g)$
Step 3: $N_2(g) + O_2(g) \rightarrow 2\ NO(g)$

The above represents a proposed mechanism for the reaction of NO_2 with CO. What are the overall products of the reaction?

(A) NO and CO_2
(B) O_2 and CO_2
(C) N_2 and NO
(D) NO only

62. The difference in energy between the transition state and the reactants is:

(A) the kinetic energy
(B) the activation energy
(C) the free energy
(D) the reaction energy

63. The table below gives the initial concentrations and rates for three experiments.

EXPERIMENT	INITIAL $[ClO_2]$ (mol L^{-1})	INITIAL $[OH^-]$ (mol L^{-1})	INITIAL RATE OF FORMATION OF ClO_2 (mol L^{-1} s^{-1})
1	0.100	0.100	2.30×10^5
2	0.200	0.100	9.20×10^5
3	0.200	0.200	1.84×10^6

The reaction is $2ClO_2(aq) + 2OH^-(aq) \rightarrow ClO_2^-(aq) + ClO_3^-(aq) + H_2O(l)$. What is the rate law for this reaction?

(A) Rate = $k[ClO_2]^2[OH^-]^2$
(B) Rate = $k[ClO_2]$
(C) Rate = $k[ClO_2]^2[OH^-]$
(D) Rate = $k[OH^-]^2$

Chapter 15

64. Which of the following CANNOT behave as both a Brønsted base and a Brønsted acid?

(A) HCO_3^-
(B) HPO_4^{2-}
(C) HSO_4^-
(D) CO_3^{2-}

65.

Acid	K_a, acid dissociation constant
H_3PO_4	7.2×10^{-3}
$H_2PO_4^-$	6.3×10^{-8}
HPO_4^{2-}	4.2×10^{-13}

Using the information from the preceding table, which of the following is the best choice for preparing a pH = 8.5 buffer?

(A) $K_2HPO_4 + K_3PO_4$
(B) $K_2HPO_4 + KH_2PO_4$
(C) K_3PO_4
(D) K_2HPO_4

66. What is the ionization constant, K_a, for a weak monoprotic acid if a 0.5 molar solution has a pH of 5.0?

(A) 3×10^{-4}
(B) 2×10^{-10}
(C) 7×10^{-8}
(D) 1×10^{-6}

67. Assuming all concentrations are 1 M, which of the following is the most acidic solution (lowest pH)?

(A) KBr (potassium bromide) and HBr (hydrobromic acid)
(B) $H_2C_2O_4$ (oxalic acid) and KHC_2O_4 (potassium hydrogen oxalate)
(C) NH_3 (ammonia) and NH_4NO_3 (ammonium nitrate)
(D) $(CH_3)_2NH$ (dimethylamine) and $HC_2H_3O_2$ (acetic acid)

68. Assuming all concentrations are equal, which of the following solutions has pH nearest 7?

(A) KBr (potassium bromide) and HBr (hydrobromic acid)
(B) $H_2C_2O_4$ (oxalic acid) and KHC_2O_4 (potassium hydrogen oxalate)
(C) NH_3 (ammonia) and NH_4NO_3 (ammonium nitrate)
(D) $(CH_3)_2NH$ (dimethylamine) and $HC_2H_3O_2$ (acetic acid)

69. Which of the following yields a buffer with a pH > 7 upon mixing equal volumes of 1 M solutions?

(A) KBr (potassium bromide) and HBr (hydrobromic acid)
(B) $H_2C_2O_4$ (oxalic acid) and KHC_2O_4 (potassium hydrogen oxalate)
(C) NH_3 (ammonia) and NH_4NO_3 (ammonium nitrate)
(D) $(CH_3)_2NH$ (dimethylamine) and $HC_2H_3O_2$ (acetic acid)

70. Which of the following will give a buffer with a pH < 7 when equal volumes of 1 M solutions of each of the components are mixed?

(A) KBr (potassium bromide) and HBr (hydrobromic acid)
(B) $H_2C_2O_4$ (oxalic acid) and KHC_2O_4 (potassium hydrogen oxalate)
(C) NH_3 (ammonia) and NH_4NO_3 (ammonium nitrate)
(D) $(CH_3)_2NH$ (dimethylamine) and $HC_2H_3O_2$ (acetic acid)

71. Determine the $OH^-(aq)$ concentration in 0.0010 M pyridine (C_5H_5N) solution. (The K_b for pyridine is 9×10^{-9}.)

(A) 5×10^{-1} M
(B) 1×10^{-3} M
(C) 3×10^{-6} M
(D) 9×10^{-9} M

72. $SnS(s) + 2H^+(aq) \rightleftharpoons Sn^{2+}(aq) + H_2S(aq)$

The successive acid dissociation constants for H_2S are 9.5×10^{-8} (K_{a1}) and 1×10^{-19} (K_{a2}). The K_{sp}, the solubility product constant, for SnS equals 1.0×10^{-25}. What is the equilibrium constant for the above reaction?

(A) $9.5 \times 10^{-8}/1.0 \times 10^{-25}$
(B) $9.5 \times 10^{-27}/1.0 \times 10^{-25}$
(C) $1.0 \times 10^{-25}/9.5 \times 10^{-27}$
(D) $1 \times 10^{-19}/1.0 \times 10^{-25}$

73. $N_2O_4(g) \rightleftharpoons 2NO_2(g)$ endothermic

An equilibrium mixture of the compounds is placed in a sealed container at 150°C. Which of the following changes may increase the amount of the product?

(A) raising the temperature of the container
(B) increasing the volume of the container and raising the temperature of the container
(C) adding 1 mole of Ar(g) to the container
(D) adding 1 mole of Ar(g) to the container and raising the temperature of the container

74. The K_{sp} for LaF_3 is 2×10^{-9}. What is the molar solubility of this compound in water?

(A) $2 \times 10^{-9}/27$

(B) $\sqrt[4]{2 \times 10^{-19}}$

(C) $\sqrt[2]{2 \times 10^{-19}}$

(D) $\sqrt[4]{2 \times 10^{-9}/27}$

75. The K_{sp} for $Mn(OH)_2$ is 1.6×10^{-13}. What is the molar solubility of this compound in water?

(A) $\sqrt[3]{4.0 \times 10^{-14}}$

(B) 1.6×10^{-13}

(C) $\sqrt[3]{4.0 \times 10^{-13}}$

(D) $\sqrt[2]{4.0 \times 10^{-14}}$

76. $FeS(s) + 2\ H^+(aq) \rightleftharpoons Fe^{2+}(aq) + H_2S(aq)$

What is the equilibrium constant for the above reaction? The successive acid dissociation constants for H_2S are 9.5×10^{-8} (K_{a1}) and 1×10^{-19} (K_{a2}). The K_{sp}, the solubility product constant, for FeS equals 5.0×10^{-18}.

(A) $9.5 \times 10^{-27}/5.0 \times 10^{-18}$
(B) $5.0 \times 10^{-18}/9.5 \times 10^{-27}$
(C) $5.0 \times 10^{-18}/9.5 \times 10^{-8}$
(D) $9.5 \times 10^{-8}/5.0 \times 10^{-18}$

77. The K_{sp} for $Cr(OH)_3$ is 1.6×10^{-30}. What is the molar solubility of this compound in water?

(A) $\sqrt[4]{1.6 \times 10^{-30}}$
(B) $\sqrt[4]{1.6 \times 10^{-30}/27}$
(C) 1.6×10^{-30}
(D) $1.6 \times 10^{-30}/27$

Chapter 16

78. $I^-(aq) + H^+(aq) + MnO_4^-(aq) \rightarrow Mn^{2+}(aq) + H_2O(l) + I_2(s)$

What is the coefficient of H^+ when the above reaction is balanced?

(A) 12
(B) 32
(C) 16
(D) 8

79. How many moles of Au will deposit on the cathode when 0.60 Faradays of electricity is passed through a 1.0 M solution of Au^{3+}?

(A) 0.60 mol
(B) 0.30 mol
(C) 0.40 mol
(D) 0.20 mol

80. $Sn^{2+}(aq) + 2\ Fe^{3+}(aq) \rightarrow Sn^{4+}(aq) + 2\ Fe^{2+}(aq)$

The reaction shown above was used in an electrolytic cell. The voltage measured for the cell was not equal to the calculated E° for the cell. Which of the following could explain this discrepancy?

(A) Both of the solutions were at 25°C instead of 0°C.
(B) The anode and cathode were different sizes.
(C) The anion in the anode compartment was chloride instead of nitrate, as in the cathode compartment.
(D) One or more of the ion concentrations was not 1 M.

Questions 81–82 refer to the following half-reaction in an electrolytic cell:

$2\ SO_4^{2-}(aq) + 10\ H^+(aq) + 8\ e^- \rightarrow$
$S_2O_3^{2-}(aq) + 5\ H_2O(l)$

81. Choose the correct statement from the following list.

(A) The sulfur is oxidized.
(B) This is the cathode reaction.
(C) The oxidation state of sulfur does not change.
(D) The H^+ serves as a catalyst.

82. If a current of 0.60 amperes is passed through the electrolytic cell for 0.75 h, how should you calculate the grams of $S_2O_3^{2-}(aq)$ formed?

(A) (0.60) (0.75) (3,600) (112)/(96,500) (8)
(B) (0.60) (0.75) (3,600) (112)/(96,500) (10)
(C) (0.60) (0.75) (60) (32)/(96,500) (8)
(D) (0.60) (0.75) (3,600) (112)/(10)

83. $2\ BrO_3^-(aq) + 12\ H^+(aq) + 10\ e^- \rightarrow Br_2(aq) + 6\ H_2O(l)$

If a current of 5.0 A is passed through the electrolytic cell for 0.50 h, how should you calculate the number of grams of Br_2 that will form?

(A) (5.0) (0.50) (3,600) (159.8)/(10)
(B) (5.0) (0.50) (3,600) (159.8)/(96,500) (10)
(C) (5.0) (0.50) (60) (159.8)/(96,500) (10)
(D) (5.0) (0.50) (3,600) (79.9)/(96,500) (10)

84. $2\,IO_3^-(aq) + 6\,H_2O(l) + 10\,e^- \rightarrow I_2(s) + 12\,OH^-(aq)$

Using the above reaction, if a current of 7.50 A is passed through the electrolytic cell for 0.45 h, how should you calculate the grams of I_2 that will form?

(A) (7.50) (0.45) (3,600) (253.8)/(10)
(B) (7.50) (0.45) (3,600) (126.9)/(96,500) (10)
(C) (7.50) (0.45) (60) (253.8)/(96,500) (10)
(D) (7.50) (0.45) (3,600) (253.8)/(96,500) (10)

Chapter 17

85. When $^{226}_{88}Ra$ decays, it emits two α particles, then a β, particle, followed by an α particle. The resulting nucleus is:

(A) $^{212}_{83}$ Bi
(B) $^{222}_{86}$ Rn
(C) $^{214}_{82}$ Pb
(D) $^{214}_{83}$ Bi

86. Which of the following lists the types of radiation in the correct order of increasing penetrating power?

(A) α, γ, β
(B) β, α, γ
(C) α, β, γ
(D) β, γ, α

87. What is the missing product in the following nuclear reaction?

$$^{236}_{92}U \rightarrow 4\,^{1}_{0}n + ^{136}_{53}I + \_\_\_\_$$

(A) $^{90}_{39}$ Y
(B) $^{96}_{38}$ Sr
(C) $^{96}_{39}$ Y
(D) $^{98}_{40}$ Zr

88. If 75% of a sample of pure $^{3}_{1}H$ decays in 24.6 yr, what is the half-life of $^{3}_{1}H$?

(A) 24.6 yr
(B) 18.4 yr
(C) 12.3 yr
(D) 6.15 yr

Chapter 18

89. Alkenes are hydrocarbons with the general formula C_nH_{2n}. If a 1.40 g sample of any alkene is combusted in excess oxygen, how many moles of water will form?

(A) 0.2
(B) 0.1
(C) 1.5
(D) 0.7

90. What type of compound is shown?

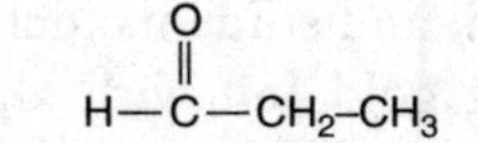

(A) an alcohol
(B) an aldehyde
(C) a ketone
(D) an ester

Chapter 19

Questions on this chapter are incorporated into the chapters concerning the specific experiments.

› Answers and Explanations

Chapter 5

1. D—The others (nonmetals) form anions.

2. B—Decreasing radii is related to increasing charges, or for going up a column (with equal charges), or moving toward the right in a period of the periodic table. This explanation will not be sufficient for the free-response portion of the test, where it is necessary to address such factors as the effective nuclear charge.

3. C—The element that is farthest away from F on the periodic table.

4. A—Hexaammine = $(NH_3)_6$; cobalt(III) = Co^{3+}; and nitrate = NO_3^-.

5. D—Hexaammine = $(NH_3)_6$; chromium(III) = Cr^{3+}; chloride = Cl^-

6. D—Rutherford, and students, determined this by bombarding gold foil with alpha particles and detecting the deflection of some of the particles.

Chapter 6

7. A—The balanced chemical equation is:

$3\ Mn(OH)_2(s) + 2\ H_3AsO_4(aq) \rightarrow$
$Mn_3(AsO_4)_2(s) + 6\ H_2O(l)$

8. D—$Ca(OH)_2$, NaOH, and Na_2CO_3 are strong electrolytes (strong bases or soluble salts) and should be separated. You should know all the strong bases and that sodium compounds are soluble. Cancel all spectator ions (Na^+ and OH^-).

9. C—The hydroxide is low because it combined with some of the iron, so Fe^{2+} will be low. There is no other ion that the hydroxide ion could combine with to form a precipitate. The nitrate is double the potassium because there are two moles of nitrate per mole of iron(II) nitrate instead of one ion per mole, as in potassium hydroxide.

10. C—Copper is blue, not red, and carbonate and aluminum are colorless. Iron slowly hydrolyzes (reacts with water) to form solid $Fe(OH)_3$ (rust).

Chapter 7

11. C—(0.1000 mol $Cr_2O_7^{2-}$/1,000 mL)(45.20 mL)(6 mol Fe^{2+}/1 mol $Cr_2O_7^{2-}$)(1/75.00 mL)(1,000 mL/L)

12. C—Either calculate the percent Mn in each oxide: (A) 77.4%; (B) 69.6%; (C) 49.5%; (D) 63.2% or determine the empirical formula from the percent manganese and the percent oxygen (= 100.0 – 49.5).

13. A—$H_2C_2O_4$ is the limiting reagent as the amount is less than the stoichiometric ratio indicates. The calculation is $(2.5\text{ mol } H_2C_2O_4)\left(\dfrac{2\text{ mol } MnSO_4}{5\text{ mol } H_2C_2O_4}\right)$.

14. A—The coefficients in the balanced equation are 2, 2, and 3. Therefore, (1.0 mol $KClO_3$) (3 mol O_2/2 mol $KClO_3$) = 1.5 mol.

Chapter 8

15. A—(0.400 mol Ba)(1 mol H_2/1 mol Ba)(22.4 L/mol). Note that the 22.4 L/mol only works at STP.

16. D—This is an application of Charles's law, which relates volume to temperature. There is a direct relationship between volume and the absolute temperature. Doubling either volume or temperature, with moles and pressure remaining constant, doubles the other.

17. D—The average kinetic energy of a gas depends upon the temperature. Since the temperature of the two gases is the same, the average kinetic energy of the gases is the same. The moles are the same, so the number of particles and the volumes must be the same. Density is mass over volume, and since the balloons have the same volume, the one with more mass will have the higher density.

18. A—These are the basic differences between ideal and real gases.

19. D—It is necessary to first write the balanced chemical equation:

$$2\ Al(s) + 6\ HCl(g) \rightarrow 2\ AlCl_3(s) + 3\ H_2(g)$$

(13.5 g Al)(1 mol Al/27.0 g Al)(3 mol H_2/2 mol Al)(22.4 L/mol H_2)

20. A—The mole fraction of CO times the total pressure yields the partial pressure. The mole fraction of CO is the moles of CO (4.0) divided by the total moles (10.0).

21. A—$n = PV/RT = (0.993 \text{ atm})(0.237 \text{ L}) / (0.0821 \text{ L atm/K mol})(373 \text{ K}) = 7.69 \times 10^{-3}$ mol

molar mass $= 0.548 \text{ g}/7.69 \times 10^{-3}$ mol $= 71.3$ g/mol

This example illustrates the importance of rounding in calculations where no calculator is available. The answers are not close together; therefore, a rough calculation will lead to the correct answer. Also, you should notice the answer D is impossible for any substance.

22. A—For the rate to be one-half, the molar mass of the other gas must be the square of the molar mass of helium ($4^2 = 16$).

23. A—The average kinetic energy of the molecules depends on the temperature. The correct answer involves a temperature difference (333 K – 303 K). Do not forget that ALL gas law calculations require Kelvin temperatures.

Chapter 9

24. D—This is the definition of the ionization energy.

25. C—This is a basic postulate of kinetic molecular theory.

26. A—This is one of the properties of free energy.

27. B—This is the definition of the lattice energy.

28. B—This is an application of Hess's law.

$\frac{1}{2}[2ClF(g) + O_2(g) \rightarrow Cl_2O(g) + OF_2(g)]$ $\frac{1}{2}(167.5 \text{ kJ})$

$\frac{1}{2}[2 F_2(g) + O_2(g) \rightarrow 2 OF_2(g)]$ $\frac{1}{2}(-43.5 \text{ kJ})$

$\frac{1}{2}[Cl_2O(g) + 3 OF_2(g) \rightarrow 2 ClF_3(l) + 2 O_2(g)]$ $-\frac{1}{2}(394.1 \text{ kJ})$

$ClF(g) + F_2(g) \rightarrow ClF_3(l)$ -135.1 kJ

As always, rounding and estimating will save time.

29. C—The one with the greatest increase in the moles of gas.

30. D—Nonspontaneous means $\Delta G > 0$. For a reaction to become spontaneous at lower temperature ($\Delta G < 0$) means $\Delta H < 0$ and $\Delta S < 0$.

Chapter 10

31. C—Atoms with only completely filled shells or subshells are not paramagnetic, they are diamagnetic. From the choices in this problem, these are: Be, Mg, He, Kr, and Zn; therefore, any answer containing one of these cannot be the correct choice. It is not necessary to work through a possible solution until encountering a diamagnetic species. Also, it might be helpful to look on the periodic table.

32. C—Transition metal ions are, in general, s^0 and p^0 or p^6 with the possibility of having one or more electrons in the d orbitals. C could be Cr^{3+}.

33. B—The noble gases, except helium, are ns^2np^6. In this case, $n = 4$, and the gas is krypton.

34. D—Halogens are ns^2np^5. In this case, $n = 2$, and the halogen is F.

35. A—The 1p orbital does not exist.

36. D—This is a statement of the uncertainty principle.

37. C—According to Hund's rule, the nitrogen 2p electrons enter the 2p orbitals individually (with spins parallel).

38. A—The Pauli exclusion principle states this limitation for all orbitals.

39. B—The d orbitals are less effectively shielded than the s orbitals. Due to this difference, the s orbitals have lower energy.

40. C—Mg becomes Mg^{2+}. The element is N, which can become N^{3-}.

Chapter 11

41. C—The iodine has five bonding pairs and one lone pair.

42. B—This is the only one with only single bonds. The other molecules have double or triple bonds. All double and triple bonds are a combination of σ and π bonds.

43. C—Use VSEPR; only the tetrahedral CF_4 is nonpolar. The other materials form a square pyramidal

(IF_5), T-shaped (BrF_3), and irregular tetrahedral (SF_4), and, therefore, are polar.

44. C—The only ionic bonds are present in the sodium compounds (eliminating B and D). The nitride ion has no internal bonding (eliminating A), but the nitrate ion has both σ and π bonds.

45. C—Draw the Lewis structures. The number of unshared pairs: (A) 1; (B) 0; (C) 3; (D) 0.

46. C—Draw the Lewis structure; the carbon on the left in the formula is sp^3, and the other is sp^2.

Chapter 12

47. D—Both graphite and diamond are covalent network solids.

48. A—Calcium is a metal, and answer A applies to metallic bonding.

49. C—Calcium carbonate is an ionic compound.

50. B—Sulfur dioxide consists of polar molecules.

51. D—This is the definition of the critical point.

52. D—This is a consequence of metallic bonding as the atoms can easily move past each other without breaking any bonds.

53. C—The carbonyl, C=O, and —OH groups are capable of participating in hydrogen bonds.

54. C—The more —OH groups, the more hydrogen bonding, and the more soluble in water (where hydrogen bonding also occurs).

55. D—The solid begins to melt at A and finishes melting at B.

56. B—The gas-liquid line always has a positive slope, which eliminates A. Answer B negates C; therefore, both cannot be correct. The triple point is not the same as the freezing point.

Chapter 13

57. A—Molarity is moles per liter, and the moles are already known; therefore, only the volume is necessary to complete the determination.

58. D—(0.800 L)(0.50 mol Br^-/L) = 0.40 mol needed.

(0.800 L)(0.20 mol Br^-/L) = 0.16 mol present.

[(0.40 – 0.16) mol Br^- to be added] (1 mol $MgBr_2$/2 mol Br^-)

59. D—(0.5000 L)(5.00 mol/L)(63.0 g/mol) = 158 g As always, estimate the answer by rounding the values.

60. C—Equimolar gives a mole fraction of 0.5. 0.5 × 480 mm Hg + 0.5 × 50 mm Hg = 265 mm Hg (total vapor pressure) mole fraction ethyl ether = (0.5 × 480 mm Hg)/265 mm Hg.

Chapter 14

61. A—Add the equations and cancel anything that appears on both sides of the reaction arrows.

62. B—This is the definition of the activation energy.

63. C—The table shows second order in chlorine dioxide (comparing experiments 1 and 2), because doubling the ClO_2 concentration quadruples (2^2) the rate. The reaction is first order in the hydroxide ion (comparing experiments 2 and 3), because doubling the OH^- concentration doubles (2^1) the rate. When making this determination, make sure there is only one concentration changing, i.e., do not compare experiments 1 and 3.

Chapter 15

64. D—To be an acid, the species must have an H^+ to donate, and to be a base, the species must be able to accept an H^+. The carbonate ion has no H^+ to donate to be an acid.

65. B—Start with the acid with a pK_a as near 8.5 ($K = 10^{-8.5}$) as possible ($H_2PO_4^-$). To go to a higher pH, add the acid (conjugate base) with the smaller K_a (higher pK_a).

66. B—This is an approximation. At pH = 5, $[H^+] = 10^{-5}$ M; therefore, $K_a = (10^{-5})^2/0.5$.

67. A—HBr is a strong acid, and with equal concentrations and no base present, it will give the lowest pH.

68. D—The weak acid and weak base give a nearly neutral solution, as they will tend to neutralize each other.

69. C—Only B and C are buffers. B is acidic (pH < 7) and C is basic (pH > 7).

70. B—Only B and C are buffers. B is acidic (pH < 7) and C is basic (pH > 7).

71. C—$[OH^-] = (0.0010 \times 9 \times 10^{-9})^{1/2} = (9 \times 10^{-12})^{1/2}$ Estimate—the square root of 10^{-12} will be 10^{-6}.

72. C—$K = K_{sp}/K_{a1}K_{a2}$

In this case, the key is setting up the calculation but not doing the calculation.

73. B—Adding Ar yields no change, as it is not part of the equilibrium. Increasing the temperature of an endothermic equilibrium will increase the amount of product.

74. D—$K_{sp} = [La^{3+}][F^-]^3 = [x][3x]^3 = 27x^4$. Solve for x. It is only necessary to set up the problem. This requires a knowledge of what the equilibrium is ($LaF_3(s) \rightleftharpoons La^{3+}(aq) + 3\ F^-(aq)$) and how to write the equilibrium expression ($K_{sp} = [La^{3+}][F^-]^3$).

75. A—The equilibrium constant expression for the dissolving of manganese(II) hydroxide is:

$$K_{sp} = [Mn^{2+}][OH^-]^2 = 1.6 \times 10^{-13}$$

If s is used to indicate the molar solubility, the equilibrium expression becomes:

$$K_{sp} = (s)(2s)^2 = 4s^3 = 1.6 \times 10^{-13}$$

This rearranges to: $s = \sqrt[3]{K/4}$

76. B—$K = K_{sp}/K_{a1}K_{a2} = 5.0 \times 10^{-18}/(9.5 \times 10^{-8})(1 \times 10^{-19})$

77. B—$K_{sp} = [Cr^{3+}][OH^-]^3 = [x][3x]^3 = 27x^4 = 1.6 \times 10^{-30}$. Solve for x.

Chapter 16

78. C—The balanced equation is

$$10\ I^-(aq) + 16\ H^+(aq) + 2\ MnO_4^-(aq) \rightarrow 2\ Mn^{2+}(aq) + 8\ H_2O(l) + 5\ I_2(s)$$

79. D—It is only necessary to know the mole ratio for the reaction ($Au^{3+}(aq) + 3\ e^- \rightarrow Au(s)$), which gives (0.60 F)(1 mol Au/3 F) = 0.20 moles.

80. D—The cell must be nonstandard. This could be due to variations in temperature (not 25°C) or concentrations (1 M) that are not standard.

81. B—A reduction is shown. Reductions take place at the cathode.

82. A—Use dimensional analysis: (0.60 coul/s)(0.75 h)(3,600 s/h)(112 g $S_2O_3^{2-}$/mol $S_2O_3^{2-}$)/(96,500 coul/F)(8 F/mol $S_2O_3^{2-}$)

83. B—Recall that 5.0 amp is 5.0 C/s. The calculation would be:

$$\left(\frac{5.0\ \text{C}}{\text{s}}\right)\left(\frac{3{,}600\ \text{s}}{\text{h}}\right)(0.50\ \text{h})\left(\frac{1\ \text{F}}{96{,}500\ \text{C}}\right) \times \left(\frac{1\ \text{mol Br}_2}{10\ \text{F}}\right)\left(\frac{159.8\ \text{g Br}_2}{1\ \text{mol Br}_2}\right)$$

84. D—Dimensional analysis:

(7.50 coul/s)(0.45 h)(3,600 s/h)(253.8 g I_2/mol I_2)/(96,500 coul/F)(10 F)

Chapter 17

85. D—The mass of an alpha particle is 4 and the mass of a beta particle is negligible. The mass number (superscript) should be 226 − (4 + 4 + 0 + 4) = 214. The charge on an alpha particle is +2 and the charge on the beta particle is −1; therefore, the atomic number (subscript) should be 88 − (2 + 2 − 1 + 2) = 83.

86. C—Alpha particles are the least penetrating, and gamma rays are the most penetrating.

87. C—Mass difference = 236 − 4(1) − 136 = 96.

Atomic number difference = 92 − 4(0) − 53 = 39.

88. C—After one half-life, 50% would remain. After another half-life this would be reduced by 1/2 to 25%. The total amount decayed is 75%. Thus, 24.6 years must be two half-lives of 12.3 years each.

Chapter 18

89. B—The general formula simplifies to CH_2, which has a molar mass of 14 g/mol. This leads to (1.40 g)(1 mol/14 g).

90. B.

Chapter 19

Questions on this chapter are incorporated into the chapters concerning the specific experiments.

Scoring and Interpretation

Now that you have finished and scored the diagnostic exam, it is time for you to learn what it all means. First, note any area where you had difficulty. This should not be limited to unfamiliar material. You should do this even if you got the correct answer. Determine where this material is covered in the book. Plan to spend additional time on the chapter in question. There is material you may not recognize because you have not gotten that far in class.

There are no free-response questions on this diagnostic exam; such questions are not useful at this point. There will be examples of free-response questions later in this book. We will use the multiple-choice questions to provide an estimate of your preparation. This is a simplified approach based on these questions. Do not try to do more than use these results as a general guide.

Raw Score (number right)	Approximate AP Score
55–91	5
41–54	4
30–40	3
18–29	2
0–17	1

If you did better than you expected—great! Be careful not to become overconfident. Much more will need to be done before you take the AP Chemistry exam.

If you did not do as well as you would have liked, don't panic. There is plenty of time for you to prepare for the exam. This is a guide to allow you to know which path you need to follow.

No matter what your results were, you are about to begin your 5 steps to a 5.

Good Luck!

Develop Strategies for Success

CHAPTER 4

How to Approach Each Question Type

IN THIS CHAPTER

Summary: Use these question-answering strategies to raise your AP score.

Key Ideas

Multiple-Choice Questions

- Read the question carefully.
- Try to answer the question yourself before reading the answer choices.
- Drawing a picture can help.
- Don't spend too much time on any one question.
- In-depth calculations are not necessary; approximate the answer by rounding.

Free-Response Questions

- Write clearly and legibly.
- Be consistent from one part of your answer to another.
- Draw a graph if one is required.
- If the question can be answered with one word or number, don't write more.
- If a question asks "how," tell "why" as well.

Multiple-Choice Questions

Because you are a seasoned student accustomed to the educational testing machine, you have surely participated in more standardized tests than you care to count. You probably know some students who always seem to ace the multiple-choice questions, and some students who would rather set themselves on fire than sit for another round of "bubble trouble." We hope that, with a little background and a few tips, you might improve your scores on this important component of the AP Chemistry exam.

First, the background. Every multiple-choice question has three important parts:

1. The **stem** is the basis for the actual question. Sometimes this comes in the form of a fill-in-the-blank statement, rather than a question.

   Example: The mass number of an atom is the sum of the atomic number and ________.

   Example: What two factors lead to real gases deviating from the predictions of Kinetic Molecular Theory?

2. The **correct answer option**. Obviously, this is the one selection that best completes the statement or responds to the question in the stem. Because you have purchased this book, you will select this option many, many times.

3. **Distracter options**. Just as it sounds, these are the incorrect answers intended to distract anyone who decided not to purchase this book. You can locate this person in the exam room by searching for the individual who is repeatedly smacking his or her forehead on the desktop.

Students who do well on multiple-choice exams are so well prepared that they can easily find the correct answer, but other students do well because they are perceptive enough to identify and avoid the distracters. Much research has been done on how best to study for, and complete, multiple-choice questions. You can find some of this research by using your favorite Internet search engine, but here are a few tips that many chemistry students find useful.

1. *Let's be careful out there*. You must carefully read the question. This sounds obvious, but you would be surprised how tricky those test developers can be. For example, rushing past, and failing to see, the use of a negative, can throw a student.

   Example: Which of the following is *not* true of the halogens?
   a. They are nonmetals.
   b. They form monatomic anions with a −1 charge.
   c. In their standard states they may exist as solids, liquids, or gases.
   d. All may adopt positive oxidation states.
   e. They are next to the noble gases on the periodic table.

   A student who is going too fast, and ignores the negative *not*, might select option (a), because it is true and it was the first option that the student saw.

   You should be very careful about the wording. It is easy to skip over small words like "not," "least," or "most." You must make sure you are answering the correct question. Many students make this type of mistake—do not add your name to the list.

2. *See the answer, be the answer*. Many people find success when they carefully read the question and, before looking at the alternatives, visualize the correct answer. This allows the person to narrow the search for the correct option and identify the distracters. Of course, this visualization tip is most useful for students who have used this book to thoroughly review the chemistry content.

Example: When Robert Boyle investigated gases, he found the relationship between pressure and volume to be _________.

Before you even look at the options, you should know what the answer is. Find that option, and then quickly confirm to yourself that the others are indeed wrong.

3. *Never say never.* Words like "never" and "always" are absolute qualifiers. If these words are in one of the choices, it is rarely the correct choice.

   Example: Which of the following is true about a real gas?
   a. There are never any interactions between the particles.
   b. The particles present always have negligible volumes.

   If you can think of any situation where the statements in (a) and (b) are untrue, then you have discovered distracters and can eliminate these as valid choices.

4. *Easy is as easy does.* It's exam day and you're all geared up to set this very difficult test on its ear. Question number one looks like a no-brainer. Of course! The answer is 7, choice c. Rather than smiling at the satisfaction that you knew the answer, you doubt yourself. Could it be that easy? Sometimes they *are* just that easy.

5. *Sometimes, a blind squirrel finds an acorn.* Should you guess? Try to eliminate one or more answers before you guess. Then pick what you think is the best answer. You are not penalized for guessing, so don't leave an answer blank.

6. *Draw it, nail it.* Many questions are easy to answer if you do a quick sketch in the margins of your test book. Hey, you paid for that test book; you might as well use it.

   Example: The rate of the reverse reaction will be slower than the rate of the forward reaction if the relative energies of the reactants and products are:

	Reactant	*Product*
a.	High	Equal to the reactants
b.	Low	Equal to the reactants
c.	High	Higher than the reactants
d.	Low	Higher than the reactants
e.	High	Lower than the reactants

   These types of question are particularly difficult, because the answer requires two ingredients. The graph that you sketch in the margin will speak for itself.

7. *Come back, Lassie, come back!* Pace yourself. If you do not immediately know the answer to a question—skip it. You can come back to it later. You have approximately 90 seconds per question. You can get a good grade on the test even if you do not finish all the questions. If you spend too much time on a question you may get it correct; however, if you go on you might get several questions correct in the same amount of time. The more questions you read, the more likely you are to find the ones for which you know the answers. You can help yourself on this timing by practice.

Times are given for the various tests in this book; if you try to adhere strictly to these times, you will learn how to pace yourself automatically.

8. *Timing is everything, kid.* You have about 90 seconds for each of the 60 questions. Keep an eye on your watch as you pass the halfway point. If you are running out of time and you have a few questions left, skim them for the easy (and quick) ones so that the rest of your scarce time can be devoted to those that need a little extra reading or thought.
9. *Think!* But do not try to outthink the test. The multiple-choice questions are straightforward—do not over-analyze them. If you find yourself doing this, pick the simplest answer. If you know the answer to a "difficult" question—give yourself credit for preparing well; do not think that it is too easy and that you missed something. There are easy questions and difficult questions on the exam.
10. *Change is good?* You should change answers only as a last resort. You can mark your test so you can come back to a questionable problem later. When you come back to a problem, make sure you have a definite reason for changing the answer.

Other things to keep in mind:

- Take the extra half of a second required to fill in the bubbles clearly.
- Don't smudge anything with sloppy erasures. If your eraser is smudgy, ask the proctor for another.
- Absolutely, positively, check that you are bubbling the same line on the answer sheet as the question you are answering. I suggest that every time you turn the page you double-check that you are still lined up correctly.

Free-Response Questions

You will have 90 minutes to complete Section II, the free-response part of the AP Chemistry exam. There will be a total of seven free-response questions (FRQs) of two different types. Three questions will be of the multipart type. Plan on spending a maximum of 20–25 minutes per question on these three. You will be given some information (the question stem), and then you will have several questions to answer related to that stem. These questions will be, for the most part, unrelated to each other. You might have a lab question, an equilibrium constant question, and so on. But all of these questions will be related to the original stem. The other type of free-response question will be of the single-part type. There will be four of these. Plan on allowing 3–10 minutes per question.

There are a number of kinds of questions that are fair game in the free-response section (Section II). One category is quantitative. You might be asked to analyze a graph or a set of data, and answer questions associated with this data. In many cases you will be required to perform appropriate calculations.

Another category of questions will be ones that refer to a laboratory setting/experiment. These lab questions tend to fall into two types: analysis of observations/data or the design of experiments. In the first type you might be given a set of data, for instance, kinetics data, and then be required to determine the order of reaction and/or the rate constant using that set of data. In the second type you might be asked to design a laboratory procedure given a set of equipment/reagents to accomplish a certain task, such as separation of certain metal ions in a mixture. You must use the equipment given, but you do not have to use all of the equipment.

The third category of questions on the exam involves questions related to representations of atoms or molecules. These representations might include such things as Lewis structures, ball-and-stick models, or space-filling models. You might be asked to take one and convert it to another or to choose a particular representation that is the most useful in describing certain observations.

Your score on the free-response questions amounts to one-half of your grade and, as long-time readers of essays, we assure you that there is no other way to score highly than to know your stuff. While you can guess on a multiple-choice question and have a 1/5 chance of getting the correct answer, there is no room for guessing in this section. There are, however, some tips that you can use to enhance your FRQ scores.

1. *Easy to read—easy to grade.* Organize your responses around the separate parts of the question and clearly label each part of your response. In other words, do not hide your answer; make it easy to find and easy to read. It helps you and it helps the reader to see where you are going. Trust me: helping the reader can never hurt. Which leads me to a related tip . . . Write in English, not Sanskrit! Even the most levelheaded and unbiased reader has trouble keeping his or her patience while struggling with bad handwriting. (We have actually seen readers waste almost 10 minutes using the Rosetta stone to decipher a paragraph of text that was obviously written by a time-traveling student from the Egyptian Empire.)
2. *Consistently wrong can be good.* The free-response questions are written in several parts. If you are looking at an eight-part question, it can be scary. However, these questions are graded so that you can salvage several points even if you do not correctly answer the first part. The key thing for you to know is that you must be consistent, even if it is consistently wrong. For example, you may be asked to draw a graph showing a phase diagram. Following sections may ask you to label the triple point, critical point, normal boiling point, and vapor pressure—each determined by the appearance of your graph. So let's say you draw your graph, but you label it incorrectly. Obviously, you are not going to receive that point. However, if you proceed by labeling the other points correctly in your *incorrect* quantity, you would be surprised how forgiving the grading rubric can be.
3. *Have the last laugh with a well-drawn graph.* There are some points that require an explanation (i.e., "Describe how . . ."). Not all free-response questions require a graph, but a garbled paragraph of explanation can be saved with a perfect graph that tells the reader you know the answer to the question. This does not work in reverse . . .
4. *If I say draw, you had better draw.* There are what readers call "graphing points" and these cannot be earned with a well-written paragraph. For example, if you are asked to draw a Lewis structure, certain points will be awarded for the picture, and only the picture. A delightfully written and entirely accurate paragraph of text will not earn the graphing points. You also need to label graphs clearly. You might think that a downward-sloping line is obviously a decrease, but some of those graphing points will not be awarded if lines and points are not clearly, and accurately, identified.
5. *Give the answer, not a dissertation.* There are some parts of a question where you are asked to simply "identify" something. This type of question requires a quick piece of analysis that can literally be answered in one word or number. That point will be given if you provide that one word or number whether it is the only word you write, or the fortieth. For example, you may be given a table that shows how a reaction rate varies with concentration. Suppose the correct rate is 2. The point is given if you say "2," "two," and maybe even "ii." If you write a novel concluding with the word "two," you will get the point, but you have wasted precious time. This brings me to . . .

6. *Welcome to the magic kingdom.* If you surround the right answer to a question with a paragraph of chemical wrongness, you will usually get the point, so long as you say the magic word. The only exception is a direct contradiction of the right answer. For example, suppose that when asked to "identify" the maximum concentration, you spend a paragraph describing how the temperature may change the solubility and the gases are more soluble under increased pressure, and then say the answer is two. You get the point! You said the "two" and "two" was the magic word. However, if you say that the answer is two, but that it is also four, but on Mondays, it is six, you have contradicted yourself and the point will not be given.
7. *"How" really means "how" and "why."* Questions that ask how one variable is affected by another—and these questions are legion—require an explanation, even if the question doesn't seem to specifically ask how *and* why. For example, you might be asked to explain how effective nuclear charge affects the atomic radius. If you say that the "atomic radius decreases," you may have received only one of two possible points. If you say that this is "because effective nuclear charge has increased," you can earn the second point.
8. *Read the question carefully.* The free-response questions tend to be multipart questions. If you do not fully understand one part of the question, you should go on to the next part. The parts tend to be stand-alone. If you make a mistake in one part, you will not be penalized for the same mistake a second time.
9. *Budget your time carefully.* Spend 1–2 minutes reading the question and mentally outlining your response. You should then spend the next 3–5 minutes outlining your response. Finally, you should spend about 15 minutes answering the question. A common mistake is to overdo the answer. The question is worth a limited number of points. If your answer is twice as long, you will not get more points. You will lose time you could spend on the remainder of the test. Make sure your answers go directly to the point. There should be no deviations or extraneous material in your answer.
10. *Make sure you spend some time on each section.* Grading of the free-response questions normally involves a maximum of one to three points for each part. You will receive only a set maximum number of points. Make sure you make an attempt to answer each part. You cannot compensate for leaving one part blank by doubling the length of the answer to another part.

    You should make sure the grader is able to find the answer to each part. This will help to ensure that you get all the points you deserve. There will be at least a full page for your answer. There will also be questions with multiple pages available for the answer. You are not expected to use all of these pages. In some cases, the extra pages are there simply because of the physical length of the test. The booklet has a certain number of pages.
11. *Outlines are very useful.* They not only organize your answer, but they also can point to parts of the question you may need to reread. Your outline does not need to be detailed: just a few keywords to organize your thoughts. As you make the outline, refer back to the question; this will take care of any loose ends. You do not want to miss any important points. You can use your outline to write a well-organized answer to the question. The grader is not marking on how well you wrote your answer, but a well-written response makes it easier for the grader to understand your answer and to give you all the points you deserve.

12. *Grading depends on what you get right in your answer.* If you say something that is wrong, it is not counted against you. Always try to say something. This will give you a chance for some partial credit. Do not try too hard and negate something you have already said. The grader needs to know what you mean; if you say something and negate it later, there will be doubt.

13. *Do not try to outthink the test.* There will always be an answer. For example, in the reaction question, "no reaction" will not be a choice. If you find yourself doing this, pick the simplest answer. If you know the answer to a "difficult" question—give yourself credit for preparing well; do not think that it is too easy, and that you missed something. There are easy questions and difficult questions on the exam.

    Questions concerning experiments will be incorporated into both the multiple-choice and free-response questions. This means that you will need to have a better understanding of the experiments in order to discuss not only the experiment itself, but also the underlying chemical concepts.

14. *Be familiar with all the suggested experiments.* It may be that you did not perform a certain experiment, so carefully review any that are unfamiliar in Chapter 19. Discuss these experiments with your teacher.

15. *Be familiar with the equipment.* Not only be familiar with the name of the equipment used in the experiment, but how it is used properly. For example, the correct use of a buret involves reading of the liquid meniscus.

16. *Be familiar with the basic measurements required for the experiments.* For example, in a calorimetry experiment you do not *measure* the change in temperature, you *calculate* it. You measure the initial and final temperatures.

17. *Be familiar with the basic calculations involved in each experiment.* Review the appropriate equations given on the AP exam. Know which ones may be useful in each experiment. Also, become familiar with simple calculations that might be used in each experiment. These include calculations of moles from grams, temperature conversions, and so on.

18. *Other things to keep in mind:*

    - Begin every free-response question with a reading period. Use this time well to jot down some quick notes to yourself, so that when you actually begin to respond, you will have a nice start.
    - The questions are written in logical order. If you find yourself explaining part **C** before responding to part **B**, back up and work through the logical progression of topics.
    - Abbreviations are your friends. You can save time by using commonly accepted abbreviations for chemical variables and graphical curves. With practice, you will get more adept at their use. There are a number of abbreviations present in the additional information supplied with the test. If you use any other abbreviations, make sure you define them.

STEP 4

Review the Knowledge You Need to Score High

Basics

IN THIS CHAPTER

Summary: This chapter on basic chemical principles should serve as a review if you have had a pre-AP chemistry course in school. We assume (and we all know about assumptions) that you know about such things as the scientific method, elements, compounds, and mixtures. We may mention elementary chemistry topics like this, but we will not spend a lot of time discussing them. When you are using this book, have your textbook handy. If we mention a topic and it doesn't sound familiar, go to your textbook and review it in depth. We will be covering topics that are on the AP exam. There is a lot of good information in your text that is not covered on the AP exam, so if you want more, read your text.

Keywords and Equations

This section of each chapter will contain the mathematical equations and constants that are supplied to you on the AP exam. We have tried to use, as much as possible, the exact format that is used on the test.

T = temperature	n = number of moles	m = mass	P = pressure
V = volume	D = density	v = velocity	M = molar mass
KE = kinetic energy	t = time (seconds)		

Boltzmann's constant, $k = 1.38 \times 10^{-23}$ J K^{-1}
electron charge, $e = -1.602 \times 10^{-19}$ coulomb
1 electron volt per atom = 96.5 kJ mol^{-1}

$D = \frac{m}{V}$ $\quad$ K = °C + 273

Avogadro's number = 6.022×10^{23} mol^{-1}

Units and Measurements

Almost all calculations in chemistry involve both a number and a unit. One without the other is useless. Every time you complete a calculation, be sure that your units have canceled and that the desired unit is written with the number.
Always show your units!

Units

The system of units used in chemistry is the SI system (Système International), which is related to the metric system. There are base units for length, mass, etc. and decimal prefixes that modify the base unit. Since most of us do not tend to think in these units, it is important to be able to convert back and forth from the English system to the SI system. These three conversions are useful ones, although knowing the others might allow you to simplify your calculations:

mass: 1 pound = 0.4536 kg (453.6 g)
volume: 1 quart = 0.9464 dm^3 (0.9464 L)
length: 1 inch = 2.54 cm (exact)

As shown above, the SI unit for volume is the cubic meter (m^3), but most chemists use the liter (L, which is equal to 1 cubic decimeter (dm^3)) or milliliter (mL). The appendixes list the SI base units and prefixes, as well as some English–SI equivalents.

We in the United States are used to thinking of temperature in Fahrenheit, but most of the rest of the world measures temperature in Celsius. On the Celsius scale water freezes at 0°C and boils at 100°C. Here are the equations needed to convert from Fahrenheit to Celsius and vice versa:

$$°C = \frac{5}{9}[°F - 32]$$

$$°F = \frac{9}{5}(°C) + 32$$

Many times, especially in working with gases, chemists use the Kelvin scale. Water freezes at 273.15 K and boils at 373.15 K. To convert from Celsius to kelvin:

$$K = °C + 273.15$$

Absolute zero is 0 K and is the point at which all molecular motion ceases.

The density of a substance is commonly calculated in chemistry. The **density (*D*)** of an object is calculated by dividing the mass of the object by its volume. (Some authors will use a lowercase d to represent the density term; be prepared for either.) Since density is independent of the quantity of matter (a big piece of gold and a little piece have the same density), it can be used for identification purposes. The most common units for density in chemistry are g/cm^3 or g/mL.

Measurements

We deal with two types of numbers in chemistry—exact and measured. Exact values are just that—exact, by definition. There is no uncertainty associated with them. There are exactly 12 items in a dozen and 144 in a gross. Measured values, like the ones you deal with in the lab, have uncertainty associated with them because of the limitations of our measuring instruments. When those measured values are used in calculations, the answer must reflect that combined uncertainty by the number of significant figures that are reported in the final answer. The more significant figures reported, the greater the certainty in the answer.

The measurements used in calculations may contain varying numbers of significant figures, so carry as many as possible until the end and then round off the final answer. The least precise measurement will determine the significant figures reported in the final answer. Determine the number of significant figures in each *measured* value (not the exact ones) and then, depending on the mathematical operations involved, round off the final answer to the correct number of significant figures. Here are the rules for determining the number of significant figures in a measured value:

1. All non-zero digits (1, 2, 3, 4, etc.) are significant.
2. Zeroes between non-zero digits are significant.
3. Zeroes to the left of the first non-zero digit are not significant.
4. Zeroes to the right of the last non-zero digit are significant if there is a decimal point present, but not significant if there is no decimal point.

Rule 4 is a convention that many of us use, but some teachers or books may use alternative methods.

By these rules, 230,500. would contain 6 significant figures, but 230,500 would contain only 4.

Another way to determine the number of significant figures in a number is to express it in scientific (exponential) notation. The number of digits shown is the number of significant figures. For example, 2.305×10^{-5} would contain 4 significant figures. You may need to review exponential notation.

In determining the number of significant figures to be expressed in the final answer, the following rules apply:

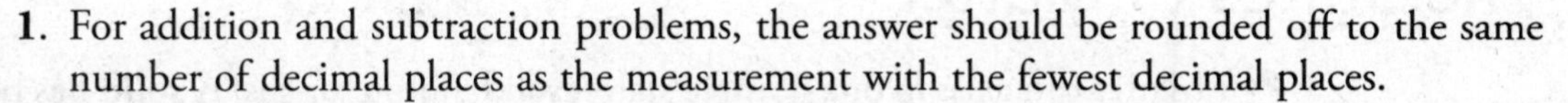

1. For addition and subtraction problems, the answer should be rounded off to the same number of decimal places as the measurement with the fewest decimal places.
2. For multiplication and division problems, round off the answer to the same number of significant figures in the measurement with the fewest significant figures.

Remember: Carry as many numbers as possible throughout the calculation and only round off the final answer.

The use of an improper number of significant figures may lower your score on the AP exam.

Dimensional Analysis—the Factor Label Method

Dimensional analysis, sometimes called the factor label (unit conversion) method, is a method for setting up mathematical problems. Mathematical operations are conducted with the units associated with the numbers, and these units are canceled until only the unit of the desired answer is left. This results in a setup for the problem. Then the mathematical operations can efficiently be conducted and the final answer calculated and rounded off to the correct number of significant figures. For example, to determine the number of centimeters in 2.3 miles:

First, write down the initial data as a fraction:

$$\frac{2.3\,\text{mi}}{1}$$

Convert from miles to feet:

$$\frac{2.3\,\text{mi}}{1} \times \frac{5{,}280\,\text{ft}}{1\,\text{mi}}$$

Convert from feet to inches:

$$\frac{2.3\,\text{mi}}{1} \times \frac{5{,}280\,\text{ft}}{1\,\text{mi}} \times \frac{12\,\text{in}}{1\,\text{ft}}$$

Finally, convert from inches to centimeters:

$$\frac{2.3\text{ mi}}{1} \times \frac{5{,}280\text{ ft}}{1\text{ mi}} \times \frac{12\text{ in}}{1\text{ ft}} \times \frac{2.54\text{ cm}}{1\text{ in}}$$

The answer will be rounded off to 2 significant figures based upon the 2.3 miles, since all the other numbers are exact:

$$\frac{2.3\text{ mi}}{1} \times \frac{5{,}280\text{ ft}}{1\text{ mi}} \times \frac{12\text{ in}}{1\text{ ft}} \times \frac{2.54\text{ cm}}{1\text{ in}} = 3.7 \times 10^5\,\text{cm}$$

Sometimes on the AP exam, only setups will be given as possible answers. Write the correct setup to the problem and then see which one of the answers represents your answer.

Remember: The units must cancel!

Also: Make sure that the answer is legible and reasonable!

The States of Matter

Matter can exist in one of three states: solid, liquid, or gas. A **solid** has both a definite shape and a definite volume. At the molecular level, the particles that make up a solid are close together and many times are locked into a very regular framework called a crystal lattice. Molecular motion exists, but it is slight.

A **liquid** has a definite volume but no definite shape. It conforms to the container in which it is placed. The particles are moving much more than in the solid. There are usually clumps of particles moving relatively freely among other clumps.

A **gas** has neither definite shape nor volume. It expands to fill the container in which it is placed. The particles move rapidly with respect to each other and act basically independently of each other.

We will indicate the state of matter that a particular substance is in by a parenthetical s, l, or g. Thus, $H_2O(s)$ would represent solid water (ice), while $H_2O(g)$ would represent gaseous water (steam). For a more detailed discussion of solids, liquids, and gases see Chapters 8 and 12.

The Structure of the Atom

Historical Development

The first modern atomic theory was developed by John Dalton and first presented in 1808. Dalton used the term **atom** (first used by Democritus) to describe the tiny, indivisible particles of an element. Dalton also thought that atoms of an element are the same and atoms of different elements are different. In 1897, J. J. Thompson discovered the existence of the first subatomic particle, the **electron**, by using magnetic and electric fields. In 1909, Robert Millikan measured the charge on the electron in his oil drop experiment (electron charge = -1.6022×10^{-19} coulombs), and from that he calculated the mass of the electron.

Thompson developed an atomic model, the raisin pudding model, which described the atom as being a diffuse positively charged sphere with electrons scattered throughout.

Ernest Rutherford, in 1910, was investigating atomic structure by shooting positively charged alpha particles at a thin gold foil. Most of the particles passed through with no deflection, a few were slightly deflected, and every once in a while an alpha particle was deflected back towards the alpha source. Rutherford concluded from this scattering experiment that the atom was mostly empty space where the electrons were, and that there was a dense core of positive charge at the center of the atom that contained most of the atom's mass. He called that dense core the **nucleus**.

Subatomic Particles

Our modern theory of the atom describes it as an electrically neutral sphere with a tiny nucleus at the center, which holds the positively charged protons and the neutral neutrons. The negatively charged electrons move around the nucleus in complex paths, all of which comprise the **electron cloud**. Table 5.1 summarizes the properties of the three fundamental subatomic particles:

Table 5.1 The Three Fundamental Subatomic Particles

NAME	SYMBOL	CHARGE	MASS (AMU)	MASS (G)	LOCATION
proton	p^+	1+	1.007	1.673×10^{-24}	nucleus
neutron	n^0	0	1.009	1.675×10^{-24}	nucleus
electron	e^-	1–	5.486×10^{-4}	9.109×10^{-28}	outside nucleus

Many teachers and books omit the charges on the symbols for the proton and neutron.

The **amu (atomic mass unit)** is commonly used for the mass of subatomic particles and atoms. An amu is 1/12 the mass of a carbon-12 atom, which contains 6 protons and 6 neutrons (C-12).

Since the atom itself is neutral, the number of electrons must equal the number of protons. However, the number of neutrons in an atom may vary. Atoms of the same element (same number of protons) that have differing numbers of neutrons are called **isotopes**. A specific isotope of an element can be represented by the following symbolization:

$$^{A}_{Z}X$$

X represents the element symbol taken from the periodic table. *Z* is the **atomic number** of the element, the number of protons in the nucleus. *A* is the **mass number**, the sum of the protons and neutrons. By subtracting the atomic number (p) from the mass number (p + n), the number of neutrons may be determined. For example, $^{238}_{92}U$ (U-238) contains 92 protons, 92 electrons, and (238 – 92) 146 neutrons.

Electron Shells, Subshells, and Orbitals

According to the latest atomic model, the electrons in an atom are located in various energy levels or shells that are located at different distances from the nucleus. The lower the number of the shell, the closer to the nucleus the electrons are found. Within the shells, the electrons are grouped in **subshells** of slightly different energies. The number associated with the shell is equal to the number of subshells found at that energy level. For example, energy

Table 5.2 Summary of Atomic Shell, Subshells, and Orbitals for Shells 1–4

SHELL (ENERGY LEVEL)	SUBSHELL	NUMBER OF ORBITALS	ELECTRON CAPACITY
1	s	1	2 total
2	s	1	2
	p	3	6
			8 total
3	s	1	2
	p	3	6
	d	5	10
			18 total
4	s	1	2
	p	3	6
	d	5	10
	f	7	14
			32 total

level 2 (shell 2) has two subshells. The subshells are denoted by the symbols s, p, d, f, etc. and correspond to differently shaped volumes of space in which the probability of finding the electrons is high. The electrons in a particular subshell may be distributed among volumes of space of equal energies called **orbitals.** There is one orbital for an s subshell, three for a p, five for a d, seven for an f, etc. Only two electrons may occupy an orbital. Table 5.2 summarizes the shells, subshells, and orbitals in an atom. Chapter 10 on Spectroscopy, Light, and Electrons has a discussion of the origin of this system.

Energy-Level Diagrams

The information above can be shown in graph form as an energy-level diagram, as shown in Figure 5.1:

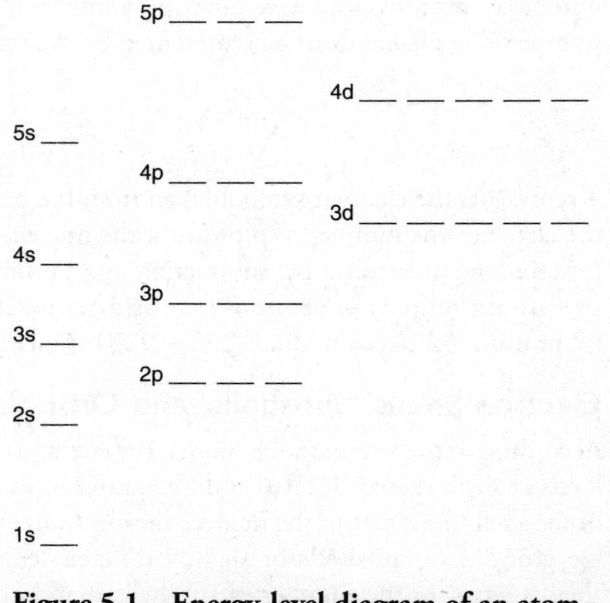

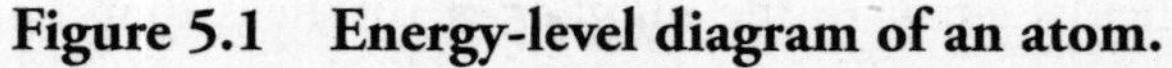

Figure 5.1 Energy-level diagram of an atom.

Be sure to fill the lowest energy levels first (**Aufbau principle**) when using the diagram in Figure 5.1. In filling orbitals having equal energy, electrons are added to the orbitals to half fill them all before any pairing occurs (**Hund's rule**). Sometimes it is difficult to remember the relative energy position of the orbitals. Notice that the 4s fills before the 3d. Figure 5.2 may help you remember the pattern in filling. Study the pattern and be able to reproduce it during the exam.

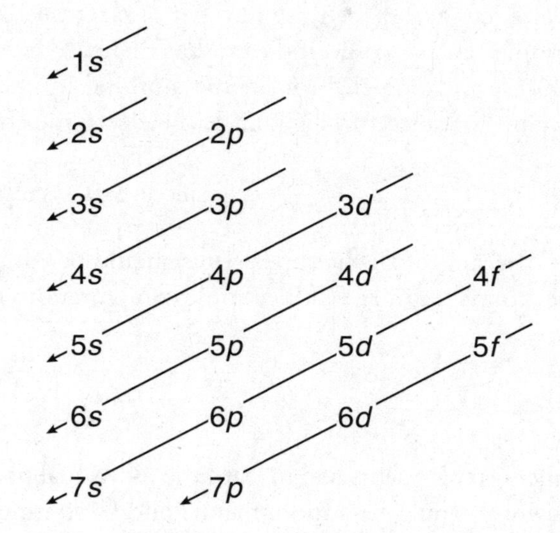

Figure 5.2 Orbital filling pattern.

Following these rules, the energy-level diagram for silicon (Z = 14) can be written as shown in Figure 5.3.

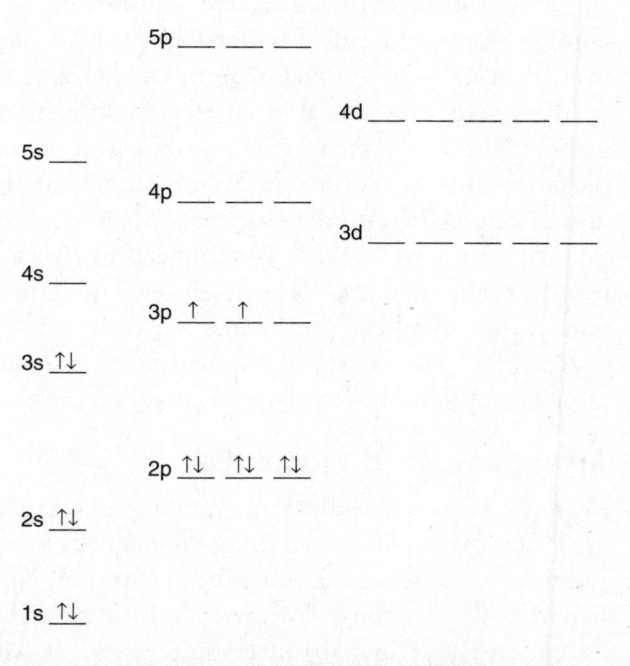

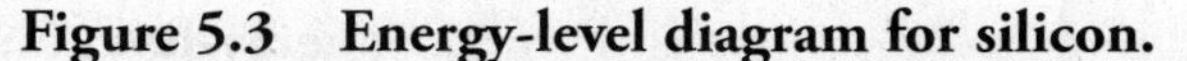

Figure 5.3 Energy-level diagram for silicon.

Although this filling pattern conveys a lot of information, it is bulky. A shorthand method for giving the same information has been developed—the electronic configuration.

Electronic Configurations

The **electronic configuration** is a condensed way of representing the pattern of electrons in an atom. Using the Aufbau build-up pattern that was used in writing the energy-level diagram, consecutively write the number of the shell (energy level), the type of orbital (s, p, d, etc.), and then the number of electrons in that orbital shown as a superscript. For example, $1s^2 2s^1$ would indicate that there are two electrons in the s-orbital in energy level (shell) 1, and one electron in the s-orbital in energy level 2. Looking at the energy-level diagram for silicon in Figure 5.3, the electronic configuration would be written as:

$$\text{silicon}: 1s^2 2s^2 2p^6 3s^2 3p^2$$

The sum of all the superscripts should be equal to the number of electrons in the atom (the atomic number, Z). Electronic configurations can also be written for cations and anions.

Periodic Table

If chemistry students had to learn the individual properties of the 100+ elements that are now known, it would be a monumental and frustrating task. Early scientists had to do just that. Then several scientists began to notice trends in the properties of the elements and began grouping them in various ways. In 1871, a Russian chemist, Dmitri Mendeleev, introduced the first modern periodic table. He arranged the elements in terms of increasing atomic mass. He then arranged columns so that elements that had similar properties were in the same column. Mendeleev was able to predict the existence and properties of elements that were then unknown. Later, when they were discovered, Mendeleev's predictions were remarkably accurate. Later the periodic table was rearranged to sequence the elements by increasing atomic number, not mass. The result is the modern periodic table shown in Figure 5.4.

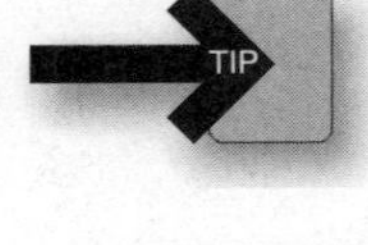

This is not the periodic table supplied on the AP exam. The one in this book has family and period labels. Become familiar with these labels so that you can effectively use the unlabeled one. You may wish to add labels to the one supplied with the AP exam.

Each square on this table represents a different element and contains three bits of information. The first is the element symbol. You should become familiar with the symbols of the commonly used elements. Second, the square lists the atomic number of the element, usually centered above the element. This integer represents the number of protons in the element's nucleus. The atomic number will always be a whole number. Third, the square lists the element's mass, normally centered underneath the element symbol. This number is not a whole number, because it is the weighted average (taking into consideration abundance) of all the masses of the naturally occurring isotopes of that element. The mass number can never be less than the atomic number.

Arrangement of Elements

There are a number of different groupings of elements on the periodic table that may be utilized. One system involves putting the elements into three main groups—metals, nonmetals, and metalloids (semimetals). Look at Figure 5.4. Notice the heavy, stair-stepped line starting at boron (B) and going downward and to the right. The elements to the left of that line (except for H, Ge, and Sb) are classified as metals. **Metals** are normally solids (mercury being an exception), shiny, and good conductors of heat and electricity. They can be hammered

The periodic table

Key: Element Name / **Atomic number** / **Symbol** / Atomic weight (mean relative mass)

1	2		3	4	5	6	7	8	9	10	11	12	13	14	15	16	17	18
Hydrogen 1 **H** 1.008																		Helium 2 **He** 4.002602(2)
Lithium 3 **Li** 6.94	Beryllium 4 **Be** 9.012182(3)												Boron 5 **B** 10.81	Carbon 6 **C** 12.011	Nitrogen 7 **N** 14.007	Oxygen 8 **O** 15.999	Fluorine 9 **F** 18.9984032(5)	Neon 10 **Ne** 20.1797(6)
Sodium 11 **Na** 22.98976928(2)	Magnesium 12 **Mg** 24.3050(6)												Aluminium 13 **Al** 26.9815386(2)	Silicon 14 **Si** 28.085	Phosphorus 15 **P** 30.973762(2)	Sulfur 16 **S** 32.06	Chlorine 17 **Cl** 35.45	Argon 18 **Ar** 39.948(1)
Potassium 19 **K** 39.0983(1)	Calcium 20 **Ca** 40.078(4)		Scandium 21 **Sc** 44.955912(6)	Titanium 22 **Ti** 47.867(1)	Vanadium 23 **V** 50.9415(1)	Chromium 24 **Cr** 51.9961(6)	Manganese 25 **Mn** 54.938045(5)	Iron 26 **Fe** 55.845(2)	Cobalt 27 **Co** 58.933195(5)	Nickel 28 **Ni** 58.6934(4)	Copper 29 **Cu** 63.546(3)	Zinc 30 **Zn** 65.38(2)	Gallium 31 **Ga** 69.723(1)	Germanium 32 **Ge** 72.63(1)	Arsenic 33 **As** 74.92160(2)	Selenium 34 **Se** 78.96(3)	Bromine 35 **Br** 79.904(1)	Krypton 36 **Kr** 83.798(2)
Rubidium 37 **Rb** 85.4678(3)	Strontium 38 **Sr** 87.62(1)		Yttrium 39 **Y** 88.90585(2)	Zirconium 40 **Zr** 91.224(2)	Niobium 41 **Nb** 92.90638(2)	Molybdenum 42 **Mo** 95.96(2)	Technetium 43 **Tc** [97.91]	Ruthenium 44 **Ru** 101.07(2)	Rhodium 45 **Rh** 102.90550(2)	Palladium 46 **Pd** 106.42(1)	Silver 47 **Ag** 107.8682(2)	Cadmium 48 **Cd** 112.411(8)	Indium 49 **In** 114.818(3)	Tin 50 **Sn** 118.710(7)	Antimony 51 **Sb** 121.760(1)	Tellurium 52 **Te** 127.60(3)	Iodine 53 **I** 126.90447(3)	Xenon 54 **Xe** 131.293(6)
Caesium 55 **Cs** 132.9054519(2)	Barium 56 **Ba** 137.327(7)	57–70 *	Lutetium 71 **Lu** 174.9668(1)	Hafnium 72 **Hf** 178.49(2)	Tantalum 73 **Ta** 180.94788(2)	Tungsten 74 **W** 183.84(1)	Rhenium 75 **Re** 186.207(1)	Osmium 76 **Os** 190.23(3)	Iridium 77 **Ir** 192.217(3)	Platinum 78 **Pt** 195.084(9)	Gold 79 **Au** 196.966569(4)	Mercury 80 **Hg** 200.59(2)	Thallium 81 **Tl** 204.38	Lead 82 **Pb** 207.2(1)	Bismuth 83 **Bi** 208.98040(1)	Polonium 84 **Po** [209]	Astatine 85 **At** [210]	Radon 86 **Rn** [222]
Francium 87 **Fr** [223.02]	Radium 88 **Ra** [226.03]	89–102 **	Lawrencium 103 **Lr** [262.11]	Rutherfordium 104 **Rf** [265.12]	Dubnium 105 **Db** [268.13]	Seaborgium 106 **Sg** [271.13]	Bohrium 107 **Bh** [270]	Hassium 108 **Hs** [277.15]	Meitnerium 109 **Mt** [276.15]	Darmstadtium 110 **Ds** [281.16]	Roentgenium 111 **Rg** [280.16]	Copernicium 112 **Cn** [285.17]	Ununtrium 113 **Uut** [284.18]	Ununquadium 114 **Uuq** [289.19]	Ununpentium 115 **Uup** [288.19]	Ununhexium 116 **Uuh** [293]	Ununseptium 117 **Uus** [294]	Ununoctium 118 **Uuo** [294]

*lanthanoids	Lanthanum 57 **La** 138.90547(7)	Cerium 58 **Ce** 140.116(1)	Praseodymium 59 **Pr** 140.90765(2)	Neodymium 60 **Nd** 144.242(3)	Promethium 61 **Pm** [144.91]	Samarium 62 **Sm** 150.36(2)	Europium 63 **Eu** 151.964(1)	Gadolinium 64 **Gd** 157.25(3)	Terbium 65 **Tb** 158.92535(2)	Dysprosium 66 **Dy** 162.500(1)	Holmium 67 **Ho** 164.93032(2)	Erbium 68 **Er** 167.259(3)	Thulium 69 **Tm** 168.93421(2)	Ytterbium 70 **Yb** 173.054(5)
**actinoids	Actinium 89 **Ac** [227.03]	Thorium 90 **Th** 232.03806(2)	Protactinium 91 **Pa** 231.03588(2)	Uranium 92 **U** 238.02891(3)	Neptunium 93 **Np** [237.05]	Plutonium 94 **Pu** [244.06]	Americium 95 **Am** [243.06]	Curium 96 **Cm** [247.07]	Berkelium 97 **Bk** [247.07]	Californium 98 **Cf** [251.08]	Einsteinium 99 **Es** [252.08]	Fermium 100 **Fm** [257.10]	Mendelevium 101 **Md** [258.10]	Nobelium 102 **No** [259.10]

Figure 5.4 The periodic table.

into thin sheets (malleable) and extruded into wires (ductile). Chemically, metals tend to lose electrons in reactions, to form cations.

Elements bordering the stair-stepped line (B, Si, Ge, As, Sb, Te) are classified as metalloids. **Metalloids** have properties of both metals and nonmetals. Their unusual electrical properties make them valuable in the semiconductor and computer industry.

The rest of the elements, to the right of the metalloids, are called nonmetals. **Nonmetals** have properties that are often the opposite of metals. Some are gases, are poor conductors of heat and electricity, are neither malleable nor ductile, and tend to gain electrons in their chemical reactions to form anions.

Another way to group the elements on the periodic table is in terms of periods and groups (families). **Periods** are the horizontal rows, which have consecutive atomic numbers. The periods are numbered from 1 to 7. Elements in the same period do not have similar properties in terms of reactions.

The vertical rows on the periodic table are called **groups** or **families**. They may be labeled in one of two ways. An older and still widely used system is to label each group with a Roman numeral and a letter, A or B. The groups that are labeled with an A are called the main-group elements, while the B groups are called the **transition elements**. Two other horizontal groups, the **inner transition elements**, have been pulled out of the main body of the periodic table. The Roman numeral at the top of the main-group families indicates the number of **valence** (outermost shell) **electrons** in that element. **Valence electrons** are normally considered to be only the s and p electrons in the outermost energy level. The transition elements (B groups) are filling d-orbitals, while the inner transition elements are filling f-orbitals.

Four main-group families are given special names, which you should remember:

- IA group (Group 1) — alkali metals
- IIA group (Group 2) — alkaline earth metals
- VIIA group (Group 17) — halogens
- VIIIA group (Group 18) — noble gases

Another way to label the groups is to consecutively number the groups from left to right, 1–18. This method is newer than the other labeling method, and it has not gained wide use. Most teachers and chemists still prefer and use the older method.

Trends in Periodic Properties

Trends are useful on the multiple-choice portion of the AP exam, but simply stating a trend will **not** be sufficient on the free-response portion of the exam. You must give the reason behind the trend. For example, "higher on the periodic table" is a trend, but not a reason.

The overall attraction an electron experiences is due to the **effective nuclear charge**. This attraction is related to the positive nuclear charge interacting with the negative electrons. Electrons between the nucleus and the electron under consideration interfere with, or shield, that electron from the full nuclear charge. This shielding lessens the nuclear charge. Within a period, the shielding is nearly constant; however, the effective nuclear charge will increase with an increasing number of protons (atomic number). Within the same family or group, as the atomic number increases so does the shielding, resulting in a relatively constant effective nuclear charge.

The size of an atom is generally determined by the number of energy levels occupied by electrons. This means that as we move from top to bottom within a group, the size of the atom increases due to the increased number of shells containing electrons. As we move from left to right within a period (within the same valence shell), the atomic size decreases somewhat owing to the increased effective nuclear charge for the electrons. This increased attraction is related to the increasing number of protons within the nucleus. The size of a cation is smaller than the neutral atom, because in many cases an entire energy shell has been removed, while an anion is larger than the corresponding neutral atom since the nuclear attraction is being distributed over additional electrons. As the number of electrons changes so will the electron–electron repulsion. The greater the electron–electron repulsion, the larger the species becomes, and vice versa.

The **ionization energy (IE)** is the energy needed to completely remove an electron from an atom. It may be expressed in terms of 1 atom or a mole of atoms. Energy is required in this process in order to overcome the attraction of the nucleus for the electrons. There are two factors affecting the magnitude of the ionization energy. One is the size of the atom. The closer the electrons are to the nucleus, the more energy is needed to overcome the effective nuclear charge.

Therefore, ionization energy tends to decrease from top to bottom within a group, since the valence electrons (the first ones to be lost) are farther away from the nucleus.

The other factor is the magnitude of the effective nuclear charge. The greater the effective nuclear charge, the more energy is required to remove the electron. Since the effective nuclear charge increases from left to right within a period, the ionization energies will also increase from left to right. The increased effective nuclear charge results in the atom becoming slightly smaller, which also leads to a greater nuclear attraction for the electrons.

The ionization energy for the removal of a second electron is greater in all cases than the first, because the electron is being pulled away from a positively charged ion and the attraction is greater than from a neutral atom.

The **electron affinity (EA)** is the energy change that results from adding an electron to an atom or ion. The trends in electron affinity are not quite as regular as size or ionization energy. In general, electron affinity increases from left to right within a period (owing to the increased effective nuclear charge), and decreases from top to bottom within a group owing to increased atomic or ionic size. Noble gases are an exception—they have no EA.

Do not forget that the trends mentioned in this section may help you on the multiple-choice portion of the AP exam. However, it is the underlying reasons that you need for the free-response portion.

Oxidation Numbers

Oxidation numbers are bookkeeping numbers that allow chemists to do things like balance redox equations. Don't confuse oxidation numbers with the charge on an ion. Oxidation numbers are assigned to elements in their natural state or in compounds using the following rules:

- The oxidation number of an element in its elemental form (i.e., H_2, Au, Ag, N_2) is zero.
- The oxidation number of a monoatomic ion is equal to the charge on the ion. The oxidation number of Mg^{2+} is +2. Note that the charge is written with number first, then sign; for oxidation numbers it is sign, then number.
- The sum of all the oxidation numbers of all the elements in a neutral molecule is zero. The sum of all the oxidation numbers in a polyatomic ion is equal to the charge on the ion.
- The alkali metal ions have an oxidation number of +1 in all their compounds.
- The alkaline earth metals have an oxidation number of +2 in all their compounds.
- The oxidation number of hydrogen in compounds is +1, except it is −1 when combined with metals or boron in binary compounds.
- The oxidation number of halogens in their compounds is −1 except when combined with another halogen above them on the periodic table, or with oxygen.
- The oxidation number of oxygen is −2 in compounds, except for peroxides, in which it is −1.

Determine the oxidation number of sulfur in sulfuric acid, H_2SO_4. The sum of all the oxidation numbers must equal zero, since this is a neutral compound. The oxidation numbers of hydrogen (+1) and oxygen (−2) are known, so the oxidation number of sulfur can be determined:

$$\begin{array}{c} 2(+1) + ? + 4(-2) = 0 \\ H_2SO_4 \end{array}$$

The oxidation number of sulfur in this compound must be +6.

Nomenclature Overview

This overview covers some of the rules for naming simple inorganic compounds. There are additional rules, and some exceptions to these rules. The first part of this overview discusses the rules for deriving a name from a chemical formula. In many cases, the formula may be determined from the name by reversing this process. The second part examines situations in which additional information is needed to generate a formula from the name of a compound. The transition metals present some additional problems; therefore, there is a section covering transition metal nomenclature and coordination compounds.

Binary Compounds

Binary compounds are compounds that consist of only two elements. Some binary compounds have special names, and these special names supersede any of the rules given below. H_2O is water, NH_3 is ammonia, and CH_4 is methane. All other binary compounds have a name with a suffix *ide*. Binary compounds may be subdivided into metal type, nonmetal type, and acid type.

(a) Metal type These binary compounds begin with metals. The metal is given first in the formula. In general, metals are the elements on the left-hand side of the periodic table, and the nonmetals are on the right-hand side. Hydrogen, a nonmetal, is an exception to this generalization.

First name the metal, then name the nonmetal with the suffix *ide*. Examples:

Formula	Name
Na_2O	sodium oxide
$MgCl_2$	magnesium chloride

The ammonium ion (NH_4^+) is often treated as a metal, and its compounds are named under this rule. Thus, NH_4Cl is named ammonium chloride.

(b) Nonmetal type These binary compounds have formulas that begin with a nonmetal. Prefixes are used to indicate the number of each atom present. No prefixes are used for hydrogen. Naming the compounds can best be explained using the following examples:

Formula	Name
CO	carbon monoxide
SO_3	sulfur trioxide
P_4O_{10}	tetraphosphorus decoxide

Carbon monoxide is one of the very few cases where the prefix *mono* is used. In general, you should not use *mono* in any other compound.

Some of the prefixes used to denote the numbers of atoms in a compound are listed below:

Number of atoms	Prefix
1	mono
2	di
3	tri
4	tetra
5	penta
6	hexa
7	hepta
8	octa
9	nona
10	deca

On many occasions the terminal *a* or *o* is dropped for oxides, so they read as pentoxide, heptoxide, or monoxide.

In normal nomenclature, the nonmetal prefixes are not used if a metal is present. One of the few exceptions to this is MnO_2, sometimes called manganese dioxide.

(c) Acid type These binary compounds have formulas that begin with hydrogen. If the compound is not in solution, the naming is similar to that of the metal type.

If the compound is dissolved in H_2O, indicated by (*aq*), the compound takes on the prefix *hydro* and the suffix *ic*. If the compound is not in solution, the state of matter should be shown as follows:

$$HCl(g), HF(l)$$

If the formula has no designation of phase or water, either name may be used. Examples for naming these compounds are:

Formula	Name
$HCl(g)$	hydrogen chloride
$H_2S(g)$	hydrogen sulfide
$HCl(aq)$	*hydro*chlor*ic* acid
$H_2S(aq)$	*hydro*sulfur*ic* acid
HCl	hydrogen chloride or *hydro*chlor*ic* acid
H_2S	hydrogen sulfide or *hydro*sulfur*ic* acid

HCN (hydrocyanic acid) is named using these rules. However, in this case, it does not matter if the phase or water is indicated.

Ternary Compounds

Ternary compounds are those containing three or more elements. If the first element in the formula is hydrogen, it is usually classified as an acid. If the formula contains oxygen in addition to the hydrogen, the compound is usually classified as an oxyacid. In general, if the first element in the formula is not hydrogen, the compound is classified as a salt.

Ternary acids are usually named with the suffixes *ic* or *ous*. The exceptions are the acids derived from ions with an *ide* suffix (see HCN in the preceding section). These acids undergo many reactions to form salts, compounds of a metal, and the ion of an acid. The ions from the acids H_2SO_4 and HNO_3 are SO_4^{2-}, NO_3^-. If an acid name has the suffix *ic*, the ion of this acid has a name with the suffix *ate*. If an acid name has the suffix *ous*, the ion has a name with the suffix *ite*. Salts have the same suffixes as the suffixes of the ions. The difference between the acid with a suffix *ic* and the acid with the suffix *ous* can many times be determined by visual inspection of the formula. The acid with the suffix *ous* usually has one fewer oxygen atom than the acid with the suffix *ic*. Examples:

Formula	Name of the acid	Formula	Name of the acid
H_2SO_4	sulfuric acid	HNO_3	nitric acid
H_2SO_3	sulfurous acid	HNO_2	nitrous acid

When the ternary compound is not an acid, the first element is usually a metal. In these cases, the name of the compound is simply the name of the metal followed by the name of the ion. The ammonium ion is treated as a metal in these cases.

The following are examples:

Acid formula	Acid name	Ion name	Salt formula	Salt name
H_2SO_4	sulfuric acid	sulfate ion	Na_2SO_4	sodium sulfate
H_2SO_3	sulfurous acid	sulfite ion	Na_2SO_3	sodium sulfite
HNO_3	nitric acid	nitrate ion	KNO_3	potassium nitrate
HNO_2	nitrous acid	nitrite ion	KNO_2	potassium nitrite
H_3PO_4	phosphoric acid	phosphate ion	$(NH_4)_3PO_4$	ammonium phosphate

Writing Formulas

To write the formula from the name of a binary compound containing only nonmetals, simply write the symbols for the separate atoms with the prefixes converted to subscripts.

In all compounds, the total charge must be zero. There are NO exceptions. Thus, to determine the formula in those cases where no prefixes are given, it is necessary to have some idea what the individual charges are. The species with the positive charge is listed and named first; this is followed by the species with the negative charge. Subscripts may be needed to make sure the sum of the charges (valances) will equal zero. Examples:

1. Magnesium oxide

$$Mg^{2+}O^{2-} = +2 - 2 = 0$$

This gives MgO.

2. Sodium oxide

$$Na^{1+}O^{2-} = +1 - 2 = -1 \text{ thus a subscript is needed}$$

$$Na_2^{2(1+)}O^{2-} = 2(+1) - 2 = 0$$

This gives Na_2O.

3. Aluminum oxide

$$Al^{3+}O^{2-} = +3 - 2 = +1 \text{ thus a subscript is needed}$$

$$Al_2^{2(3+)}O_3^{3(-2)} = 2(+3) + 3(-2) = 0$$

This gives Al_2O_3.

If a polyatomic ion must be increased to achieve zero charge, parentheses should be used. An example of this is shown as:

$$NH_4^+SO_4^{2-} = +1 - 2 = -1$$

$$(NH_4)_2^{2(1+)}SO_4^{2-} = 2(+1) - 2 = 0$$

This gives $(NH_4)_2SO_4$.

One way of predicting the values of the subscripts is to crisscross the valences. This is not a rule of nomenclature, but for practice purposes in this exercise it will be referred to as the crisscross rule. It works most of the time and therefore is worth considering. Example:

$Al^{3+}O^{2-}$ crisscross the 2 from the oxygen charge to the aluminum and the 3 from the aluminum charge to the oxygen

$Al_2^{3+}O_3^{2-}$

If the crisscross rule is applied, you should reduce the formula if possible. For example:

$Mn^{4+}O^{2}$ crisscrosses to Mn_2O_4, which reduces to MnO_2

If a formula is given, the crisscross rule can be reversed to give the valences:

$$Al_2O_3$$

$$Al_2^{3+}O_3^{2-}$$

As a first approximation, the valences of the representative elements can be predicted from their position on the periodic table. Hydrogen and the metals have positive charges beginning with +1 on the left and increasing by one as you proceed to the right on the periodic table (skipping the transition metals). Nonmetals begin with 0 in the rightmost column of the periodic table and decrease by 1 as you move to the left on the periodic table. Metalloids may be treated as metals or nonmetals. Examples are:

$$Na^{+}\ Al^{3+}\ Pb^{4+}\ N^{3-}\ Se^{2-}\ I^{-}$$
$$Na^{+}\ Mg^{2+}\ Al^{3+}\ Si^{4+}\ P^{3-}\ S^{2-}\ Cl^{-}\ Au^{0}$$

Transition Metals

Many transition metals and the group of six elements centered around lead on the periodic table commonly have more than one valence. The valence of these metals in a compound must be known before the compound can be named. Modern nomenclature rules indicate the valence of one of these metals with a Roman numeral suffix (Stock notation). Older nomenclature rules used different suffixes to indicate the charge. Examples:

1. $FeCl_3$
   $Fe^{3+}Cl_3{}^{1-}$ (crisscross rule)

   The compound is named iron(III) chloride or ferric chloride.

2. $FeCl_2$

   If chloride is −1, two chloride ions are −2. Fe has a valence of +2, to give a total charge of zero. The name is iron(II) chloride or ferrous chloride.

3. MnO_2

   Mn^{4+} (found previously)
   The name would be manganese(IV) oxide, although it is often named manganese dioxide.

The Roman numeral suffix is part of the name of the metal. Thus iron(III) is one word. Stock notation should be used for all metals that have a variable valence. This includes almost all the transition elements and the elements immediately around lead on the periodic table. Stock notation is often omitted for Zn, Cd, and Ag, as they do not have variable valences.

The valences of some common metals and acids are listed in the appendixes.

Coordination Compounds

Coordination compounds contain a complex. In general, a complex may be recognized because it is enclosed in square brackets [ ]. The square brackets are omitted when the actual structure of the complex is uncertain.

A complex is composed of a central atom, normally a metal, surrounded by atoms or groups of atoms called ligands. One way of forming a complex is illustrated below:

$$Ni^{2+} + 6\ H_2O \rightarrow [Ni(H_2O)_6]^{2+}$$

In this reaction the metal behaves as a Lewis acid and accepts a pair of electrons from the Lewis base (ligand). In this case the ligand is water, with the oxygen atom donating one of its lone pairs to the nickel. The oxygen atom is called the donor atom. In this complex, there are six donor atoms.

A complex may be ionic or neutral. An ionic complex is called a complex ion. A neutral complex is a type of coordination compound. The only difference in naming coordination compounds or complex ions is that anionic complex ions have an *ate* suffix.

A coordination compound may contain more than one complex ion or material that is not part of the complex, but it must have an overall neutral charge. Examples of coordination compounds are: $[Pt(NH_3)_2Cl_2]$, $K_2[Mn(C_2O_4)_3]$, and $[Ni(H_2O)_6]SO_4$.

When writing formulas the metal (central atom) is *always* listed first within the brackets. However, when writing names the metal name is *always* given last. Any material not listed within the brackets is named separately.

Examples:

$[Ru(NH_3)_5(N_2)]Cl_2$	coordination compound
$[Ru(NH_3)_5(N_2)]^{2+}$	complex ion (cationic)
$[PtNH_3Cl_2(C_5H_5N)]$	coordination compound
$[IF_6]^-$	complex ion (anionic) (the name must end in *-ate*)
$K[IF_6]$	coordination compound (same *-ate* ending)

If everything in the formula is enclosed within one set of brackets, the entire name will be one word. If there is material outside the brackets, this outside material is named separately.

Just as with simpler compounds, cations are always named before anions. Thus, a cationic complex would be the first word in the name, and an anionic complex would be the last word in a name (with an *ate* ending).

Examples:

$[Ni(H_2O)_4Cl_2]$	tetraaquadichloronickel(II)
$[Co(NH_3)_6]Cl_3$	hexaamminecobalt(III) chloride
$K_2[PtCl_4]$	potassium tetrachloroplatinate(II)

When naming a complex, or when writing the formula for a complex, the ligands are listed alphabetically. Again, do not forget that metals are first in the formula and last in the name.

The names of anionic ligands always end in an *o*. Neutral ligands are basically unchanged. Two common exceptions in the case of neutral ligands are NH_3 = ammine (note the double *m*), and H_2O = aqua. Other common ligands and their names are listed in the appendixes.

Multiple identical ligands have prefixes added to designate the number of such ligands:

2	di-	5	penta-	8	octa-
3	tri-	6	hexa-	9	nona-
4	tetra-	7	hepta-	10	deca-

Examples:

$[Co(NH_3)_6]Cl_3$	hexaamminecobalt(III) chloride
$[Cr(NO)_4]$	tetranitrosylchromium(0)

If the ligand name contains a prefix or begins with a vowel (except ammine and aqua), alternate prefixes should be used:

2	bis-	5	pentakis-	8	octakis-
3	tris-	6	hexakis-	9	nonakis-
4	tetrakis-	7	heptakis-	10	decakis-

When using the alternate prefixes, it is common practice to enclose the name of the ligand within parentheses. Either type of prefix is added after the ligands have been alphabetized.

Examples:

$[Cr(en)_3]Cl_3$	Tris(ethylenediamine)chromium(III) chloride
$K_2[Ge(C_2O_4)_3]$	Potassium tris(oxalato)germanate

Anionic complexes always have names ending in *ate*. This will require a change in the name of the metal. Thus, aluminum would become aluminate, and zinc would become zincate. The only exceptions to this are some of the metals whose symbols are based on Latin or Greek names. These exceptions are:

Metal (Greek or Latin name)	Symbol	Anionic name
copper (cuprum)	Cu	cuprate
silver (argentum)	Ag	argentate
gold (aurum)	Au	aurate
iron (ferrum)	Fe	ferrate
tin (stannum)	Sn	stannate
lead (plumbum)	Pb	plumbate

Examples:

$K[Au(CN)_4]$	potassium tetracyanoaurate(III)
$(NH_4)_2[PbCl_6]$	ammonium hexachloroplumbate(IV)

If the metal ion may exist in more than one oxidation state, this oxidation state should be listed, in Roman numerals, *immediately* after the name of the metal ion. The Roman numeral is enclosed in parentheses and is considered part of the same word, and not a separate grouping. If the metal occurs in only one oxidation state, no such indicator is used. This notation is the Stock system discussed earlier.

Experiments

Experiments involving the basic material covered in this chapter have been placed in the in-depth chapters throughout the remainder of this book.

Common Mistakes to Avoid

Between the two of us, we have almost 60 years of teaching experience. We've seen a lot of student mistakes. We will try to to steer you clear of the most common ones.

1. Always show your units in mathematical problems.
2. In the conversion from °F to °C, be sure to subtract 32 from the Fahrenheit temperature first, then multiply by 5/9.
3. In the conversion from °C to °F, be sure to multiply the Celsius temperature by 9/5, then add 32.
4. There is no degree sign used for kelvin.
5. Only consider measured values for significant figures.
6. When considering whether or not zeroes to the right of the last non-zero digit are significant, pay attention to whether or not there is a decimal point.
7. Round off only your final answer, not intermediate calculations.
8. In working problems, be sure that your units cancel.
9. If you are solving for cm, for example, be sure you end up with cm and not 1/cm.
10. Make sure your answer is a reasonable one.
11. Don't confuse the mass number (A) with the atomic number (Z).
12. When determining valence electrons, only the s and p electrons are considered.
13. Don't put more than 2 electrons in any individual orbital.

14. Always fill lowest energy levels first.
15. Half fill orbitals of equal energy before pairing up the electrons.
16. In writing the electronic configuration of an atom, make sure you use the correct filling order.
17. Don't confuse the periods with the groups on the periodic table.
18. Don't confuse ionization energy with electron affinity.
19. Don't confuse oxidation numbers with ionic charge.
20. In naming compounds, don't confuse metal and nonmetal type binary compounds. Prefixes are used only with nonmetal types.
21. Be careful when using the crisscross rule to reduce the subscripts to their lowest whole-number ratio.
22. Be sure to report the proper number of significant figures.
23. Simply knowing a periodic trend will allow you to pick the correct multiple-choice answer, but be prepared to explain the trend in free-response questions.

› Review Questions

Here are questions you can use to review the content of this chapter and practice for the AP Chemistry exam. First are 25 multiple-choice questions similar to what you will encounter in Section I of the AP Chemistry exam. Following those is a four-part free-response question like the ones in Section II of the exam. To make these questions an even more authentic practice for the actual exam, time yourself following the instructions provided.

Multiple-Choice Questions

Answer the following questions in 30 minutes. You may not use a calculator. You may use the periodic table and the equation sheet at the back of this book.

1. In most of its compounds, this element exists as a monatomic cation.

(A) O
(B) Cl
(C) Na
(D) N

2. This element may form a compound with the formula $CaXO_4$.

(A) Se
(B) Cl
(C) P
(D) Na

3. Which of the following elements may occur in the greatest number of different oxidation states?

(A) C
(B) F
(C) O
(D) Ca

4. Choose the group that does NOT contain isotopes of the same element.

		Number of protons	Number of neutrons
(A)	Atom I	18	18
	Atom II	18	19
(B)	Atom I	25	30
	Atom II	25	31
(C)	Atom I	37	42
	Atom II	37	41
(D)	Atom I	82	126
	Atom II	81	126

5. Which of the following groups has the species correctly listed in order of increasing radius?

(A) Mg^{2+}, Ca^{2+}, Ba^{2+}
(B) K^+, Na^+, Li^+
(C) Br^-, Cl^-, F^-
(D) Na, Mg, Al

6. Which of the following elements has the lowest electronegativity?

(A) C
(B) K
(C) Al
(D) I

7. Choose the ion with the largest ionic radius.

(A) F^-
(B) Al^{3+}
(C) K^+
(D) I^-

8. What is the name of the energy change when a gaseous atom, in the ground state, adds an electron?

(A) ionization energy
(B) sublimation energy
(C) atomization energy
(D) electron affinity

9. The following ionization energies are reported for element *X*. (All the values are in kJ/mol.)

First	Second	Third	Fourth	Fifth
500	4,560	6,910	9,540	13,400

Based on the above information, the most likely identity of *X* is:

(A) Mg
(B) Cl
(C) Al
(D) Na

10. In general, as the atomic numbers increase within a period, the atomic radius:

(A) decreases
(B) increases
(C) first decreases and then increases
(D) does not change

11. Which of the following elements is a reactive gas?

(A) chlorine
(B) gold
(C) sodium
(D) radon

12. Which of the following elements is an unreactive metal?

(A) chlorine
(B) gold
(C) sodium
(D) radon

13. Which of the following represents the correct formula for potassium trisoxalatoferrate(III)?

(A) $P_3[Fe(C_2O_4)_3]$
(B) $K_3[Fe(C_2O_4)_3]$
(C) $KFe_3(C_2O_4)_3$
(D) $K_3[Fe_3(C_2O_4)_3]$

14. Which of the following substances will produce a colorless aqueous solution?

(A) $Zn(NO_3)_2$
(B) $CuSO_4$
(C) $K_2Cr_2O_7$
(D) $Co(NO_3)_2$

15. This element is a liquid at room temperature.

(A) Hg
(B) Th
(C) Na
(D) Cl

16. Which of the following elements is present in chlorophyll?

(A) K
(B) Ga
(C) Al
(D) Mg

17. What is the symbol for the element that forms a protective oxide coating?

(A) K
(B) Ga
(C) Al
(D) Mg

18. Which of the following elements is important in the semiconductor industry to improve the conductivity of germanium, Ge?

(A) K
(B) Ga
(C) Al
(D) Mg

19. Which of the following aqueous solutions is blue?

(A) $CuSO_4$
(B) $Cr_2(SO_4)_3$
(C) $NiSO_4$
(D) $ZnSO_4$

20. In order to separate two substances by fractional crystallization, the two substances must differ in which of the following?

(A) solubility
(B) specific gravity
(C) vapor pressure
(D) viscosity

21. In a flame test, copper compounds impart which of the following colors to a flame?

(A) red
(B) orange
(C) blue to green
(D) violet

22. What should you do if you spill sulfuric acid on the countertop?

(A) Neutralize the acid with vinegar.
(B) Sprinkle solid NaOH on the spill.
(C) Neutralize the acid with $NaHCO_3$ solution.
(D) Neutralize the acid with an Epsom salt ($MgSO_4$) solution.

23. Which of the following can be achieved by using a visible-light spectrophotometer?

(A) Run a flame test to determine if Na^+ or K^+ is in a solution.
(B) Find the concentration of a $KMnO_4$ solution.
(C) Detect the presence of isolated double bonds.
(D) Measure the strength of a covalent bond.

24. You have an aqueous solution of NaCl. The simplest method for the separation of NaCl from the solution is:

(A) evaporation of the solution to dryness
(B) centrifuging the solution
(C) filtration of the solution
(D) electrolysis of the solution

25. The determination that atoms have small, dense nuclei is attributed to:

(A) Rutherford
(B) Becquerel
(C) Einstein
(D) Dalton

Answers and Explanations for the Multiple-Choice Questions

1. C—All the other elements are nonmetals. Nonmetals usually form monatomic anions.

2. A—The element cannot be a metal (Na). A nonmetal that can have a +6 oxidation state is necessary. P has a maximum of +5. Cl may be +5 or +7. Se, in column 16, can easily be +6.

3. A—Based on their positions on the periodic table:

C	+4 to – 4
F	– 1 and 0 (element)
O	– 2 to 0
Ca	+2 and 0

4. D—Isotopes MUST have the same number of protons. Different isotopes of an element have different numbers of neutrons.

5. A—All the others are in decreasing order. Ions in the same column and with the same charge increase in size when going down a column the same as atoms. Atoms in the same row increase in size toward the left side. This argument is not sufficient on the free-response portion of the exam.

6. B—In general, the element farthest from F on the periodic table will have the lowest electronegativity. There are exceptions, but you normally do not need to concern yourself with exceptions.

7. D—The very large iodine atom (near the bottom of the periodic table) gains an electron to make it even larger. This reasoning is not sufficient on the free-response portion of the exam.

8. D—The definition of electron affinity is the energy change when a ground-state gaseous atom adds an electron.

9. D—The more electrons removed, the higher the values should be. The large increase between the first and second ionization energies indicates a change in electron shell. The element, *X*, has only 1 valence electron. This is true for Na. For the other elements the numbers of valence electrons are: Mg – 2; Cl – 7; and Al – 3.

10. A—The increase in the number of protons in the nucleus has a greater attraction (greater effective nuclear charge) for the electrons being added in the same energy level. Thus, the electrons are pulled closer to the nucleus and the size slightly decreases. This thought process should be used on the free-response portion of the AP exam; however, simply remembering that radii decrease across a period is sufficient for most multiple-choice questions.

11. A—The only other gas is radon, and it is inert.

12. B—Sodium is a metal on the left side of the periodic table. Metals on the left side of the periodic table are very reactive. Radon is not a metal.

13. B—Ferrate(III) means Fe^{3+}, while trisoxalato means $(C_2O_4)_3^{6-}$; three potassium atoms are needed to balance the charge.

14. A—B is blue; C is orange; and D is pink to red.

15. A—Chlorine is a gas; all the others are solid metals.

16. D—Magnesium is present in chlorophyll.

17. C—Aluminum forms a protective oxide coating.

18. B—Gallium, adjacent to Ge on the periodic table, is one of the elements that will improve the conductivity of germanium.

19. A—B is purple; C is green; and D is colorless.

20. A—Fractional crystallization works because the less soluble material separates first.

21. C—A could be Li or Sr; B is Ca; and D is K.

22. C—Adding a weak base solution, such as $NaHCO_3$, which will not only neutralize the acid but will help to disperse the heat, is the best choice.

23. B—A solution containing a colored substance is necessary.

24. A—Separation of materials in solution is normally not simple; therefore, removal of the solvent through evaporation is the best choice.

25. A—Rutherford, and his students, demonstrated the existence of the nucleus.

Free-Response Question

Both authors have been AP free-response graders for years. Here is a free-response question for practice.

You have 10 minutes to do the following question. You may use a calculator and the tables in the back of the book.

Question 1

Use the periodic table and other information concerning bonding and electronic structure to explain the following observations.

(a) The radii of the iron cations are less than that of an iron atom, and Fe^{3+} is smaller than Fe^{2+}.

(b) When moving across the periodic table from Li to Be to B, the first ionization energy increases from Li to Be, then drops for B. The first ionization energy of B is greater than that of Li.

(c) The electron affinity of F is higher than the electron affinity of O.

(d) The following observations have been made about the lattice energy and ionic radii of the compounds listed below. Compare NaF to CaO, and then compare CaO to BaO. All of the solids adopt the same crystal structure.

Compound	Ionic radius of cation (pm)	Ionic radius of anion (pm)	Lattice energy (kJ/mole)
NaF	116	119	911
CaO	114	126	3,566
BaO	149	126	3,202

Answer and Explanation for the Free-Response Question

Notice that all the answers are very short. Do not try to fill all the space provided on the exam. You score points by saying specific things, not by the bulk of material. The graders look for certain keywords or phrases. The answers should not contain statements that contradict each other; otherwise, there may be a penalty. Contradictions most commonly occur when the student tries to say too much. On the AP exam, the different parts of the free-response questions tend to be more diverse than this one, as this question focuses on this chapter, whereas the AP free-response questions focus on the entire course.

(a) The observed trend of radii is $Fe > Fe^{2+} > Fe^{3+}$. There is an increase in the effective nuclear charge in this series. As electrons are removed, the repulsion between the remaining electrons decreases. The larger the effective nuclear charge, the greater the attraction of the electrons toward the nucleus and the smaller the atom or ion becomes.

Give yourself 1 point for "effective nuclear charge" and 1 point for the effective nuclear charge, and give yourself 1 point for the remainder of the discussion.

(b) When moving across a period on the periodic table, the value of the effective nuclear charge increases with atomic number. This causes a general increase from Li to Be to B. DO NOT use the argument that ionization energies increase to the right on the periodic table, unless you also discuss effective nuclear charge.

The even higher value of Be (greater than B) is due to the increased stability of the electron configuration of Be. Beryllium has a filled s-subshell. Filled subshells have an increased stability, and additional energy is required to pull an electron away.

This effective nuclear charge argument is worth 1 point. Give yourself 0 points if you say that the ionization energy increases to the right on the periodic table. This is an observation; it is not an explanation. Give yourself 1 more point for the filled subshell discussion.

(c) The effective nuclear charge in F is greater than the effective nuclear charge in O. This causes a greater attraction of the electrons. DO NOT use the argument that electron affinity increases to the right on the periodic table, unless you also discuss effective nuclear charge.

You get 1 point for this answer.

(d) Because all these solids adopt the same structure, the structure is irrelevant. The sizes of the anions are similar; thus, anion size arguments are not important. Two factors, other than structure and anion size, are important here. The two compounds with the highest lattice energies contain divalent ions (+2 or −2), while NaF contains univalent ions (+1 or −1). The higher the charge is, the greater the attraction between the ions is. The lattice energy increases as the attraction increases.

You get 1 point for correctly discussing the charges. The difference between the CaO and BaO values is because the larger the ion is, the lower the attraction is (greater separation). The lower attraction leads to a lower lattice energy. This size argument will get you 1 point.

Total your points. The maximum is 7.

❯ Rapid Review

Here is a brief review of the most important points in the chapter. If something sounds unfamiliar, study it in the chapter and your textbook.

- Know the metric measurement system and some metric/English conversions.
- Know how to convert from any one of the Fahrenheit/Celsius/Kelvin temperature scales to the other two.
- The density of a substance is mass per unit volume.
- Know how to determine the number of significant figures in a number, the rules for how many significant figures are to be shown in the final answer, and the round-off rules.
- Know how to set up problems using the factor label method.
- Know the differences between a solid, a liquid, and a gas at both the macroscopic and microscopic levels.
- Know what part Dalton, Thompson, Millikan, and Rutherford had in the development of the atomic model.
- Know the three basic subatomic particles—proton, neutron, and electron—their symbols, mass in amu, and their location.
- Isotopes are atoms of the same element that have differing numbers of neutrons.
- Electrons are located in major energy levels called shells. Shells are divided into subshells, and there are orbitals for each subshell.
- Know the electron capacity of each orbital (always 2).
- Be able to write both the energy-level diagram and the electronic configuration of an atom or ion by applying both the Aufbau build-up principle and Hund's rule.
- Know how the modern periodic table was developed, including the differences between Mendeleev's table and the current table.

- Periods are the horizontal rows on the periodic table; the elements have properties unlike the other members of the period.
- Groups or families are the vertical rows on the periodic table; the elements have similar properties.
- Know the properties of metals, nonmetals, and metalloids and which elements on the periodic table belong to each group.
- Valence electrons are outer-shell electrons.
- The IA family is known as the alkali metals; the IIAs are the alkaline earth metals; the VIIAs are the halogens; and the VIIIAs are the noble gases.
- Know why atoms get larger as we go from top to bottom in a group and slightly smaller as we move from left to right on the periodic table. Remember that on the free-response section, simply quoting a trend is not sufficient in answering the question. This is true for all trends.
- Ionization energy is the energy it takes to remove an electron from a gaseous atom or ion. It decreases from top to bottom and increases from left to right on the periodic table. Much the same trend is noted for electron affinity, the energy change that takes place when an electron is added to a gaseous atom or ion. The trends depend on the size of the atom or ion and its effective nuclear charge.
- Oxidation numbers are bookkeeping numbers. Know the rules for assigning oxidation numbers.
- Be able to name binary metal type and nonmetal type compounds, as well as ternary compounds, oxyacids, simple coordination compounds, etc.

CHAPTER 6

Reactions and Periodicity

IN THIS CHAPTER

Summary: Chemistry is the world of chemical reactions. Chemical reactions power our society, our environment, and our bodies. Some chemical species called **reactants** get converted into different substances called **products**. During this process there are energy changes that take place. It takes energy to break old bonds. Energy is released when new bonds are formed. Does it take more energy to break the bonds than is released in the formation of the new bonds? If so, energy will have to be constantly supplied to convert the reactants into products. This type of reaction is said to be **endothermic**, absorbing energy. If more energy is released than is needed to break the old bonds, then the reaction is said to be **exothermic**, releasing energy. The chemical reactions that provide the energy for our world are exothermic reactions. In Chapter 9 Thermodynamics, you can read in more depth about the energy changes that occur during reactions.

Reactions occur because of collisions. One chemical species collides with another at the right place and transfers enough energy, and a chemical reaction occurs. Such reactions can be very fast or very slow. In Chapter 14 on Kinetics, you can study how reactions occur and the factors that affect the speed of reactions. But in this chapter we will review the balancing of chemical equations, discuss the general types of chemical reactions, and describe why these reactions occur.

Keywords and Equations

There are no keywords or equations listed on the AP exam that are specific to this chapter.

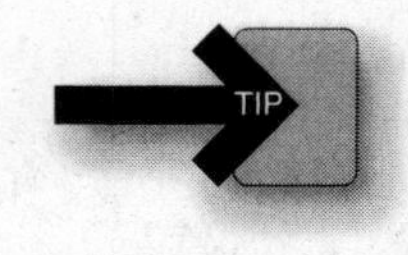

Reactions questions will always appear in the free-response section of the AP exam.

AP Exam Format

Beginning with the 2007 AP exam, the treatment of chemical reactions was changed from previous years. In the past, the free-response questions concerning chemical reactions simply involved the formulas of the reactants and products chosen from a series of reactions. You were not expected to write balanced chemical equations. However, under the current AP Chemistry exam format you no longer are able to choose from a list of reactions. You are expected to write a balanced chemical equation for every reaction given and answer one or more questions about each reaction. If the reaction occurs in aqueous solution, you will have to write the net ionic equation for the process. For the reactions question of the AP exam, you will be expected not only to balance the equation, but to have an understanding of why the reaction occurs. The reactions and concepts described may also appear in other parts of the AP exam, such as the multiple-choice sections. Again, you will need to have an understanding of why a particular reaction occurs. As you study this chapter, pay particular attention to the explanations that accompany the reactions and equations. You will be expected to demonstrate your understanding on the AP exam.

General Aspects of Chemical Reactions and Equations

Balancing Chemical Equations

The authors hope that, because you are preparing to take the AP exam, you have already been exposed to the balancing of chemical equations. We will quickly review this topic and point out some specific aspects of balancing equations as the different types of chemical reactions are discussed.

A balanced chemical equation provides many types of information. It shows which chemical species are the reactants and which species are the products. It may also indicate in which state of matter the reactants and products exist. Special conditions of temperature, catalysts, etc. may be placed over or under the reaction arrow. And, very important, the coefficients (the integers in front of the chemical species) indicate the number of each reactant that is used and the number of each product that is formed. These coefficients may stand for individual atoms/molecules or they may represent large numbers of them called moles (see Chapter 7 Stoichiometry for a discussion of moles). The basic idea behind the balancing of equations is the **Law of Conservation of Matter**, which says that in ordinary chemical reactions matter is neither created nor destroyed. The number of each type of reactant atom has to equal the number of each type of product atom. This requires adjusting the reactant and product coefficients—balancing the equation. When finished, the coefficients should be in the lowest possible whole-number ratio.

Most equations are balanced by inspection. This means basically a trial-and-error, methodical approach to adjusting the coefficients. One procedure that works well is to balance the homonuclear (same nucleus) molecule last. Chemical species that fall into this category include the diatomic elements, which you should know: H_2, O_2, N_2, F_2, Cl_2, Br_2, and I_2. This is especially useful when balancing combustion reactions. If a problem states that oxygen gas was used, then knowing that oxygen exists as the diatomic element is absolutely necessary in balancing the equation correctly.

Periodic Relationships

The periodic table can give us many clues as to the type of reaction that is taking place. One general rule, covered in more detail in Chapter 11 Bonding, is that nonmetals react with other nonmetals to form covalent compounds, and that metals react with nonmetals to form ionic compounds. If the reaction that is producing the ionic compound is occurring in solution, you will be expected to write the net ionic equation for the reaction. Also, because of the wonderful arrangement of the periodic table, the members of a family or group (a vertical grouping) all react essentially in the same fashion. Many times, in reactions involving the loss of electrons (oxidation), as we proceed from top to bottom in a family the reaction rate (speed) increases. Conversely, in reactions involving the gain of electrons (reduction) the reaction rate increases as we move from the bottom of a family to the top. Recall also that the noble gases (VIIIA) undergo very few reactions. Other specific periodic aspects will be discussed in the various reaction sections.

General Properties of Aqueous Solutions

Many of the reactions that you will study occur in aqueous solution. Water is called the universal solvent, because it dissolves so many substances. It readily dissolves ionic compounds as well as polar covalent compounds, because of its polar nature. Ionic compounds that dissolve in water (dissociate) form **electrolyte** solutions, which conduct electrical current owing to the presence of ions. The ions can attract the polar water molecules and form a bound layer of water molecules around themselves. This process is called **solvation**. Refer to Chapter 13 Solutions and Colligative Properties for an in-depth discussion of solvation.

Even though many ionic compounds dissolve in water, many others do not. If the attraction of the oppositely charged ions in the solid for each other is greater than the attraction of the polar water molecules for the ions, then the salt will not dissolve to an appreciable amount. If solutions containing ions such as these are mixed, precipitation will occur, because the strong attraction of the ions for each other overcomes the weaker attraction for the water molecules.

As mentioned before, certain covalent compounds, like alcohols, readily dissolve in water because they are polar. Since water is polar, and these covalent compounds are also polar, water will act as a solvent for them (general rule of solubility: "Like dissolves like"). Compounds like alcohols are **nonelectrolytes**—substances that do not conduct an electrical current when dissolved in water. However, certain covalent compounds, like acids, will **ionize** in water, that is, form ions:

$$\mathrm{HCl(aq) \rightarrow H^+(aq) + Cl^-(aq)}$$

There are several ways of representing reactions that occur in water. Suppose, for example, that we were writing the equation to describe the mixing of a lead(II) nitrate solution with a sodium sulfate solution and showing the resulting formation of solid lead(II) sulfate. One type of equation that can be written is the **molecular equation**, in which both the reactants and products are shown in the undissociated form:

$$\mathrm{Pb(NO_3)_2(aq) + Na_2SO_4(aq) \rightarrow PbSO_4(s) + 2NaNO_3(aq)}$$

Molecular equations are quite useful when doing reaction stoichiometry problems (see Chapter 7).

Showing the soluble reactants and products in the form of ions yields the **ionic equation** (sometimes called the total ionic equation):

$$Pb^{2+}(aq) + 2NO_3^-(aq) + 2Na^+(aq) + SO_4^{2-}(aq) \rightarrow PbSO_4(s) + 2Na^+(aq) + 2NO_3^-(aq)$$

Writing the equation in the ionic form shows clearly which species are really reacting and which are not. In the example above, Na^+ and NO_3^- appear on both sides of the equation. They do not react, but are simply there in order to maintain electrical neutrality of the solution. Ions like this, which are not actually involved in the chemical reaction taking place, are called **spectator ions.**

The **net ionic equation** is written by dropping out the spectator ions and showing only those chemical species that are involved in the chemical reaction:

$$Pb^{2+}(aq) + SO_4^{2-}(aq) \rightarrow PbSO_4(s)$$

This net ionic equation focuses only on the substances that are actually involved in the reaction. It indicates that an aqueous solution containing Pb^{2+} (any solution, not just $Pb(NO_3)_2(aq)$) will react with any solution containing the sulfate ion to form insoluble lead(II) sulfate. If this equation form is used, the spectator ions involved will not be known, but in most cases this is not a particular problem, since the focus is really the general reaction, and not the specific one. You will be expected to write the balanced net ionic equation for many of the reactions on the test.

Precipitation Reactions

Precipitation reactions involve the formation of an insoluble compound, a **precipitate,** from the mixing of two soluble compounds. Precipitation reactions normally occur in aqueous solution. The example above that was used to illustrate molecular equations, ionic equations, etc., was a precipitation reaction. A solid, lead(II) sulfate, was formed from the mixing of the two aqueous solutions. In order to predict whether or not precipitation will occur if two solutions are mixed, you must:

1. Learn to write the correct chemical formulas from the names; on the AP exam names are frequently given instead of formulas in the reaction section.
2. Be able to write the reactants and products in their ionic form, as in the ionic equation example above. Be sure, however, that you do not try to break apart molecular compounds such as most organic compounds, or insoluble species.
3. Know and be able to apply the following solubility rules by combining the cation of one reactant with the anion of the other in the correct formula ratio, and determining the solubility of the proposed product. Then do the same thing for the other anion/cation combination.
4. On the AP exam, you will be expected to explain why a substance is soluble/insoluble. Simply quoting the solubility rule is not sufficient.

Learn the following solubility rule:

All sodium, potassium, ammonium, and nitrate salts are soluble in water.

ENRICHMENT

Although not required for the AP exam, the following solubility rules are very useful:

- All salts containing acetate (CH_3COO^-), and perchlorates (ClO_4^-) are *soluble.*
- All chlorides (Cl^-), bromides (Br^-), and iodides (I^-) are *soluble,* except those of Cu^+, Ag^+, Pb^{2+}, and Hg_2^{2+}.
- All salts containing sulfate (SO_4^{2-}) are *soluble,* except those of Pb^{2+}, Ca^{2+}, Sr^{2+}, and Ba^{2+}.

Salts containing the following ions are normally **insoluble**:

- Most carbonates (CO_3^{2-}) and phosphates (PO_4^{3-}) are *insoluble,* except those of Group IA and the ammonium ion.
- Most sulfides (S^{2-}) are *insoluble,* except those of Group IA and IIA and the ammonium ion.
- Most hydroxides (OH^-) are *insoluble,* except those of Group IA, calcium, and barium.
- Most oxides (O^{2-}) are *insoluble,* except for those of Group IA, and Group IIA, which react with water to form the corresponding soluble hydroxides.

Let's see how one might apply these rules. Suppose a solution of lead(II) nitrate is mixed with a solution of sodium iodide. Predict what will happen.

Write the formulas:

$$Pb(NO_3)_2(aq) + NaI(aq) \rightarrow$$

Convert to the ionic form:

$$Pb^{2+}(aq) + 2NO_3^-(aq) + Na^+(aq) + I^-(aq) \rightarrow$$

Predict the possible products by combining the cation of one reactant with the anion of the other and vice versa:

$$PbI_2 + NaNO_3$$

Apply the solubility rules to the two possible products:

$PbI_2(s)$ *Insoluble,* therefore a precipitate will form.

$NaNO_3(aq)$ *Soluble,* no precipitate will form.

Complete the chemical equation and balance it:

$$Pb(NO_3)_2(aq) + 2NaI(aq) \rightarrow PbI_2(s) + 2NaNO_3(aq)$$

$$Pb^{2+}(aq) + 2I^-(aq) \rightarrow PbI_2(s)$$

If both possible products are soluble, then the reaction would be listed as NR (No Reaction). In the reaction question part of the AP exam, there will be a possible reaction for every part of the question. If at least one insoluble product is formed, the reaction is sometimes classified as a **double displacement (replacement) or metathesis reaction**.

Oxidation–Reduction Reactions

Oxidation–reduction reactions, commonly called **redox reactions**, are an extremely important category of reaction. Redox reactions include combustion, corrosion, respiration, photosynthesis, and the reactions involved in electrochemical cells (batteries). The driving force involved in redox reactions is the exchange of electrons from a more active species to a less active one. You can predict the relative activities from a table of activities or a half-reaction table. Chapter 16 Electrochemistry goes into depth about electrochemistry and redox reactions.

The AP free-response booklet includes a table of half-reactions, which you may use for help during this part of the exam. A similar table can be found in the back of this book. Alternatively, you may wish to memorize the common oxidizing and reducing agents.

Redox is a term that stands for **red**uction and **ox**idation. **Reduction** is the gain of electrons and **oxidation** is the loss of electrons. For example, suppose a piece of zinc metal is placed in a solution containing the blue Cu^{2+} cation. Very quickly a reddish solid forms on the surface of the zinc metal. That substance is copper metal. As the copper metal is deposited, the blue color of the solution begins to fade. At the molecular level, the more active zinc metal is losing electrons to form the Zn^{2+} cation, and the Cu^{2+} ion is gaining electrons to form the less active copper metal. These two processes can be shown as:

$$Zn(s) \rightarrow Zn^{2+}(aq) + 2e^- \quad \text{(oxidation)}$$
$$Cu^{2+}(aq) + 2e^- \rightarrow Cu(s) \quad \text{(reduction)}$$

The electrons that are being lost by the zinc metal are the same electrons that are being gained by the copper(II) ion. The zinc metal is being oxidized and the copper(II) ion is being reduced. Further discussions on why reactions such as these occur can be found in the section on single-displacement reactions later in this chapter.

Something must cause the oxidation (taking the electrons) and that substance is called the **oxidizing agent** (the reactant being reduced). In the example above, the oxidizing agent is the Cu^{2+} ion. The reactant undergoing oxidation is called the **reducing agent** because it is furnishing the electrons that are being used in the reduction half-reaction. Zinc metal is the reducing agent above. The two half-reactions, oxidation and reduction, can be added together to give you the overall redox reaction. When doing this, the electrons must cancel—that is, there must be the same number of electrons lost as electrons gained:

$$Zn(s) + Cu^{2+}(aq) + 2e^- \rightarrow Zn^{2+}(aq) + 2e^- + Cu(s)$$

$$\textit{or} \quad Zn(s) + Cu^{2+}(aq) \rightarrow Zn^{2+}(aq) + Cu(s)$$

In these redox reactions there is a simultaneous loss and gain of electrons. In the oxidation reaction (commonly called a half-reaction) electrons are being lost, but in the reduction half-reaction those very same electrons are being gained. So, in redox reactions electrons are being exchanged as reactants are being converted into products. This electron exchange may be direct, as when copper metal plates out on a piece of zinc, or it may be indirect, as in an electrochemical cell (battery).

Another way to determine what is being oxidized and what is being reduced is by looking at the change in oxidation numbers of the reactant species. (See Chapter 5 Basics for a discussion of oxidation numbers and how to calculate them.) On the AP exam you may be asked to assign oxidation numbers and/or identify changes in terms of oxidation numbers. Oxidation is indicated by an increase in oxidation number. In the example above, the Zn metal went from an oxidation state of zero to +2. Reduction is indicated by a decrease in oxidation number. Cu^{2+} went from an oxidation state of +2 to zero. In order to figure out whether a particular reaction is a redox reaction, write the net ionic equation. Then determine the oxidation numbers of each element in the reaction. If one or more elements have changed oxidation number, it is a redox reaction.

There are several types of redox reaction that are given specific names. In the next few pages we will examine some of these types of redox reaction.

Combination Reactions

Combination reactions are reactions in which two or more reactants (elements or compounds) combine to form one product. Although these reactions may be of a number of different types, some types are definitely redox reactions. These include reactions of metals with nonmetals to form ionic compounds, and the reaction of nonmetals with other nonmetals to form covalent compounds.

$$2K(s) + Cl_2(g) \rightarrow 2KCl(s)$$
$$2H_2(g) + O_2(g) \rightarrow 2H_2O(l)$$

In the first reaction, we have the combination of an active metal with an active nonmetal to form a stable ionic compound. The very active oxygen reacts with hydrogen to form the stable compound water. The hydrogen and potassium are undergoing oxidation, while the oxygen and chlorine are undergoing reduction.

Decomposition Reactions

Decomposition reactions are reactions in which a compound breaks down into two or more simpler substances. Although not all decomposition reactions are redox reactions, many are. For example, the thermal decomposition reactions, such as the common laboratory experiment of generating oxygen by heating potassium chlorate, are decomposition reactions:

$$2KClO_3(s) \xrightarrow[\Delta]{} 2KCl(s) + 3O_2(g)$$

In this reaction the chlorine is going from the less stable +5 oxidation state to the more stable −1 oxidation state. While this is occurring, oxygen is being oxidized from −2 to 0.

Another example is **electrolysis**, in which an electrical current is used to decompose a compound into its elements:

$$2H_2O(l) \xrightarrow[\text{electricity}]{} 2H_2(g) + O_2(g)$$

The spontaneous reaction would be the opposite one; therefore, we must supply energy (in the form of electricity) in order to force the nonspontaneous reaction to occur.

Single Displacement Reactions

Single displacement (replacement) reactions are reactions in which atoms of an element replace the atoms of another element in a compound. All of these single replacement reactions are redox reactions, since the element (in a zero oxidation state) becomes an ion. Most single displacement reactions can be categorized into one of three types of reaction:

- A metal displacing a metal ion from solution
- A metal displacing hydrogen gas (H_2) from an acid or from water
- One halogen replacing another halogen in a compound

Remember: It is an **element** displacing another atom from a compound. The displaced atom appears as an element on the product side of the equation.

Table 6.1 Activity Series of Metals in Aqueous Solution

$Li(s)$	$\rightarrow$	$Li^{+}(aq)$	+	e^{-}	Most easily oxidized
$K(s)$	$\rightarrow$	$K^{+}(aq)$	+	e^{-}	
$Ba(s)$	$\rightarrow$	$Ba^{2+}(aq)$	+	$2\,e^{-}$	
$Sr(s)$	$\rightarrow$	$Sr^{2+}(aq)$	+	$2\,e^{-}$	
$Ca(s)$	$\rightarrow$	$Ca^{2+}(aq)$	+	$2\,e^{-}$	
$Na(s)$	$\rightarrow$	$Na^{+}(aq)$	+	e^{-}	
$Mg(s)$	$\rightarrow$	$Mg^{2+}(aq)$	+	$2\,e^{-}$	
$Al(s)$	$\rightarrow$	$Al^{3+}(aq)$	+	$3\,e^{-}$	
$Mn(s)$	$\rightarrow$	$Mn^{2+}(aq)$	+	$2\,e^{-}$	
$Zn(s)$	$\rightarrow$	$Zn^{2+}(aq)$	+	$2\,e^{-}$	
$Cr(s)$	$\rightarrow$	$Cr^{2+}(aq)$	+	$2\,e^{-}$	
$Fe(s)$	$\rightarrow$	$Fe^{2+}(aq)$	+	$2\,e^{-}$	
$Cd(s)$	$\rightarrow$	$Cd^{2+}(aq)$	+	$2\,e^{-}$	
$Co(s)$	$\rightarrow$	$Co^{2+}(aq)$	+	$2\,e^{-}$	
$V(s)$	$\rightarrow$	$V^{3+}(aq)$	+	$3\,e^{-}$	
$Ni(s)$	$\rightarrow$	$Ni^{2+}(aq)$	+	$2\,e^{-}$	
$Sn(s)$	$\rightarrow$	$Sn^{2+}(aq)$	+	$2\,e^{-}$	
$Pb(s)$	$\rightarrow$	$Pb^{2+}(aq)$	+	$2\,e^{-}$	
$H_2(g)$	$\rightarrow$	$2\,H^{+}(aq)$	+	$2\,e^{-}$	
$Cu(s)$	$\rightarrow$	$Cu^{2+}(aq)$	+	$2\,e^{-}$	
$Ag(s)$	$\rightarrow$	$Ag^{+}(aq)$	+	e^{-}	
$Hg(l)$	$\rightarrow$	$Hg^{2+}(aq)$	+	$2\,e^{-}$	
$Pd(s)$	$\rightarrow$	$Pd^{2+}(aq)$	+	$2\,e^{-}$	
$Pt(s)$	$\rightarrow$	$Pt^{2+}(aq)$	+	$2\,e^{-}$	
$Au(s)$	$\rightarrow$	$Au^{3+}(aq)$	+	$3\,e^{-}$	Least easily oxidized

Reactions will always appear in the free-response section of the AP Chemistry exam. This may not be true in the multiple-choice part.

For the first two types, a table of metals relating their ease of oxidation to each other is useful in being able to predict what displaces what. Table 6.1 shows the **activity series for metals**, which lists the metal and its oxidation in order of decreasing ease of oxidation. An alternative to the activity series is a table of half-cell potentials, as discussed in Chapter 16 Electrochemistry. In general, the more active the metal, the lower its potential.

Elements on this activity series can displace ions of metals *lower* than themselves on the list. If, for example, one placed a piece of tin metal into a solution containing $Cu(NO_3)_2(aq)$, the Sn would replace the Cu^{2+} cation:

$$Sn(s) + Cu(NO_3)_2(aq) \rightarrow Sn(NO_3)_2(aq) + Cu(s)$$
$$Sn(s) + Cu^{2+}(aq) \rightarrow Sn^{2+}(aq) + Cu(s)$$

The second equation is the net ionic form that is often required on the AP exam.

If a piece of copper metal was placed in a solution of $Sn(NO_3)_2(aq)$, there would be no reaction, since copper is lower than tin on the activity series. This table allows us to also predict that if sodium metal is placed in water, it will displace hydrogen, forming hydrogen gas:

$$2\,Na(s) + 2\,H_2O(l) \rightarrow 2\,NaOH(aq) + H_2(g)$$
$$2\,Na(s) + 2\,H_2O(l) \rightarrow 2\,Na^{+}(aq) + 2\,OH^{-}(aq) + H_2(g)$$

The Group IA and IIA elements on the activity table will displace hydrogen from water, but not the other metals shown. All the metals above hydrogen will react with acidic solutions to produce hydrogen gas:

$$Co(s) + 2\,HCl(aq) \rightarrow CoCl_2(aq) + H_2(g)$$
$$Co(s) + 2\,H^+(aq) \rightarrow Co^{2+}(aq) + H_2(g)$$

Halogen reactivity decreases as one goes from top to bottom in the periodic table, because of the decreasing electronegativity. Therefore, a separate activity series for the halogens can be developed:

$$F_2$$
$$Cl_2$$
$$Br_2$$
$$I_2$$

The above series indicates that if chlorine gas were dissolved in a KI(aq) solution, the elemental chlorine would displace the iodide ion:

$$Cl_2(aq) + 2\,KI(aq) \rightarrow 2\,KCl(aq) + I_2(s)$$
$$Cl_2(aq) + 2\,I^-(aq) \rightarrow 2\,Cl^-(aq) + I_2(s)$$

Combustion Reactions

Combustion reactions are redox reactions in which the chemical species rapidly combines with oxygen and usually emits heat and light. Reactions of this type are extremely important in our society as the sources of heat energy. Complete combustion of carbon yields carbon dioxide, and complete combustion of hydrogen yields water. The complete combustion of **hydrocarbons**, organic compounds containing only carbon and hydrogen, yields carbon dioxide and water:

$$2\,C_2H_6(g) + 7\,O_2(g) \rightarrow 4CO_2(g) + 6\,H_2O(g)$$

If the compound also contains oxygen, such as in alcohols, ethers, etc., the products are still carbon dioxide and water:

$$2\,CH_3OH(l) + 3\,O_2(g) \rightarrow 2CO_2(g) + 4\,H_2O(g)$$

If the compound contains sulfur, the complete combustion produces sulfur dioxide, SO_2:

$$2\,C_2H_6S(g) + 9\,O_2(g) \rightarrow 4\,CO_2(g) + 6\,H_2O(g) + 2\,SO_2(g)$$

If nitrogen is present, it will normally form the very stable nitrogen gas, N_2.

In all of these reactions, the driving force is the highly reactive oxygen forming a very stable compound(s). This is shown by the exothermic nature of the reaction.

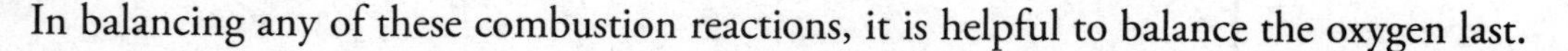
In balancing any of these combustion reactions, it is helpful to balance the oxygen last.

Coordination Compounds

When a salt is dissolved in water, the metal ions, especially transition metal ions, form a complex ion with water molecules and/or other species. A **complex ion** is composed of a metal ion bonded to two or more molecules or ions called **ligands**. These are Lewis acid–base reactions. For example, suppose $Cr(NO_3)_3$ is dissolved in water. The Cr^{3+} cation attracts water molecules to form the complex ion $Cr(H_2O)_6^{3+}$. In this complex ion, water acts as the ligand. If ammonia is added to this solution, the ammonia can displace the water molecules from the complex:

$$[Cr(H_2O)_6]^{3+}(aq) + 6\,NH_3(aq) \leftrightharpoons [Cr(NH_3)_6]^{3+}(aq) + 6\,H_2O(l)$$

In reactions involving coordination compounds, the metal acts as the Lewis acid (electron-pair acceptor), while the ligand acts as a Lewis base (electron-pair donor). In the reaction above, the ammonia ligand displaced the water ligand from the chromium complex because nitrogen is a better electron-pair donor (less electronegative) than oxygen.

The nitrogen in the ammonia and the oxygen in the water are the donor atoms. They are the atoms that actually donate the electrons to the Lewis acid. The **coordination number** is the number of donor atoms that surround the central atom. As seen above, the coordination number for Cr^{3+} is 6. Coordination numbers are usually 2, 4, or 6, but other values can be possible. Silver (Ag^+) commonly forms complexes with a coordination number of 2; zinc (Zn^{2+}), copper (Cu^{2+}), nickel (Ni^{2+}), and platinum (Pt^{2+}) commonly form complexes with a coordination number of 4; most other central ions have a coordination number of 6.

$$AgCl(s) + 2\,NH_3(aq) \rightarrow [Ag(NH_3)_2]^+(aq) + Cl^-(aq)$$

$$Zn(OH)_2(s) + 2\,OH^-(aq) \rightarrow [Zn(OH)_4]^{2-}(aq)$$

$$Fe^{3+}(aq) + 6\,CN^-(aq) \rightarrow [Fe(CN)_6]^{3-}(aq)$$

Acid–Base Reactions

Acids and bases are extremely common, as are the reactions between acids and bases. The driving force is often the hydronium ion reacting with the hydroxide ion to form water. Chapter 15 on Equilibrium describes the equilibrium reactions of acids and bases, as well as some information concerning acid–base titration. After you finish this section, you may want to review the acid–base part of the Equilibrium chapter.

Properties of Acids, Bases, and Salts

At the macroscopic level, acids taste sour, may be damaging to the skin, and react with bases to yield salts. Bases taste bitter, feel slippery, and react with acids to form salts.

At the microscopic level, **acids** are defined as proton (H^+) donors (Brønsted–Lowry theory) or electron-pair acceptors (Lewis theory). **Bases** are defined as proton (H^+) acceptors (Brønsted–Lowry theory) or electron-pair donors (Lewis theory). Consider the gas-phase reaction between hydrogen chloride and ammonia:

$$HCl(g) + {:}NH_3(g) \rightarrow HNH_3^+Cl^- \quad (\text{or } NH_4^+Cl^-)$$

HCl is the acid, because it is donating an H^+ and the H^+ will accept an electron pair from ammonia. Ammonia is the base, accepting the H^+ and furnishing an electron pair that the H^+ will bond with via coordinate covalent bonding. **Coordinate covalent bonds** are covalent bonds in which one of the atoms furnishes both of the electrons for the bond. After the bond is formed, it is identical to a covalent bond formed by donation of one electron by both of the bonding atoms.

Acids and bases may be **strong**, dissociating completely, or **weak**, partially dissociating and forming an equilibrium system.(See Chapter 15 for the details on weak acids and bases.) Strong acids include:

1. Hydrochloric, HCl

2. Hydrobromic, HBr

3. Hydroiodic, HI

4. Nitric, HNO_3

5. Chloric, $HClO_3$

6. Perchloric, $HClO_4$

7. Sulfuric, H_2SO_4

The strong acids above are all compounds that ionize completely in aqueous solution, yielding hydrogen ions and the anions from the acid.

Strong bases include:

1. Alkali metal (Group IA) hydroxides (LiOH, NaOH, KOH, RbOH, CsOH)

2. $Ca(OH)_2$, $Sr(OH)_2$, and $Ba(OH)_2$

The strong bases listed above are all compounds that dissociate completely, yielding the hydroxide ion (which is really the base, not the compound).

Unless told otherwise, assume that acids and bases not on the lists above are weak and will establish an equilibrium system when placed into water.

Some salts have acid–base properties. For example, ammonium chloride, NH_4Cl, when dissolved in water will dissociate and the ammonium ion will act as a weak acid, donating a proton. We will examine these acid–base properties in more detail in the next section.

Certain oxides can have acidic or basic properties. These properties often become evident when the oxides are dissolved in water. In most case, reactions of this type are not redox reactions.

Many oxides of metals that have a +1 or +2 charge are called basic oxides (basic anhydrides), because they will react with acids.

$$Fe_2O_3(s) + 6HCl(aq) \rightarrow 2FeCl_3(aq) + 3H_2O(l)$$
$$Fe_2O_3(s) + 6H^+(aq) \rightarrow 2Fe^{3+}(aq) + 3H_2O(l)$$

Many times they react with water to form a basic solution:

$$Na_2O(s) + H_2O(l) \rightarrow 2NaOH(aq)$$
$$Na_2O(s) + H_2O(l) \rightarrow 2Na^+(aq) + 2OH^-(aq)$$

Many nonmetal oxides are called acidic oxides (acidic anhydrides), because they react with water to form an acidic solution:

$$CO_2(g) + H_2O(l) \rightarrow H_2CO_3(aq)$$

$H_2CO_3(aq)$ is named carbonic acid and is the reason that most carbonated beverages are slightly acidic. It is also the reason that soft drinks have fizz, because carbonic acid will decompose to form carbon dioxide and water.

Acid–Base Reactions

In general, acids react with bases to form a salt and, usually, water. The salt will depend upon which acid and base are used:

$$HCl(aq) + NaOH(aq) \rightarrow H_2O(l) + NaCl(aq)$$
$$HNO_3(aq) + KOH(aq) \rightarrow H_2O(l) + KNO_3(aq)$$
$$HBr(aq) + NH_3(aq) \rightarrow NH_4Br(aq)$$

Reactions of this type are called **neutralization reactions.**

The first two neutralization equations are represented by the same net ionic equation:

$$H^+(aq) + OH^-(aq) \rightarrow H_2O(l)$$

In the third case, the net ionic equation is different:

$$H^+(aq) + NH_3(aq) \rightarrow NH_4^+(aq)$$

As mentioned previously, certain salts have acid–base properties. In general, salts containing cations of strong bases and anions of strong acids are neither acidic nor basic. They are neutral, reacting with neither acids nor bases. An example would be potassium nitrate, KNO_3. The potassium comes from the strong base KOH and the nitrate from the strong acid HNO_3.

Salts containing cations not of strong bases but with anions of strong acids behave as acidic salts. An example would be ammonium chloride, NH_4Cl.

$$2NH_4Cl(aq) + Ba(OH)_2(aq) \rightarrow BaCl_2(aq) + 2NH_3(aq) + 2H_2O(l)$$
$$NH_4^+(aq) + OH^-(aq) \rightarrow NH_3(aq) + H_2O(l)$$

Cations of strong bases and anions not of strong acids are basic salts. An example would be sodium carbonate, Na_2CO_3. It reacts with an acid to form carbonic acid, which would then decompose to carbon dioxide and water:

$$2\ HCl(aq) + Na_2CO_3(aq) \rightarrow 2\ NaCl(aq) + H_2CO_3(aq)$$
$$\downarrow$$
$$CO_2(g) + H_2O(l)$$
$$2\ H^+(aq) + CO_3^{2-} \rightarrow H_2CO_3(aq) \rightarrow CO_2(g) + H_2O(l)$$

The same type of reaction would be true for acid carbonates, such as sodium bicarbonate, $NaHCO_3$.

Another group of compounds that have acid–base properties are the hydrides of the alkali metals and of calcium, strontium, and barium. These hydrides will react with water to form the hydroxide ion and hydrogen gas:

$$NaH(s) + H_2O(l) \rightarrow NaOH(aq) + H_2(g)$$
$$NaH(s) + H_2O(l) \rightarrow Na^+(aq) + OH^-(aq) + H_2(g)$$

Note that in this case, water is behaving as H^+OH^-.

Acid–Base Titrations

A common laboratory application of acid–base reactions is a titration. A **titration** is a laboratory procedure in which a solution of known concentration is used to determine the concentration of an unknown solution. For strong acid/strong base titration systems, the net ionic equation is:

$$H^+(aq) + OH^-(aq) \rightarrow H_2O(l)$$

For example, suppose you wanted to determine the molarity of an HCl solution. You would pipet a known volume of the acid into a flask and add a couple drops of a suitable acid–base indicator. An indicator that is commonly used is phenolphthalein, which is colorless in an acidic solution and pink in a basic solution. You would then fill a buret with a strong base solution (NaOH is commonly used) of known concentration. The buret allows you to add small amounts of the base solution to the acid solution in the flask. The course of the titration can also be followed by the use of a pH meter. Initially the pH of the solution will be low, since it is an acid solution. As the base is added and neutralization of the acid takes place, the pH will slowly rise. Small amounts of the base are added until one reaches the equivalence point. The **equivalence point** is that point in the titration where the number of moles of H^+ in the acid solution has been exactly neutralized with the same number of moles of OH^-:

$$\text{moles } H^+ = \text{moles } OH^- \quad \text{at the equivalence point}$$

For the titration of a strong acid with a strong base, the pH rapidly rises in the vicinity of the equivalence point. Then, as the tiniest amount of base is added in excess, the indicator turns pink. This is called the **endpoint** of the titration. In an accurate titration the endpoint will be as close to the equivalence point as possible. For simple titrations that do not use a pH meter, it is assumed that the endpoint and the equivalence point are the same, so that:

$$\text{moles } H^+ = \text{moles } OH^- \quad \text{at the endpoint}$$

After the equivalence point has been passed, the pH is greater than 7 (basic solution) and begins to level out somewhat. Figure 6.1 shows the shape of the curve for this titration.

Reaction stoichiometry can then be used to solve for the molarity of the acid solution. See Chapter 7 Stoichiometry for a discussion of solution stoichiometry.

An unknown base can be titrated with an acid solution of known concentration. One major difference is that the pH will be greater than 7 initially and will decrease as the titration proceeds. The other major difference is that the indicator will start off pink, and the color will vanish at the endpoint.

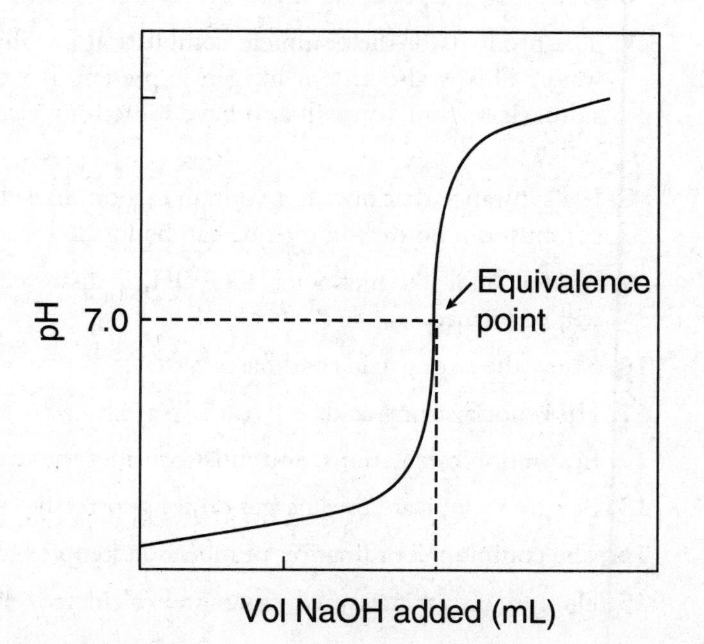

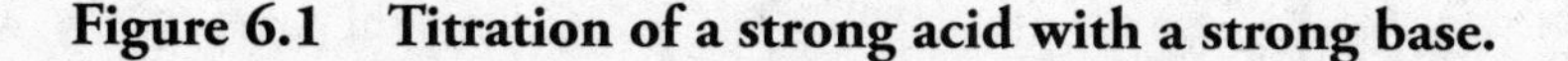

Figure 6.1 Titration of a strong acid with a strong base.

Experiments

Laboratory experiments involving reactions are usually concerned with both the reaction and the stoichiometry. You need some idea of the balanced chemical equation. In the case of an acid–base reaction, an acid reacts with a base. The acid supplies H^+ and the base accepts the H^+. If the acid is diprotic, such as H_2SO_4, it can donate two H^+.

The key to any reaction experiment is moles. The numbers of moles may be calculated from various measurements. A sample may be weighed on a balance to give the mass, and the moles calculated with the formula weight. Or the mass of a substance may be determined using a volume measurement combined with the density. The volume of a solution may be measured with a pipet, or calculated from the final and initial readings from a buret. This volume, along with the molarity, can be used to calculate the moles present. The volume, temperature, and pressure of a gas can be measured and used to calculate the moles of a gas. You must be extremely careful on the AP exam to distinguish between those values that you measure and those that you calculate.

The moles of any substance in a reaction may be converted to the moles of any other substance through a calculation using the balanced chemical equation. Other calculations are presented in Chapter 7 Stoichiometry.

Common Mistakes to Avoid

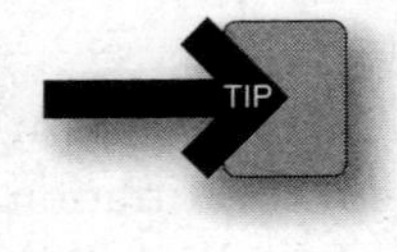

1. In balancing chemical equations **don't** change the subscripts in the chemical formula, just the coefficients.
2. Molecular compounds ionize, ionic compounds dissociate.
3. In writing ionic and net ionic equations, show the chemical species as they actually exist in solution (i.e., strong electrolytes as ions, etc.).
4. In writing ionic and net ionic equations, don't break apart covalently bonded compounds unless they are strong acids that are ionizing.
5. Know the solubility rules as guidelines, not explanations.
6. Oxidizing and reducing agents are reactants, not products.
7. The products of the complete combustion of a hydrocarbon are carbon dioxide and water. This is also true if oxygen is present as well; but if some other element, like sulfur, is present you will also have something else in addition to carbon dioxide and water.
8. If a substance that does not contain carbon, like elemental sulfur, undergoes complete combustion, no carbon dioxide can be formed.
9. If an alcohol like methanol, CH_3OH, is dissolved in water, no hydroxide ion, OH^-, will be formed.
10. Know the strong acids and bases.
11. HF is not a strong acid.
12. In titration calculations, you must consider the reaction stoichiometry.
13. Be sure to indicate the charges on ions correctly.
14. The common coordination numbers of complex ions are 2, 4, and 6.
15. Do not confuse measured values and calculated values.

Review Questions

Here are questions you can use to review the content of this chapter and practice for the AP Chemistry exam. First are 16 multiple-choice questions similar to what you will encounter in Section I of the AP Chemistry exam. Following those is a single-part free-response question like ones in Section II of the exam. To make these review questions an even more authentic practice for the actual exam, time yourself following the instructions provided.

Multiple-Choice Questions

Answer the following questions in 20 minutes. You may not use a calculator. You may use the periodic table and the equation sheet at the back of this book.

1. ___ $Fe(OH)_2(s)$ + ___ $H_3PO_4(aq)$ → ___ $Fe_3(PO_4)_2(s)$ + ___ $H_2O(l)$

After the above chemical equation is balanced, the lowest whole-number coefficient for water is:

(A) 3
(B) 1
(C) 9
(D) 6

2. This ion will generate gas bubbles upon the addition of hydrochloric acid.

(A) Cu^{2+}
(B) CO_3^{2-}
(C) Fe^{3+}
(D) Al^{3+}

3. Aqueous solutions of this ion are blue.

(A) Cu^{2+}
(B) CO_3^{2-}
(C) Fe^{3+}
(D) Al^{3+}

4. Which of the following best represents the balanced net ionic equation for the reaction of lead(II) carbonate with concentrated hydrochloric acid? In this reaction, all lead compounds are insoluble.

(A) $Pb_2CO_3 + 2\,H^+ + Cl^- \rightarrow Pb_2Cl + CO_2 + H_2O$
(B) $PbCO_3 + 2\,H^+ + 2\,Cl^- \rightarrow PbCl_2 + CO_2 + H_2O$
(C) $PbCO_3 + 2\,H^+ \rightarrow Pb^{2+} + CO_2 + H_2O$
(D) $PbCO_3 + 2\,Cl^- \rightarrow PbCl_2 + CO_3^{2-}$

5. A sample of copper metal is reacted with concentrated nitric acid in the absence of air. After the reaction, which of these final products are present?

(A) $CuNO_3$ and H_2O
(B) $Cu(NO_3)_3$, NO, and H_2O
(C) $Cu(NO_3)_2$, NO, and H_2O
(D) $CuNO_3$, H_2O, and H_2

6. Which of the following is the correct net ionic equation for the reaction of acetic acid with potassium hydroxide?

(A) $HC_2H_3O_2 + OH^- \rightarrow C_2H_3O_2^- + H_2O$
(B) $HC_2H_3O_2 + K^+ \rightarrow KC_2H_3O_2 + H^+$
(C) $HC_2H_3O_2 + KOH \rightarrow KC_2H_3O_2 + H_2O$
(D) $H^+ + OH^- \rightarrow H_2O(l)$

7. Which of the following is the correct net ionic equation for the addition of aqueous ammonia to a precipitate of silver chloride?

(A) $AgCl + 2\,NH_3 \rightarrow [Ag(NH_3)_2]^+ + Cl^-$
(B) $AgCl + 2\,NH_4^+ \rightarrow [Ag(NH_4)_2]^{3+} + Cl^-$
(C) $AgCl + NH_4^+ \rightarrow Ag^+ + NH_4Cl$
(D) $AgCl + NH_3 \rightarrow Ag^+ + NH_3Cl$

8. Potassium metal will react with water to release a gas and form a potassium compound. Which of the following is true?

(A) The final solution is basic.
(B) The gas is oxygen.
(C) The potassium compound precipitates.
(D) The potassium compound will react with strong bases.

9. A sample is tested for the presence of the Hg_2^{2+} ion. This ion, along with others, may be precipitated with chloride ion. If Hg_2^{2+} is present in the chloride precipitate, a black color will form upon treatment with aqueous ammonia. The balanced net ionic equation for the formation of this black color is:

(A) $Hg_2Cl_2 + 2\ NH_3 + 2\ H_2O \rightarrow$
$2\ Hg + 2\ NH_4^+ + 2\ Cl^- + 2\ OH^-$
(B) $Hg_2Cl_2 + 2\ NH_3 \rightarrow$
$Hg + HgNH_2Cl + NH_4^+ + Cl^-$
(C) $Hg_2Cl_2 + 2\ NH_4^+ \rightarrow 2\ Hg + 2\ NH_4Cl$
(D) $Hg_2Cl_2 + NH_4^+ \rightarrow 2\ Hg + NH_4Cl + Cl^-$

10. How many moles of $Pb(NO_3)_2$ must be added to 0.10 L of a solution that is 1.0 M in $MgCl_2$ and 1.0 M in KCl to precipitate all of the chloride ion? The compound $PbCl_2$ precipitates.

(A) 1.0 mol
(B) 0.20 mol
(C) 0.50 mol
(D) 0.15 mol

11. When 50.0 mL of 1.0 M $AgNO_3$ is added to 50.0 mL of 0.50 M HCl, a precipitate of AgCl forms. After the reaction is complete, what is the concentration of silver ions in the solution?

(A) 0.50 M
(B) 0.0 M
(C) 1.0 M
(D) 0.25 M

12. A student mixes 50.0 mL of 0.10 M $Pb(NO_3)_2$ solution with 50.0 mL of 0.10 M KCl. A white precipitate forms, and the concentration of the chloride ion becomes very small. Which of the following correctly places the concentrations of the remaining ions in order of decreasing concentration?

(A) $[NO_3^-] > [Pb^{2+}] > [K^+]$
(B) $[NO_3^-] > [K^+] > [Pb^{2+}]$
(C) $[K^+] > [NO_3^-] > [Pb^{2+}]$
(D) $[Pb^{2+}] > [NO_3^-] > [K^+]$

13. A solution is prepared for qualitative analysis. The solution contains the following ions: Co^{2+}, Pb^{2+}, and Al^{3+}. Which of the following will cause no observable reaction?

(A) Dilute $NH_3(aq)$ is added.
(B) Dilute $K_2CrO_4(aq)$ is added.
(C) Dilute $HNO_3(aq)$ is added.
(D) Dilute $K_2S(aq)$ is added.

14. Chlorine gas is bubbled through a colorless solution and the solution turns reddish. Adding a little methylene chloride to the solution extracts the color into the methylene chloride layer. Which of the following ions may be present in the original solution?

(A) Cl^-
(B) I^-
(C) SO_4^{2-}
(D) Br^-

15. The addition of excess concentrated NaOH(aq) to a 1.0 M $(NH_4)_2SO_4$ solution will result in which of the following observations?

(A) The solution becomes neutral.
(B) The formation of a brown precipitate takes place.
(C) Nothing happens because the two solutions are immiscible.
(D) The odor of ammonia will be detected.

16. ____ $C_4H_{11}N(l)$ + ____ $O_2(g) \rightarrow$ ____ $CO_2(g)$ + ____ $H_2O(l)$ + ____ $N_2(g)$

When the above equation is balanced, the lowest whole number coefficient for CO_2 is:

(A) 4
(B) 16
(C) 27
(D) 22

Answers and Explanations for the Multiple-Choice Questions

1. D—The balanced equation is:

$3Fe(OH)_3(s) + 2H_3PO_4(aq) \rightarrow$
$Fe_3(PO_4)_2(s) + 6H_2O(l)$

2. B—Carbonates produce carbon dioxide gas in the presence of an acid. None of the other ions will react with hydrochloric acid to produce a gas.

3. A—Aqueous solutions of Cu^{2+} are normally blue. Iron ions give a variety of colors but are normally colorless, or nearly so, in the absence of complexing agents. The other ions are colorless.

4. B—Lead(II) carbonate is insoluble, so its formula should be written as $PbCO_3$. Hydrochloric acid is a strong acid so it should be written as separate H^+ and Cl^- ions. Lead(II) chloride, $PbCl_2$, is insoluble, and carbonic acid, H_2CO_3, quickly decomposes to CO_2 and H_2O. Also notice that A cannot be correct because the charges do not balance.

5. C—The balanced chemical equation is:

$3\ Cu(s) + 8\ HNO_3(aq) \rightarrow 3\ Cu(NO_3)_2(aq) +$
$2\ NO(g) + 4\ H_2O(l)$

The copper is below hydrogen on the activity series, so H_2 cannot form by this acid-metal reaction. Nitric acid causes oxidation, which will oxidize copper to Cu^{2+} giving $Cu(NO_3)_2$. Some of the nitric acid reduces to NO. An oxidation and a reduction must ALWAYS be together.

6. A—Acetic acid is a weak acid; as such, it should be written as $HC_2H_3O_2$. Potassium hydroxide is a strong base so it will separate into K^+ and OH^- ions. Any potassium compound that might form is soluble and will yield K^+ ions. The potassium ions are spectator ions and are left out of the net ionic equation.

7. A—Aqueous ammonia contains primarily NH_3, which eliminates choices B and C. NH_3Cl does not exist, which eliminates choice D. The reaction produces the silver-ammonia complex, $[Ag(NH_3)_2]^+$.

8. A—The reaction of potassium to produce a potassium compound is an oxidation; therefore, there must be a reduction, and the only species available for reduction is hydrogen. The reaction is:

$2\ K(s) + 2\ H_2O(l) \rightarrow 2\ KOH(aq) + H_2(g)$

KOH is a water-soluble strong base, which will not react with other strong bases.

9. B—The black color is due to the formation of metallic mercury. Aqueous ammonia contains primarily NH_3, which eliminates choices C and D. The total charges on each side of the reaction arrow must be equal, which eliminates choice A.

10. D—The magnesium chloride gives 0.20 moles of chloride ion, and the potassium chloride gives 0.10 moles of chloride ion. A total of 0.30 moles of chloride will react with 0.15 moles of lead, because two Cl^- require one Pb^{2+}.

11. D—The HCl is the limiting reagent. The HCl will react with one-half the silver to halve the concentration. The doubling of the volume halves the concentration a second time.

12. B—All the potassium and nitrate ions remain in solution, leaving $PbCl_2$ as the only possible precipitate. Equal volumes of equal concentrations give the same number of moles of reactants; however, two nitrate ions are produced per solute formula as opposed to only one potassium ion. Initially, the lead and potassium would be equal, but some of the lead is precipitated as $PbCl_2$.

13. C—Ammonia, as a base, will precipitate the metal hydroxides since the only soluble hydroxides are the strong bases. Chromate, sulfide, and chloride ions might precipitate one or more of the ions. Nitrates, from nitric acid, are soluble; therefore, this is the solution that is least likely to cause an observable change.

14. D—Chlorine causes oxidation. It is capable of oxidizing both B and D. Answer B gives I_2, which is brownish in water and purplish in methylene chloride. Bromine solutions are reddish in both.

15. D—Excess strong base will ensure the solution to be basic and not neutral. Both ammonium and sodium salts are soluble; therefore, no precipitate can form. One aqueous solution will mix with another aqueous solution. The following acid–base reaction occurs to release ammonia gas: $NH_4^+(aq) + OH^-(aq) \rightarrow NH_3(g) + H_2O(l)$.

16. B—$4\ C_4H_{11}N(l) + 27\ O_2(g) \rightarrow$
$16\ CO_2(g) + 22\ H_2O(l) + 2\ N_2(g)$

Free-Response Question

On the new AP exams, there will be both long-answer and short-answer free-response questions. The following is an example of a short-answer question.

You have 5 minutes to do the following question. You may use a calculator and the tables in the back of the book.

Question 1

A lead(II) nitrate, $Pb(NO_3)_2$, solution is mixed with an ammonium sulfate, $(NH_4)_2SO_4$, solution and a precipitate forms. What is the precipitate and which ions, if any, are spectator ions in this reaction? Explain how you arrived at your answers.

Answer and Explanation for the Free-Response Question

There are four ions present. These ions are NH_4^+, SO_4^{2-}, Pb^{2+}, and NO_3^-. The starting materials are in solution; therefore, they are soluble. The compounds that might form in the reaction are NH_4NO_3 and $PbSO_4$. Since ammonium (and nitrate) salts are normally soluble, the precipitate must be $PbSO_4$. The spectator ions are the nitrate ions (NO_3^-) and the ammonium ions (NH_4^+), the ions not in the precipitate.

You get 1 point for correctly identifying the precipitate. You get an additional point for the identification of the spectator ions and 2 points for the explanation. There are a maximum of 4 points possible.

› Rapid Review

- Reaction questions will always appear in the free-response section of the AP exam. This may not be true in the multiple-choice part.
- Energy may be released in a reaction (exothermic) or absorbed (endothermic).
- Chemical equations are balanced by adding coefficients in front of the chemical species until the number of each type of atom is the same on both the right and left sides of the arrow.
- The coefficients in the balanced equation must be in the lowest whole-number ratio.
- Water is the universal solvent, dissolving a wide variety of both ionic and polar substances.
- Electrolytes are substances that conduct an electrical current when dissolved in water; nonelectrolytes do not.
- Most ions in solution attract and bind a layer of water molecules in a process called solvation.
- Some molecular compounds, like acids, ionize in water, forming ions.
- In the molecular equation, the reactants and products are shown in their undissociated/un-ionized form; the ionic equation shows the strong electrolytes in the form of ions; the net ionic equation drops out all spectator ions and shows only those species that are undergoing chemical change.
- Precipitation reactions form an insoluble compound, a precipitate, from the mixing of two soluble compounds.
- Learn and be able to apply the solubility rules.

- Redox reactions are reactions where oxidation and reduction take place simultaneously.
- Oxidation is the loss of electrons, and reduction is the gain of electrons.
- Combination reactions are usually redox reactions in which two or more reactants (elements or compounds) combine to form one product.
- Decomposition reactions are usually redox reactions in which a compound breaks down into two or more simpler substances.
- Single displacement reactions are redox reactions in which atoms of an element replace the atoms of another element in a compound.
- Know how to use the activity series to predict whether or not an element will displace another element.
- Combustion reactions are redox reactions in which the chemical species rapidly combine with diatomic oxygen gas, emitting heat and light. The products of the complete combustion of a hydrocarbon are carbon dioxide and water.
- Indicators are substances that exhibit different colors under acidic or basic conditions.
- Acids are proton donors (electron-pair acceptors).
- Bases are proton acceptors (electron-pair donors).
- Coordinate covalent bonds are covalent bonds in which one atom furnishes both of the electrons for the bond.
- Strong acids and bases completely ionize/dissociate, and weak acids and bases only partially ionize/dissociate.
- Know the strong acids and bases.
- Acids react with bases to form a salt and usually water in a neutralization reaction.
- Many hydrides react with water to form the hydroxide ion and hydrogen gas.
- A titration is a laboratory procedure for determining the concentration of an unknown solution using a solution of known concentration.
- The equivalence point of an acid–base titration is the point at which the moles of H^+ from the acid equals the moles of OH^- from the base. The endpoint is the point at which the indicator changes color, indicating the equivalence point.
- A complex ion is composed of a metal ion covalently bonded to two or more molecules or anions called ligands.
- The coordination number (usually 2, 4, or 6) is the number of donor atoms that can surround a metal ion in a complex.

CHAPTER 7

Stoichiometry

IN THIS CHAPTER

Summary: The previous chapter on chemical reactions discussed reactants and products in terms of individual atoms and molecules. But an industrial chemist is not interested in the number of molecules being produced; she or he is interested in kilograms or pounds or tons of products being formed per hour or day. How many kilograms of reactants will it take? How many kilograms of products will be formed? These are the questions of interest. A production chemist is interested primarily in the macroscopic world, not the microscopic one of atoms and molecules. Even a chemistry student working in the laboratory will not be weighing out individual atoms and molecules, but large numbers of them in grams. There must be a way to bridge the gap between the microscopic world of individual atoms and molecules, and the macroscopic world of grams and kilograms. There is—it is called the mole concept, and it is one of the central concepts in the world of chemistry.

Keywords and Equations

Avogadro's number = $6.022 \times 10^{23}\ mol^{-1}$

Molarity, M = moles solute per liter solution

n = moles

m = mass $\qquad n = \dfrac{m}{M}$

M = molar mass

Moles and Molar Mass

The **mole** (**mol**) is the amount of a substance that contains the same number of particles as atoms in exactly 12 grams of carbon-12. This number of particles (atoms or molecules or ions) per mole is called **Avogadro's number** and is numerically equal to 6.022×10^{23} particles. The mole is simply a term that represents a certain number of particles, like a dozen or a pair. That relates moles to the microscopic world, but what about the macroscopic world? The mole also represents a certain mass of a chemical substance. That mass is the substance's atomic or molecular mass expressed in grams. In Chapter 5, the Basics chapter, we described the atomic mass of an element in terms of atomic mass units (amu). This was the mass associated with an individual atom. Then we described how one could calculate the mass of a compound by simply adding together the masses, in amu, of the individual elements in the compound. This is still the case, but at the macroscopic level the unit of grams is used to represent the quantity of a mole. Thus, the following relationships apply:

KEY IDEA

$$6.022 \times 10^{23} \text{ particles} = 1 \text{ mol}$$
$$= \text{atomic (molecular, formula) mass in grams}$$

The mass in grams of one mole of a substance is the **molar mass**.

The relationship above gives a way of converting from grams to moles to particles, and vice versa. If you have any one of the three quantities, you can calculate the other two. This becomes extremely useful in working with chemical equations, as we will see later, because the coefficients in the balanced chemical equation are not only the number of individual atoms or molecules at the microscopic level, but also the number of moles at the macroscopic level.

How many moles are present in 1.20×10^{25} silver atoms?
Answer:

$$(1.20\times10^{25}\text{ Ag atoms})\left(\frac{1\text{ mol Ag atoms}}{6.022\times10^{23}\text{ Ag atoms}}\right)=19.9\text{ mol Ag}$$

Percent Composition and Empirical Formulas

If the formula of a compound is known, it is a fairly straightforward task to determine the percent composition of each element in the compound. For example, suppose you want to calculate the percentage of hydrogen and oxygen in water, H_2O. First calculate the molecular mass of water:

$$1 \text{ mol } H_2O = 2 \text{ mol H} + 1 \text{ mol O}$$

Substituting the masses involved:

$$1 \text{ mol } H_2O = 2\ (1.0079 \text{ g/mol}) + 16.00 \text{ g/mol} = 18.0158 \text{ g/mol}$$

(intermediate calculation—don't worry about significant figures yet)

$$\begin{aligned}\text{percentage hydrogen} &= [\text{mass H/mass } H_2O] \times 100\\ &= [2(1.0079 \text{ g/mol})/18.0158 \text{ g/mol}] \times 100\\ &= 11.19\% \text{ H}\\ \text{percentage oxygen} &= [\text{mass O/mass } H_2O] \times 100\\ &= [16.00 \text{ g/mol}/18.0158 \text{ g/mol}] \times 100\\ &= 88.81\% \text{ O}\end{aligned}$$

As a good check, add the percentages together. They should equal 100% or be very close to it.

Determine the mass percent of each of the elements in $C_6H_{12}O_6$
Formula mass (FM) = 180.158 amu

Answer:

$$\%C = \frac{(6\,C\,\text{atoms})(12.011\ \text{amu/atom})}{(180.158\ \text{amu})} \times 100\% = 40.002\%$$

$$\%H = \frac{(12\,H\,\text{atoms})(1.008\ \text{amu/atom})}{(180.158\ \text{amu})} \times 100\% = 6.714\%$$

$$\%O = \frac{(6\,O\,\text{atoms})(15.9994\ \text{amu/atom})}{(180.158\ \text{amu})} \times 100\% = 53.2846\%$$

$$\text{Total} = 100.001\%$$

The total is a check. It should be *very* close to 100%.

In the problems above, the percentage data was calculated from the chemical formula, but the empirical formula can be determined if the percent compositions of the various elements are known. The **empirical formula** tells us what elements are present in the compound and the simplest whole-number ratio of elements. The data may be in terms of percentage, or mass, or even moles. But the procedure is still the same: convert each to moles, divide each by the smallest number, then use an appropriate multiplier if needed. The empirical formula mass can then be calculated. If the actual molecular mass is known, dividing the molecular mass by the empirical formula mass gives an integer (rounded if needed) that is used to multiply each of the subscripts in the empirical formula. This gives the **molecular (actual) formula**, which tells which elements are in the compound and the actual number of each.

For example, a sample of a gas was analyzed and found to contain 2.34 g of nitrogen and 5.34 g of oxygen. The molar mass of the gas was determined to be about 90 g/mol. What are the empirical and molecular formulas of this gas?

Answer:

$$(2.34\ \text{g N})\left(\frac{1\ \text{mol N}}{14.0\ \text{g N}}\right) = 0.167\ \text{mol N} \qquad \left(\frac{0.167}{0.167}\right) = 1\ \text{N}$$

$$(5.34\ \text{g O})\left(\frac{1\ \text{mol O}}{16.0\ \text{g O}}\right) = 0.334\ \text{mol O} \qquad \left(\frac{0.334}{0.167}\right) = 2\ \text{O}$$

$$\therefore \text{Empirical Formula} = NO_2$$

The molecular formula may be determined by dividing the actual molar mass of the compound by the empirical molar mass. In this case the empirical molar mass is 46 g/mol.

Thus $\left(\frac{90\ \text{g/mol}}{46\ \text{g/mol}}\right) = 1.96$ which, to one significant figure, is 2. Therefore, the molecular formula is twice the empirical formula—N_2O_4.

Be sure to use as many significant digits as possible in the molar masses. Failure to do so may give you erroneous ratio and empirical formulas.

Reaction Stoichiometry

As we have discussed previously, the balanced chemical equation not only indicates which chemical species are the reactants and the products, but also indicates the relative ratio of reactants and products. Consider the balanced equation of the Haber process for the production of ammonia:

$$N_2(g) + 3H_2(g) \rightarrow 2\ NH_3(g)$$

This balanced equation can be read as: *1 nitrogen molecule reacts with 3 hydrogen molecules to produce 2 ammonia molecules.* But as indicated previously, the coefficients can stand not only for the number of atoms or molecules (microscopic level), they can also stand for the number of *moles* of reactants or products. The equation can also be read as: *1 mol of nitrogen molecules reacts with 3 mol of hydrogen molecules to produce 2 mol of ammonia molecules.* And if the number of moles is known, the number of grams or molecules can be calculated. This is **stoichiometry**, the calculation of the amount (mass, moles, particles) of one substance in a chemical reaction through the use of another. The coefficients in a balanced chemical equation define the mathematical relationship between the reactants and products, and allow the conversion from moles of one chemical species in the reaction to another.

Consider the Haber process above. How many moles of ammonia could be produced from the reaction of 20.0 mol of nitrogen with excess hydrogen?

Before any stoichiometry calculation can be done, you must have a balanced chemical equation!

You are starting with moles of nitrogen and want moles of ammonia, so we'll convert from moles of nitrogen to moles of ammonia by using the ratio of moles of ammonia to moles of nitrogen as defined by the balanced chemical equation:

$$\frac{20.0\ \text{mol}\ N_2}{1} \times \frac{2\ \text{mol}\ NH_3}{1\ \text{mol}\ N_2} = 40.0\ \text{mol}\ NH_3$$

The ratio of 2 mol NH_3 to 1 mol N_2 is called the stoichiometric ratio and comes from the balanced chemical equation.

Suppose you also wanted to know how many moles of hydrogen it would take to fully react with the 20.0 mol of nitrogen. Just change the stoichiometric ratio:

$$\frac{20.0\ \text{mol}\ N_2}{1} \times \frac{3\ \text{mol}\ H_2}{1\ \text{mol}\ N_2} = 60.0\ \text{mol}\ H_2$$

Notice that this new stoichiometric ratio also came from the balanced chemical equation.

Suppose instead of moles you had grams and wanted an answer in grams. How many grams of ammonia could be produced from the reaction of 85.0 g of hydrogen gas with excess nitrogen?

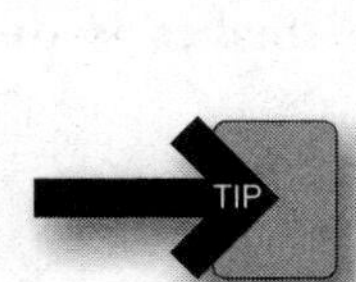

In working problems that involve something other than moles, you will still need moles. And you will need the balanced chemical equation.

In this problem we will convert from grams of hydrogen to moles of hydrogen to moles of ammonia using the correct stoichiometric ratio, and finally to grams of ammonia. And we will need the molar mass of H_2 (2.0158 g/mol) and ammonia (17.0307 g/mol):

$$\frac{85.0\ \text{g}\ H_2}{1} \times \frac{1\ \text{mol}\ H_2}{2.0158\ \text{g}} \times \frac{2\ \text{mol}\ NH_3}{3\ \text{mol}\ H_2} \times \frac{17.0307\ \text{g}}{1\ \text{mol}\ NH_3} = 478.8\ \text{g}\ NH_3$$

Actually, you could have calculated the actual number of ammonia molecules produced if you had gone from moles of ammonia to molecules (using Avogadro's number):

$$\frac{85.0\ \text{g}\ H_2}{1} \times \frac{1\ \text{mol}\ H_2}{2.0158\ \text{g}} \times \frac{2\ \text{mol}\ NH_3}{3\ \text{mol}\ H_2} \times \frac{6.022 \times 10^{23}\ \text{molecules}\ NH_3}{1\ \text{mol}\ NH_3}$$
$$= 1.693 \times 10^{25}\ \text{molecules}\ NH_3$$

In another reaction, 40.0 g of Cl_2 and excess H_2 are combined. HCl will be produced. How many grams of HCl will form?

$$H_2(g) + Cl_2(g) \rightarrow 2\ HCl(g)$$

Answer: $$(40.0\ g\ Cl_2)\left(\frac{1\ mol\ Cl_2}{70.906\ g\ Cl_2}\right)\left(\frac{2\ mol\ HCl}{1\ mol\ Cl_2}\right)\left(\frac{36.461\ g\ HCl}{1\ mol\ HCl}\right) = 41.1\ g\ HCl$$

Limiting Reactants

In the examples above, one reactant was present in excess. One reactant was completely consumed, and some of the other reactant would be left over. The reactant that is used up first is called the **limiting reactant** (L.R.). This reactant really determines the amount of product being formed. How is the limiting reactant determined? You can't assume it is the reactant in the smallest amount, since the reaction stoichiometry must be considered. There are generally two ways to determine which reactant is the limiting reactant:

1. Each reactant, in turn, is assumed to be the limiting reactant, and the amount of product that would be formed is calculated. The reactant that yields the *smallest* amount of product is the limiting reactant. The advantage of this method is that you get to practice your calculation skills; the disadvantage is that you have to do more calculations.
2. The moles of reactant per coefficient of that reactant in the balanced chemical equation is calculated. The reactant that has the smallest mole-to-coefficient ratio is the limiting reactant. This is the method that many use.

Let us consider the Haber reaction once more. Suppose that 50.0 g of nitrogen and 40.0 g of hydrogen were allowed to react. Calculate the number of grams of ammonia that could be formed.

First, write the balanced chemical equation:

$$N_2(g) + 3\ H_2(g) \rightarrow 2\ NH_3(g)$$

Next, convert the grams of each reactant to moles:

$$\frac{50.0\ g\ N_2}{1} \times \frac{1\ mol\ N_2}{28.014\ g\ N_2} = 1.7848\ mol\ N_2$$

$$\frac{40.0\ g\ H_2}{1} \times \frac{1\ mol\ H_2}{2.0158\ g\ H_2} = 19.8432\ mol\ H_2$$

Divide each by the coefficient in the balanced chemical equation. The smaller is the limiting reactant:

For N_2: 1.7848 mol N_2/1 = 1.7848 mol/coefficient *limiting reactant*
For H_2: 19.8432 mol H_2/3 = 6.6144 mol/coefficient

Finally, base the stoichiometry of the reaction on the limiting reactant:

$$\frac{50.0\ g\ N_2}{1} \times \frac{1\ mol\ N_2}{28.014\ g\ N_2} \times \frac{2\ mol\ NH_3}{1\ mol\ N_2} \times \frac{17.0307\ g}{1\ mol\ NH_3} = 60.8\ g\ NH_3$$

Any time the quantities of more than one reactant are given it is probably an L.R. problem.

Let's consider another case. To carry out the following reaction: $P_2O_5(s) + 3H_2O(l) \rightarrow 2H_3PO_4\ (aq)$ 125 g of P_2O_5 and 50.0 g of H_2O were supplied. How many grams of H_3PO_4 may be produced?

Answer:

1. Convert to moles:

$$(125 \text{ g } P_2O_5)\left(\frac{1 \text{ mol } P_2O_5}{142 \text{ g } P_2O_5}\right) = 0.880 \text{ mol } P_2O_5$$

$$(50.0 \text{ g } H_2O)\left(\frac{1 \text{ mol } H_2O}{18.0 \text{ g } H_2O}\right) = 2.78 \text{ mol } H_2O$$

2. Find the limiting reactant:

$$\frac{0.880 \text{ mol}}{1 \text{ mol}} = 0.880 \; P_2O_5$$

$$\frac{2.78 \text{ mol}}{3 \text{ mol}} = 0.927 \; H_2O$$

The 1 mol and the 3 mol come from the balanced chemical equation. The 0.880 is smaller, so this is the L.R.

3. Finish using the number of moles of the L.R.:

$$(0.880 \text{ mol } P_2O_5)\left(\frac{2 \text{ mol } H_3PO_4}{1 \text{ mol } P_2O_5}\right)\left(\frac{98.0 \text{ g } H_3PO_4}{1 \text{ mol } H_3PO_4}\right) = 172 \text{ g}$$

Percent Yield

In the preceding problems, the amount of product calculated based on the limiting-reactant concept is the maximum amount of product that could be formed from the given amount of reactants. This maximum amount of product formed is called the **theoretical yield**. However, rarely is the amount that is actually formed (the **actual yield**) the same as the theoretical yield. Normally it is less. There are many reasons for this, but the principal reason is that most reactions do not go to completion; they establish an equilibrium system (see Chapter 15 Equilibrium for a discussion of chemical equilibrium). For whatever reason, not as much as expected is formed. The efficiency of the reaction can be judged by calculating the percent yield. The **percent yield (% yield)** is the actual yield divided by the theoretical yield, and the result is multiplied by 100% to generate percentage:

$$\% \text{ yield} = \frac{\text{actual yield}}{\text{theoretical yield}} \times 100\%$$

Consider the problem in which it was calculated that 60.8 g NH_3 could be formed. Suppose that reaction was carried out, and only 52.3 g NH_3 was formed. What is the percent yield?

$$\% \text{ yield} = \frac{52.3 \text{ g}}{60.8 \text{ g}} \times 100\% = 86.0\%$$

Let's consider another percent yield problem in which a 25.0-g sample of calcium oxide is heated with excess hydrogen chloride to produce water and 37.5 g of calcium chloride. What is the percent yield of calcium chloride?

$$CaO(s) + 2HCl(g) \rightarrow CaCl_2(aq) + H_2O(l)$$

Answer:

$$(25.0 \text{ g CaO})\left(\frac{1 \text{ mol CaO}}{56.077 \text{ g CaO}}\right)\left(\frac{1 \text{ mol } CaCl_2}{1 \text{ mol CaO}}\right)\left(\frac{110.984 \text{ g } CaCl_2}{1 \text{ mol } CaCl_2}\right) = 49.478 \text{ g } CaCl_2$$

The theoretical yield is 49.5 g.

$$\frac{37.5 \text{ g } CaCl_2}{49.478 \text{ g } CaCl_2} \times 100\% = 75.8\%$$

Note: All the units except % must cancel. This includes canceling g $CaCl_2$ with g $CaCl_2$, not simply g.

Molarity and Solution Calculations

We discuss solutions further in the chapter on solutions and colligative properties, but solution stoichiometry is so common on the AP exam that we will discuss it here briefly also. **Solutions** are homogeneous mixtures composed of a **solute** (substance present in smaller amount) and a **solvent** (substance present in larger amount). If sodium chloride is dissolved in water, the NaCl is the solute and the water the solvent.

One important aspect of solutions is their **concentration**, the amount of solute dissolved in the solvent. In the chapter on solutions and colligative properties we will cover several concentration units, but for the purpose of stoichiometry, the only concentration unit we will use at this time is molarity. **Molarity (M)** is defined as the moles of solute per liter of solution:

$$M = \text{mol solute/L solution}$$

Let's start with a simple example of calculating molarity. A solution of NaCl contains 39.12 g of this compound in 100.0 mL of solution. Calculate the molarity of NaCl.

Answer:

$$\frac{(39.12 \text{ g NaCl})\dfrac{[1 \text{ mol NaCl}]}{[58.45 \text{ g NaCl}]}}{(100.0 \text{ mL})\dfrac{[1 \text{ L}]}{[1000 \text{ mL}]}} = 6.693 \text{ M NaCl}$$

Knowing the volume of the solution and the molarity allows you to calculate the moles or grams of solute present.

Next, let's see how we can use molarity to calculate moles. How many moles of ammonium ions are in 0.100 L of a 0.20 M ammonium sulfate solution?

Answer:

$$\left[\frac{0.20 \text{ mol } (NH_4)_2SO_4}{L}\right]\left[\frac{2 \text{ mol } NH_4^+}{1 \text{ mol } (NH_4)_2SO_4}\right](0.100 \text{ L}) = 0.040 \text{ mol } NH_4^+$$

Stoichiometry problems (including limiting-reactant problems) involving solutions can be worked in the same fashion as before, except that the volume and molarity of the solution must first be converted to moles.

If 35.00 mL of a 0.1500 M KOH solution is required to titrate 40.00 mL of a phosphoric acid solution, what is the concentration of the acid? The reaction is:

$$2KOH\ (aq) + H_3PO_4\ (aq) \rightarrow K_2HPO_4\ (aq) + 2H_2O\ (l)$$

Answer:

$$\frac{(35.00 \text{ mL})\dfrac{(0.1500 \text{ mol KOH})(1 \text{ mol } H_3PO_4)}{(1{,}000 \text{ mL})(2 \text{ mol KOH})}}{(40.00 \text{ mL})\dfrac{(1 \text{ L})}{(1{,}000 \text{ mL})}} = 0.06562 \text{ M } H_3PO_4$$

Experiments

Stoichiometry experiments must involve moles. They nearly always use a balanced chemical equation. Measurements include initial and final masses, and initial and final volumes. Calculations may include the difference between the initial and final values. Using the formula mass and the mass in grams, moles may be calculated. Moles may also be calculated from the volume of a solution and its molarity.

Once the moles have been calculated (they are never measured), the experiment will be based on further calculations using these moles.

Common Mistakes to Avoid

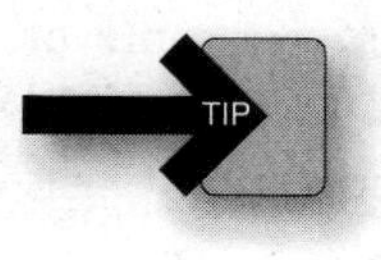

1. Avogadro's number is 6.022×10^{23} (not 10^{-23}).
2. Be sure to know the difference between molecules and moles.
3. In empirical formula problems, be sure to get the lowest ratio of whole numbers.
4. In stoichiometry problems, be sure to use the *balanced chemical equation.*
5. The stoichiometric ratio comes from the *balanced chemical equation.*
6. When in doubt, convert to moles.
7. In limiting-reactant problems, don't consider just the number of grams or even moles to determine the limiting reactant—use the mol/coefficient ratio.
8. The limiting reactant is a reactant, a chemical species to the left of the reactant arrow.
9. *Use the balanced chemical equation.*
10. Percent yield is actual yield of a substance divided by the theoretical yield of the same substance multiplied by 100%.
11. Molarity is moles of *solute* per liter of *solution*, not solvent.
12. Be careful when using Avogadro's number—use it when you need or have the number of atoms, ions, or molecules.

› Review Questions

Use these questions to review the content of this chapter and practice for the AP Chemistry exam. First are 16 multiple-choice questions similar to what you will encounter in Section I of the AP Chemistry exam. Following those is a multipart free-response question like the ones in Section II of the exam. To make these questions an even more authentic practice for the actual exam, time yourself following the instructions provided.

Multiple-Choice Questions

Answer the following questions in 20 minutes. You may not use a calculator. You may use the periodic table and the equation sheet at the back of this book.

1. How many milliliters of 0.100 M H_2SO_4 are required to neutralize 50.0 mL of 0.200 M KOH?

(A) 25.0 mL
(B) 30.0 mL
(C) 20.0 mL
(D) 50.0 mL

2. A sample of oxalic acid, $H_2C_2O_4$, is titrated with standard sodium hydroxide, NaOH, solution. A total of 45.20 mL of 0.1200 M NaOH is required to neutralize completely 20.00 mL of the acid. What is the concentration of the acid?

(A) 0.2712 M
(B) 0.1200 M
(C) 0.1356 M
(D) 0.2400 M

3. A solution is prepared by mixing 50.0 mL of 0.20 M arsenic acid, H_3AsO_4, and 50.0 mL of 0.20 M sodium hydroxide, NaOH. Which anion is present in the highest concentration?

(A) $HAsO_4^{2-}$
(B) OH^-
(C) $H_2AsO_4^-$
(D) Na^+

4. $14\ H^+(aq) + 6\ Fe^{2+}(aq) + Cr_2O_7^{2-}(aq) \rightarrow 2\ Cr^{3+}(aq) + 6\ Fe^{3+}(aq) + 7\ H_2O(l)$
This reaction is used in the titration of an iron solution. What is the concentration of the iron solution if it takes 45.20 mL of 0.1000 M $Cr_2O_7^{2-}$ solution to titrate 50.00 mL of an acidified iron solution?

(A) 0.5424 M
(B) 0.1000 M
(C) 1.085 M
(D) 0.4520 M

5. Manganese, Mn, forms a number of oxides. A particular oxide is 63.2% Mn. What is the simplest formula for this oxide?

(A) MnO
(B) Mn_2O_3
(C) Mn_3O_4
(D) MnO_2

6. Vanadium forms a number of oxides. In which of the following oxides is the vanadium-to-oxygen mass ratio 2.39:1.00?

(A) VO
(B) V_2O_3
(C) V_3O_4
(D) VO_2

7. How many grams of nitrogen are in 25.0 g of $(NH_4)_2SO_4$?

(A) 5.30 g
(B) 1.30 g
(C) 0.190 g
(D) 2.65 g

8. Sodium sulfate forms a number of hydrates. A sample of a hydrate is heated until all the water is removed. What is the formula of the original hydrate if it loses 43% of its mass when heated?

(A) $Na_2SO_4 \cdot H_2O$
(B) $Na_2SO_4 \cdot 2H_2O$
(C) $Na_2SO_4 \cdot 6H_2O$
(D) $Na_2SO_4 \cdot 8H_2O$

9. $3\ Cu(s) + 8\ HNO_3(aq) \rightarrow 3\ Cu(NO_3)_2(aq) + 2\ NO(g) + 4\ H_2O(l)$
Copper metal reacts with nitric acid according to the above equation. A 0.30 mol sample of copper metal and 10.0 mL of 12 M nitric acid are mixed in a flask. How many moles of NO gas will form?

(A) 0.060 mol
(B) 0.030 mol
(C) 0.010 mol
(D) 0.20 mol

10. Gold(III) oxide, Au_2O_3, can be decomposed to gold metal, Au, plus oxygen gas, O_2. How many moles of oxygen gas will form when 221 g of solid gold(III) oxide is decomposed? The formula mass of gold(III) oxide is 442.

(A) 0.250 mol
(B) 0.500 mol
(C) 0.750 mol
(D) 1.00 mol

11. __$C_4H_{11}N(l)$ + __$O_2(g)$ → __$CO_2(g)$ + __ $H_2O(l)$ + __ $N_2(g)$
When the above equation is balanced, the lowest whole number coefficient for O_2 is:

(A) 4
(B) 16
(C) 22
(D) 27

12. $2\ KMnO_4(aq) + 5\ H_2C_2O_4(aq) + 3\ H_2SO_4(aq) \rightarrow K_2SO_4(aq) + 2\ MnSO_4(aq) + 10\ CO_2(g) + 8\ H_2O(l)$

How many moles of $MnSO_4$ are produced when 1.0 mol of $KMnO_4$, 5.0 mol of $H_2C_2O_4$, and 3.0 mol of H_2SO_4 are mixed?

(A) 4.0 mol
(B) 5.0 mol
(C) 2.0 mol
(D) 1.0 mol

13. When the following equation is balanced, it is found that 1.00 mol of C_8H_{18} reacts with how many moles of O_2?

$$\_C_8H_{18}(g) + \_O_2(g) \rightarrow \_CO_2(g) + \_H_2O(g)$$

(A) 12.5 mol
(B) 10.0 mol
(C) 25.0 mol
(D) 37.5 mol

14. $Ca(s) + 2\ H_2O(l) \rightarrow Ca(OH)_2(aq) + H_2(g)$

Calcium reacts with water according to the above reaction. What volume of hydrogen gas, at standard temperature and pressure, is produced from 0.200 mol of calcium?

(A) 5.60 L
(B) 4.48 L
(C) 3.36 L
(D) 1.12 L

15. $2CrO_4^{2-}(aq) + 3SnO_2^{2-}(aq) + H_2O(l) \rightarrow 2\ CrO_2^{-}(aq) + 3\ SnO_3^{2-}(aq) + 2\ OH^{-}(aq)$

How many moles of OH^{-} form when 50.0 mL of 0.100 M CrO_4^{2-} is added to a flask containing 50.0 mL of 0.100 M SnO_2^{2-}?

(A) 0.100 mol
(B) 6.66×10^{-3} mol
(C) 3.33×10^{-3} mol
(D) 5.00×10^{-3} mol

16. A solution containing 0.20 mol of KBr and 0.20 mol of $MgBr_2$ in 2.0 liters of water is provided. How many moles of $Pb(NO_3)_2$ must be added to precipitate all the bromide as insoluble $PbBr_2$?

(A) 0.10 mol
(B) 0.50 mol
(C) 0.60 mol
(D) 0.30 mol

Answers and Explanations for the Multiple-Choice Questions

1. D—The reaction is $H_2SO_4(aq) + 2\ KOH(aq) \rightarrow K_2SO_4(aq) + 2\ H_2O(l)$

$$(50.0\text{ mL base})\left(\frac{0.200\text{ mol base}}{1{,}000\text{ mL base}}\right)\left(\frac{1\text{ mol acid}}{2\text{ mol base}}\right)\left(\frac{1{,}000\text{ mL acid}}{0.100\text{ mol acid}}\right) = 50.0\text{ mL}$$

2. C—The reaction is $H_2C_2O_4(aq) + 2\ NaOH(aq) \rightarrow Na_2C_2O_4(aq) + 2\ H_2O(l)$

$$(45.20\text{ mL base})\left(\frac{0.1200\text{ mol base}}{1{,}000\text{ mL base}}\right)\left(\frac{1\text{ mol acid}}{2\text{ mol base}}\right)\left(\frac{1}{20.00\text{ mL}}\right)\left(\frac{1{,}000\text{ mL}}{\text{L}}\right)$$
$$= 0.1356\text{ M acid}$$

As always, round the values to get an estimate and pick the closest answer.

3. C—Moles acid = (50.0 mL) (0.20 mol acid/1,000 mL) = 0.0100 mol
Moles base = (50.0 mL) (0.20 mol base/1,000 mL) = 0.0100 mol

There is sufficient base to react completely with only one of the ionizable hydrogen ions from the acid. This leaves $H_2AsO_4^-$. Answer D cannot be correct because it is a cation.

4. A—

$$(45.20\text{ mL } Cr_2O_7^{2-})\left(\frac{0.1000\text{ mol } Cr_2O_7^{2-}}{1{,}000\text{ mL } Cr_2O_7^{2-}}\right)\left(\frac{6\text{ mol } Fe^{2+}}{1\text{ mol } Cr_2O_7^{2-}}\right)\left(\frac{1}{50.00\text{ mL}}\right)\left(\frac{1{,}000\text{ mL}}{\text{L}}\right)$$
$$= 0.5425\text{ M}$$

This is a perfect example of where simplification is important. Change the above calculation to

$$(45.20\ \cancel{\text{mL } Cr_2O_7^{2-}})\left(\frac{0.1000\ \cancel{\text{mol } Cr_2O_7^{2-}}}{1{,}000\ \cancel{\text{mL } Cr_2O_7^{2-}}}\right)\left(\frac{6\text{ mol } Fe^{2+}}{1\ \cancel{\text{mol } Cr_2O_7^{2-}}}\right)\left(\frac{1}{50.00\ \cancel{\text{mL}}}\right)\left(\frac{1{,}000\ \cancel{\text{mL}}}{\text{L}}\right)$$
$$= 0.5425\text{ M}$$

This becomes

$$(45.20)\left(\frac{0.1000}{\cancel{1{,}000}}\right)\left(\frac{6\text{ mol } Fe^{2+}}{1}\right)\left(\frac{1}{50.00}\right)\left(\frac{\cancel{1{,}000}}{\text{L}}\right) = 0.5425\text{ M}$$

Next round and simplify to

$$(\cancel{50})\left(\frac{0.1000}{1}\right)\left(\frac{6\text{ mol } Fe^{2+}}{1}\right)\left(\frac{1}{\cancel{50.00}}\right)\left(\frac{1}{\text{L}}\right) = 0.6\text{ M}$$

Since the 45.20 was rounded up, the answer is slightly high; therefore, pick the closest answer that is lower.

5. D—63.2% Mn leaves 36.8% O (assuming 100 grams of sample) 63.2/54.94 = 1.15 mole Mn and 36.8/16.0 = 2.30 mole O

Thus, there is 1 Mn/2 O.

6. C—V: 2.39/50.94 = 0.0469 and for O:1.00/16.0 = 0.0625
0.0469/0.0469 = 1   0.0625/0.0469 = 1.33

Multiplying both by 3 gives: 3 V and 4 O.

7. A—(25.0 g $(NH_4)_2SO_4$)(1 mol$(NH_4)_2SO_4$/132 g) =
(2 mol N/1 mol$(NH_4)_2SO_4$)(14.0 g N/1 mol N) = 5.30 g

8. C—[(6 mol H_2O)(18 g/mol H_2O)]/(250 g $Na_2SO_4 \cdot 6\ H_2O$) × 100% = 43%

9. B—Calculate the moles of acid to compare to the moles of Cu:

$$(10.0\text{ mL})(12\text{ mol}/1{,}000\text{ mL}) = 0.12\text{ mol}$$

The acid is the limiting reactant, because 0.30 mol of copper requires 0.80 mol of acid. Use the limiting reactant to calculate the moles of NO formed.

$$(0.12\text{ mol acid})(2\text{ mol NO}/8\text{ mol acid}) = 0.030\text{ mol}$$

10. C—The balanced chemical equation is:

$$2\ Au_2O_3 \rightarrow 4\ Au + 3\ O_2$$

(221 g Au_2O_3)(1 mol Au_2O_3/442 g Au_2O_3)(3 mol O_2/2 mol Au_2O_3) = 0.750 mol O_2

Note the 2:1 relationship between the formula mass and the mass of reactant.

11. D—The balanced equation is:

$$4\ C_4H_{11}N(l) + 27\ O_2(g) \rightarrow 16\ CO_2(g) + 22\ H_2O(l) + 2\ N_2(g)$$

12. D—The $KMnO_4$ is the limiting reagent. Each mole of $KMnO_4$ will produce a mole of $MnSO_4$.

13. A—The balanced equation is:

$$2\ C_8H_{18}(g) + 25\ O_2(g) \rightarrow 16\ CO_2(g) + 18\ H_2O(g)$$
$$(1.00\text{ mol } C_8H_{18})(25\text{ mol } O_2/2\text{ mol } C_8H_{18}) = 12.5\text{ mol } O_2$$

14. B—(0.200 mol Ca)(1 mol H_2/1 mol Ca)(22.4 L at STP/1 mol H_2) = 4.48 L

Be careful to only use 22.4 when at STP.

15. C—There are 5.00×10^{-3} mol of CrO_4^{2-} and an equal number of moles of SnO_2^{2-}. Thus, SnO_2^{2-} is the limiting reactant (larger coefficient in the balanced reaction).

$$(5.00 \times 10^{-3}\text{ mol } SnO_2^{2-})(2\text{ mol } OH^-/3\text{ mol } SnO_2^{2-}) = 3.33 \times 10^{-3}\text{ mol } OH^-$$

16. D—The volume of water is irrelevant. 0.20 mol of KBr will require 0.10 mol of $Pb(NO_3)_2$ and 0.20 mol of $MgBr_2$ will require 0.20 mol of $Pb(NO_3)_2$. Total the two yields.

Free-Response Question

You have 15 minutes to answer the following question. You may use a calculator and the tables in the back of the book.

Question 1

The analysis of a sample of a monoprotic acid found that the sample contained 40.0% C and 6.71% H. The remainder of the sample was oxygen.

(a) Determine the empirical formula of the acid.

(b) A 0.2720 g sample of the acid, HA, was titrated with standard sodium hydroxide, NaOH, solution. Determine the molecular weight of the acid if the sample required 45.00 mL of 0.1000 M NaOH for the titration.

(c) A second sample was placed in a flask. The flask was placed in a hot water bath until the sample vaporized. It was found that 1.18 g of vapor occupied 300.0 mL at 100°C and 1.00 atmospheres. Determine the molecular weight of the acid.

(d) Using your answer from part **a**, determine the molecular formula for part **b** and for part **c**.

(e) Account for any differences in the molecular formulas determined in part **d**.

Answer and Explanation for the Free-Response Question

(a) The percent oxygen (53.3%) is determined by subtracting the carbon and the hydrogen from 100%. Assuming there are 100 grams of sample gives the grams of each element as being numerically equivalent to the percent. Dividing the grams by the molar mass of each element gives the moles of each.

For C: 40.0/12.01 = 3.33	Divide each	C = 1
For H: 6.71/1.008 = 6.66	of these by	H = 2
For O: 53.3/16.00 = 3.33	the smallest (3.33)	O = 1

This gives the empirical formula: CH_2O.

You get 1 point for correctly determining any of the elements and 1 point for getting the complete empirical formula correct.

(b) Using HA to represent the monoprotic acid, the balanced equation for the titration reaction is:

$$HA(aq) + NaOH(aq) \rightarrow NaA(aq) + H_2O(l)$$

The moles of acid may then be calculated:

$$(45.00\text{ mL NaOH})\left(\frac{0.1000\text{ M NaOH}}{1000\text{ mL}}\right)\left(\frac{1\text{ mol HA}}{1\text{ mol NaOH}}\right) = 4.500 \times 10^{-3}\text{ mol HA}$$

The molecular mass is:

$$0.2720\text{ g}/4.500 \times 10^{-3}\text{ mol} = 60.44\text{ g/mol}$$

You get 1 point for the correct number of moles of HA (or NaOH) and 1 point for the correct final answer.

(c) There are several methods to solve this problem. One way is to use the ideal gas equation as done here. The equation and the value of R are in the exam booklet. First find the moles: $n = PV/RT$. Do not forget, you MUST change temperature to kelvin.

$$n = (1.00\text{ atm})(300.0\text{ mL})(1\text{ L}/1{,}000\text{ mL})/(0.0821\text{ L atm/mol K})(373\text{ K})$$

$$n = 9.80 \times 10^{-3}\text{ mol}$$

The molecular mass is $1.18\text{ g}/9.80 \times 10^{-3}\text{ mol} = 120\text{ g/mol}$

You get 1 point for getting any part of the calculation correct and 1 point for getting the correct final answer.

(d) The approximate formula mass from the empirical (CH_2O) formula is:

$$12 + 2(1) + 16 = 30\text{ g/mol}$$

For part **b**: (60.44 g/mol)/(30 g/mol) = 2

$$\text{Molecular formula} = 2 \times \text{Empirical formula} = C_2H_4O_2$$

For part **c**: (120 g/mol)/(30 g/mol) = 4

$$\text{Molecular formula} = 4 \times \text{Empirical formula} = C_4H_8O_4$$

You get 1 point for each correct molecular formula. If you got the wrong answer in part **a**, you can still get credit for one or both of the molecular formulas if you used the part **a** value correctly.

(e) The one formula is double the formula of the other. Thus, the smaller molecule dimerizes to produce the larger molecule.

You get 1 point if you "combined" two of the smaller molecules.

Total your points. There are 9 points possible.

› Rapid Review

- The mole is the amount of substance that contains the same number of particles as exactly 12 g of carbon-12.
- Avogadro's number is the number of particles per mole, 6.022×10^{23} particles.
- A mole is also the formula (atomic, molecular) mass expressed in grams.
- If you have any one of the three—moles, grams, or particles—you can calculate the others.
- The empirical formula indicates which elements are present and the lowest whole-number ratio.
- The molecular formula tells which elements are present and the actual number of each.
- Be able to calculate the empirical formula from percent composition data or quantities from chemical analysis.
- Stoichiometry is the calculation of the amount of one substance in a chemical equation by using another one.
- *Always use the balanced chemical equation* in reaction stoichiometry problems.
- Be able to convert from moles of one substance to moles of another, using the stoichiometric ratio derived from the *balanced chemical equation.*
- In working problems that involve a quantity other than moles, sooner or later it will be necessary to convert to moles.
- The limiting reactant is the reactant that is used up first.
- Be able to calculate the limiting reactant by the use of the mol/coefficient ratio.
- Percent yield is the actual yield (how much was actually formed in the reaction) divided by the theoretical yield (the maximum possible amount of product formed) times 100%.
- A solution is a homogeneous mixture composed of a solute (species present in smaller amount) and a solvent (species present in larger amount).
- Molarity is the number of moles of solute per liter of solution. Don't confuse molarity, M or [ ], with moles, *n* or mol.
- Be able to work reaction stoichiometry problems using molarity.
- *Always use the balanced chemical equation* in reaction stoichiometry problems.

CHAPTER 8

Gases

IN THIS CHAPTER

Summary: Of the three states of matter—gases, liquids, and solids—gases are probably the best understood and have the best descriptive model. While studying gases in this chapter you will consider four main physical properties—volume, pressure, temperature, and amount—and their interrelationships. These relationships, commonly called gas laws, show up quite often on the AP exam, so you will spend quite a bit of time working problems in this chapter. But before we start looking at the gas laws, let's look at the Kinetic Molecular Theory of Gases, the extremely useful model that scientists use to represent the gaseous state.

Keywords and Equations

Gas constant, $R = 0.0821$ L atm mol^{-1} K^{-1}

$$\frac{r_1}{r_2} = \sqrt{\frac{M_2}{M_1}}$$

u_{rms} = root mean square speed
r = rate of effusion
STP = 0.000°C and 1.000 atm
$PV = nRT$

$$\left(P + \frac{n^2 a}{V^2}\right)(V - nb) = nRT$$

$P_A = P_{total} \times X_A$, where $X_A = \frac{\text{moles A}}{\text{total moles}}$

$P_{total} = P_A + P_B + P_C + \ldots$

$$\frac{P_1V_1}{T_1} = \frac{P_2V_2}{T_2}$$

$$u_{rms} = \sqrt{\frac{3kT}{m}} = \sqrt{\frac{3RT}{M}}$$

KE per molecule $= 1/2\ mv^2$

KE per mole $= \frac{3}{2}RT$

1 atm = 760 mm Hg
= 760 torr

$$n = \frac{m}{M}$$

Kinetic Molecular Theory

The **Kinetic Molecular Theory** attempts to represent the properties of gases by modeling the gas particles themselves at the microscopic level. There are five main postulates of the Kinetic Molecular Theory:

1. Gases are composed of very small particles, either molecules or atoms.
2. The gas particles are tiny in comparison to the distances between them, so we assume that the volume of the gas particles themselves is negligible.
3. These gas particles are in constant motion, moving in straight lines in a random fashion and colliding with each other and the inside walls of the container. The collisions with the inside container walls comprise the pressure of the gas.
4. The gas particles are assumed to neither attract nor repel each other. They may collide with each other, but if they do, the collisions are assumed to be elastic. No kinetic energy is lost, only transferred from one gas molecule to another.
5. The *average* kinetic energy of the gas is proportional to the Kelvin temperature.

A gas that obeys these five postulates is an **ideal gas**. However, just as there are no ideal students, there are no ideal gases: only gases that approach ideal behavior. We know that real gas particles do occupy a certain finite volume, and we know that there are interactions between real gas particles. These factors cause real gases to deviate a little from the ideal behavior of the Kinetic Molecular Theory. But a non-polar gas at a low pressure and high temperature would come pretty close to ideal behavior. Later in this chapter, we'll show how to modify our equations to account for non-ideal behavior.

Before we leave the Kinetic Molecular Theory (KMT) and start examining the gas law relationships, let's quantify a couple of the postulates of the KMT. Postulate 3 qualitatively describes the motion of the gas particles. The average velocity of the gas particles is called the **root mean square speed** and is given the symbol u_{rms}. This is a special type of average speed.

It is the speed of a gas particle having the average kinetic energy of the gas particles. Mathematically it can be represented as:

$$u_{rms} = \sqrt{\frac{3kT}{m}} = \sqrt{\frac{3RT}{M}}$$

where R is the molar gas constant (we'll talk more about it in the section dealing with the ideal gas equation), T is the **Kelvin** temperature and $\boldsymbol{M}$ is the molar mass of the gas. These root mean square speeds are very high. Hydrogen gas, H_2, at 20°C has a value of approximately 2,000 m/s.

Postulate 5 relates the average kinetic energy of the gas particles to the Kelvin temperature. Mathematically we can represent the average kinetic energy per molecule as:

$$KE \text{ per molecule} = 1/2\ mv^2$$

where m is the mass of the molecule and v is its velocity.

The average kinetic energy per mol of gas is represented by:

$$KE \text{ per mol} = 3/2\ RT$$

where R again is the ideal gas constant and T is the Kelvin temperature. This shows the direct relationship between the average kinetic energy of the gas particles and the Kelvin temperature.

Gas Law Relationships

The gas laws relate the physical properties of volume, pressure, temperature, and moles (amount) to each other. First we will examine the individual gas law relationships. You will need to know these relations for the AP exam, but the use of the individual equation is not required. Then we will combine the relationships into a single equation that you will need to be able to apply. But first, we need to describe a few things concerning pressure.

Pressure

When we use the word **pressure**, we may be referring to the pressure of a gas inside a container or to atmospheric pressure, the pressure due to the weight of the atmosphere above us. These two different types of pressure are measured in slightly different ways. Atmospheric pressure is measured using a **barometer** (Figure 8.1).

An evacuated hollow tube sealed at one end is filled with mercury, and then the open end is immersed in a pool of mercury. Gravity will tend to pull the liquid level inside the tube down, while the weight of the atmospheric gases on the surface of the mercury pool will tend to force the liquid up into the tube. These two opposing forces will quickly balance each other, and the column of mercury inside the tube will stabilize. The height of the column of mercury above the surface of the mercury pool is called the atmospheric pressure. At sea level the column averages 760 mm high. This pressure is also called 1 atmosphere (atm). Commonly, the unit torr is used for pressure, where 1 torr = 1 mm Hg, so that atmospheric pressure at sea level equals 760 torr. The SI unit of pressure is the pascal (Pa), so that 1 atm = 760 mm Hg = 760 torr = 101,325 Pa (101.325 kPa). In the United States pounds per square inch (psi) is sometimes used, so that 1 atm = 14.69 psi.

To measure the gas pressure inside a container, a **manometer** (Figure 8.2) is used. As in the barometer, the pressure of the gas is balanced against a column of mercury.

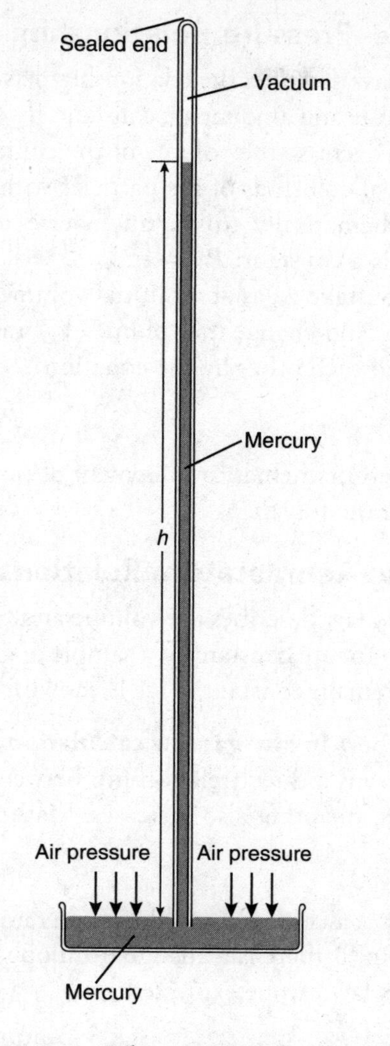

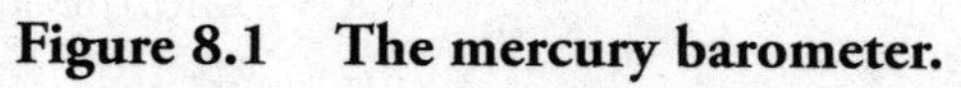

Figure 8.1 The mercury barometer.

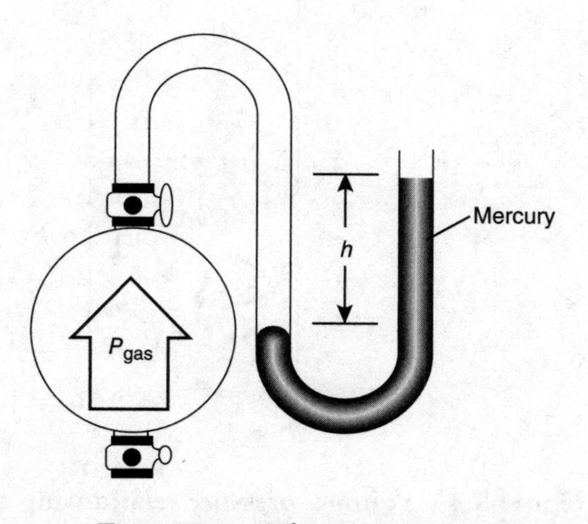

Figure 8.2 The manometer.

Volume–Pressure Relationship: Boyle's Law

Boyle's law describes the relationship between the volume and the pressure of a gas when the temperature and amount are constant. If you have a container like the one shown in Figure 8.3 and you decrease the volume of the container, the pressure of the gas increases because the number of collisions of gas particles with the container's inside walls increases.

Mathematically this is an inverse relationship, so the product of the pressure and volume is a constant: $PV = k_b$.

If you take a gas at an initial volume (V_1) and pressure (P_1) (amount and temperature constant) and change the volume (V_2) and pressure (P_2), you can relate the two sets of conditions to each other by the equation:

$$P_1V_1 = P_2V_2$$

In this mathematical statement of Boyle's law, if you know any three quantities, you can calculate the fourth.

Volume–Temperature Relationship: Charles's Law

Charles's law describes the volume and temperature relationship of a gas when the pressure and amount are constant. If a sample of gas is heated, the volume must increase for the pressure to remain constant. This is shown in Figure 8.4.

Remember: In any gas law calculation, you must express the temperature in kelvin.

There is a direct relationship between the Kelvin temperature and the volume: as one increases, the other also increases. Mathematically, Charles's law can be represented as:

$$V/T = k_c$$

where k_c is a constant and the temperature is expressed in kelvin.

Again, if there is a change from one set of volume–temperature conditions to another, Charles's law can be expressed as:

$$V_1/T_1 = V_2/T_2$$

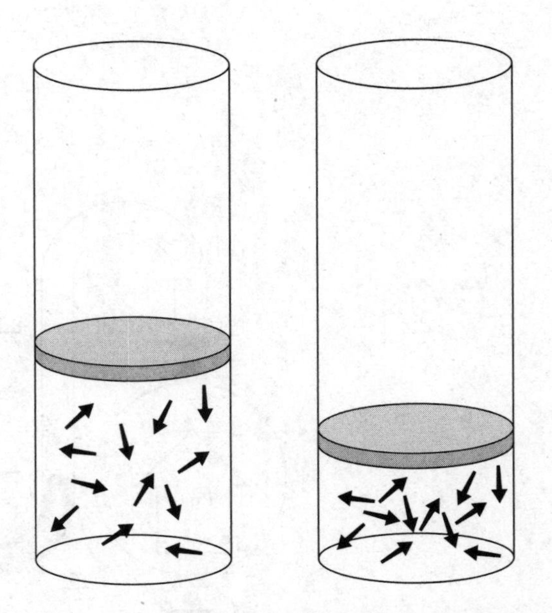

Figure 8.3 Volume–pressure relationship for gases. As the volume decreases, the number of collisions increase.

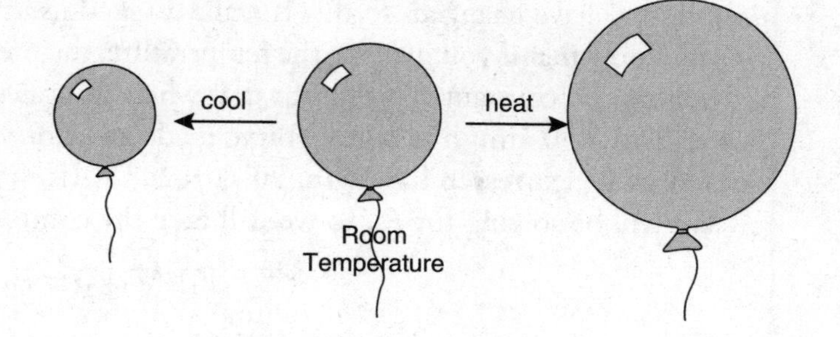

Figure 8.4 Volume–temperature relationship for gases.

Pressure–Temperature Relationship: Gay-Lussac's Law

Gay-Lussac's law describes the relationship between the pressure of a gas and its Kelvin temperature if the volume and amount are held constant. Figure 8.5 represents the process of heating a given amount of gas at a constant volume.

As the gas is heated, the particles move with greater kinetic energy, striking the inside walls of the container more often and with greater force. This causes the pressure of the gas to increase. The relationship between the Kelvin temperature and the pressure is a direct one:

$$P/T = k_g \quad \text{or} \quad P_1/T_1 = P_2/T_2$$

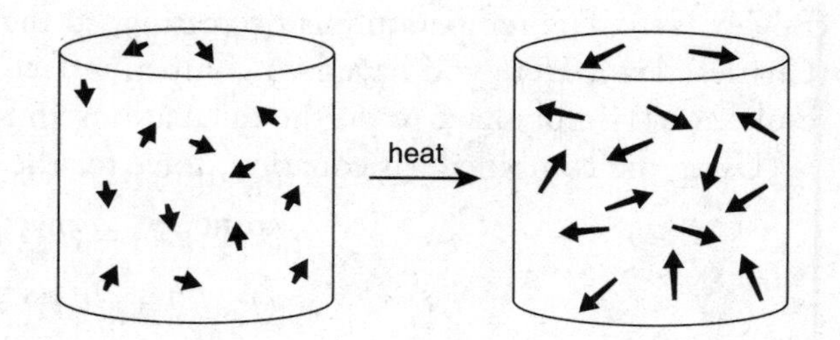

Figure 8.5 Pressure–temperature relationship for gases. As the temperature increases, the gas particles have greater kinetic energy (longer arrows) and collisions are more frequent and forceful.

Combined Gas Law

In the discussion of Boyle's, Charles's, and Gay-Lussac's laws we held two of the four variables constant, changed the third, and looked at its effect on the fourth variable. If we keep the number of moles of gas constant—that is, no gas can get in or out—then we can combine these three gas laws into one, the **combined gas law**, which can be expressed as:

$$(P_1V_1)/T_1 = (P_2V_2)/T_2$$

Again, remember: In any gas law calculation, you must express the temperature in kelvin.

In this equation there are six unknowns; given any five, you should be able to solve for the sixth.

For example, suppose a 5.0-L bottle of gas with a pressure of 2.50 atm at 20°C is heated to 80°C. We can calculate the new pressure using the combined gas law. Before we start working mathematically, however, let's do some reasoning. The volume of the bottle hasn't changed, and neither has the number of moles of gas inside. Only the temperature and pressure have changed, so this is really a Gay-Lussac's law problem. From Gay-Lussac's law you know that if you increase the temperature, the pressure should increase if the amount and volume are constant. This means that when you calculate the new pressure, it should be greater than 2.50 atm; if it is less, you've made an error. Also, **remember that the temperatures must be expressed in kelvin**. 20°C = 293 K (K = °C + 273) and 80°C = 353 K.

We will be solving for P_2, so we will take the combined gas law and rearrange for P_2:

$$(T_2P_1V_1)/(T_1V_2) = P_2$$

Substituting in the values:

$$(353\ \text{K})(2.50\ \text{atm})(5.0\ \text{L})/(293\ \text{K})(5.0\ \text{L}) = P_2$$

$$3.0\ \text{atm} = P_2$$

The new pressure is greater than the original pressure, making the answer a reasonable one. Note that all the units canceled except atm, which is the unit that you wanted.

Let's look at a situation in which two conditions change. Suppose a balloon has a volume at sea level of 10.0 L at 760.0 torr and 20°C (293 K). The balloon is released and rises to an altitude where the pressure is 450.0 torr and the temperature is −10°C (263 K). You want to calculate the new volume of the balloon. You know that you have to express the temperature in K in the calculations. It is perfectly fine to leave the pressures in torr. It really doesn't matter what pressure and volume units you use, as long as they are consistent in the problem. The pressure is decreasing, so that should cause the volume to increase (Boyle's law). The temperature is decreasing, so that should cause the volume to decrease (Charles's law). Here you have two competing factors, so it is difficult to predict the end result. You'll simply have to do the calculations and see.

Using the combined gas equation, solve for the new volume (V_2):

$$(P_1V_1)/T_1 = (P_2V_2)/T_2$$

$$(P_1V_1T_2)/(P_2T_1) = V_2$$

Now substitute the known quantities into the equation. (You could substitute the knowns into the combined gas equation first, and then solve for the volume. Do it whichever way is easier for you.)

$$(760.0\ \text{torr})(10.0\ \text{L})(263\ \text{K})/(450.0\ \text{torr})(293\ \text{K}) = V_2$$

$$15.2\ \text{L} = V_2$$

Note that the units canceled, leaving the desired volume unit of liters. Overall, the volume did increase, so in this case the pressure decrease had a greater effect than the temperature decrease. This seems reasonable, looking at the numbers. There is a relatively small change in the Kelvin temperature (293 K versus 263 K) compared to a much larger change in the pressure (760.0 torr versus 450.0 torr).

Volume–Amount Relationship: Avogadro's Law

In all the gas law problems so far, the amount of gas has been constant. But what if the amount changes? That is where Avogadro's law comes into play.

If a container is kept at constant pressure and temperature, and you increase the number of gas particles in that container, the volume will have to increase in order to keep the pressure constant. This means that there is a direct relationship between the volume and the number of moles of gas (n). This is **Avogadro's law** and mathematically it looks like this:

$$V/n = k_a \quad \text{or} \quad V_1/n_1 = V_2/n_2$$

We could work this into the combined gas law, but more commonly the amount of gas is related to the other physical properties through another relationship that Avogadro developed:

1 mol of any gas occupies 22.4 L at STP
[Standard Temperature and Pressure of 0°C (273 K) and 1 atm]

The combined gas law and Avogadro's relationship can then be combined into the ideal gas equation, which incorporates the pressure, volume, temperature, and amount relationships of a gas.

Ideal Gas Equation

The **ideal gas equation** has the mathematical form of $PV = nRT$, where:

P = pressure of the gas in atm, torr, mm Hg, Pa, etc.
V = volume of the gas in L, mL, etc.
n = number of moles of gas
R = ideal gas constant: 0.0821 L·atm/K·mol
T = Kelvin temperature

This is the value for R if the volume is expressed in liters, the pressure in atmospheres, and the temperature in kelvin (naturally). You could calculate another ideal gas constant based on different units of pressure and volume, but the simplest thing to do is to use the 0.0821 and convert the given volume to liters and the pressure to atm. And remember that you **must express the temperature in kelvin.**

Let's see how we might use the ideal gas equation. Suppose you want to know what volume 20.0 g of hydrogen gas would occupy at 27°C and 0.950 atm. You have the pressure in atm, you can get the temperature in kelvin (27°C + 273 = 300. K), but you will need to convert the grams of hydrogen gas to moles of hydrogen gas before you can use the ideal gas equation. Also, remember that hydrogen gas is diatomic, H_2.

First you'll convert the 20.0 g to moles:

$$(20.0\ \text{g/l}) \times (1\ \text{mol}\ H_2/2.016\ \text{g}) = 9.921\ \text{mol}\ H_2$$

(We're not worried about significant figures at this point, since this is an intermediate calculation.)

Now you can solve the ideal gas equation for the unknown quantity, the volume:

$$PV = nRT$$

$$V = nRT/P$$

Finally, plug in the numerical values for the different known quantities:

$$V = (9.921\ \text{mol})(0.0821\ \text{L atm/K mol})\ (300.\text{K})/0.950\ \text{atm}$$

$$V = 257\ \text{L}$$

Is the answer reasonable? You have almost 10 mol of gas. It would occupy about 224 L at STP (10 mol × 22.4 L/mol) by Avogadro's relationship. The pressure is slightly less than standard pressure of 1 atm, which would tend to increase the volume (Boyle's law), and temperature is greater than standard temperature of 0°C, which would also increase the volume (Charles's law). So you might expect a volume greater than 224 L, and that is exactly what you found.

Remember, the final thing you do when working any type of chemistry problem is answer the question: Is the answer reasonable?

Dalton's Law of Partial Pressures

Dalton's law says that in a mixture of gases (A + B + C . . .) the total pressure is simply the sum of the partial pressures (the pressures associated with each individual gas). Mathematically, Dalton's law looks like this:

$$P_{\text{Total}} = P_A + P_B + P_C + \cdots$$

Commonly Dalton's law is used in calculations involving the collection of a gas over water, as in the displacement of water by oxygen gas. In this situation there is a gas mixture: O_2 and water vapor, $H_2O(g)$. The total pressure in this case is usually atmospheric pressure, and the partial pressure of the water vapor is determined by looking up the vapor pressure of water at the temperature of the water in a reference book. Simple subtraction generates the partial pressure of the oxygen.

If you know how many moles of each gas are in the mixture and the total pressure, you can calculate the partial pressure of each gas by multiplying the total pressure by the mole fraction of each gas:

$$P_A = (P_{\text{Total}})(X_A)$$

where X_A = mole fraction of gas A. The mole fraction of gas A would be equal to the moles of gas A divided by the total moles of gas in the mixture.

Graham's Law of Diffusion and Effusion

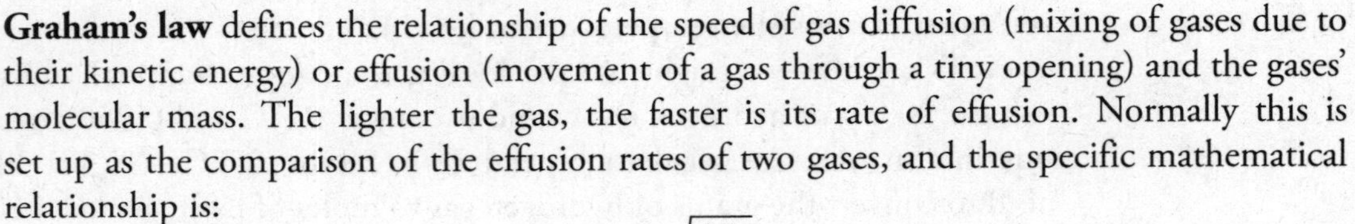

Graham's law defines the relationship of the speed of gas diffusion (mixing of gases due to their kinetic energy) or effusion (movement of a gas through a tiny opening) and the gases' molecular mass. The lighter the gas, the faster is its rate of effusion. Normally this is set up as the comparison of the effusion rates of two gases, and the specific mathematical relationship is:

$$\frac{r_1}{r_2} = \sqrt{\frac{M_2}{M_1}}$$

where r_1 and r_2 are the rates of effusion/diffusion of gases 1 and 2 respectively, and M_2 and M_1 are the molecular masses of gases 2 and 1 respectively. **Note that this is an inverse relationship.**

For example, suppose you wanted to calculate the ratio of effusion rates for hydrogen and nitrogen gases. Remember that both are diatomic, so the molecular mass of H_2 is 2.016 g/mol and the molecular mass of N_2 would be 28.02 g/mol. Substituting into the Graham's law equation:

$$\frac{r_{H_2}}{r_{N_2}} = \sqrt{\frac{M_{N_2}}{M_{H_2}}}$$

$$r_{H_2}/r_{N_2} = (28.02\ \text{g/mol}/2.016\ \text{g/mol})^{1/2} = (13.899)^{1/2} = 3.728$$

Hydrogen gas would effuse through a pinhole 3.728 times as fast as nitrogen gas. The answer is reasonable, since the lower the molecular mass, the faster the gas is moving. Sometimes we measure the effusion rates of a known gas and an unknown gas, and use Graham's law to calculate the molecular mass of the unknown gas.

Gas Stoichiometry

KEY IDEA

The gas law relationships can be used in reaction stoichiometry problems. For example, suppose you have a mixture of $KClO_3$ and NaCl, and you want to determine how many grams of $KClO_3$ are present. You take the mixture and heat it. The $KClO_3$ decomposes according to the equation:

$$2\ KClO_3(s) \rightarrow 2\ KCl(s) + 3\ O_2(g)$$

The oxygen gas that is formed is collected by displacement of water. It occupies a volume of 542 mL at 27°C. The atmospheric pressure is 755.0 torr. The vapor pressure of water at 27°C is 26.7 torr.

First, you need to determine the pressure of just the oxygen gas. It was collected over water, so the total pressure of 755.0 torr is the sum of the partial pressures of the oxygen and the water vapor:

$$P_{Total} = P_{O_2} + P_{H_2O} \text{(Dalton's law)}$$

The partial pressure of water vapor at 27°C is 26.7 torr, so the partial pressure of the oxygen can be calculated by:

$$P_{O_2} = P_{Total} - P_{H_2O} = 755.0 \text{ torr} - 26.7 \text{ torr} = 728.3 \text{ torr}$$

At this point you have 542 mL of oxygen gas at 728.3 torr and 300. K (27°C + 273). From this data you can use the ideal gas equation to calculate the number of moles of oxygen gas produced:

$$PV = nRT$$
$$PV/RT = n$$

You will need to convert the pressure from torr to atm:

$$(728.3 \text{ torr}) \times (1 \text{ atm}/760.0 \text{ torr}) = 0.9583 \text{ atm}$$

and express the volume in liters: 542 mL = 0.542 L

Now you can substitute the quantities into the ideal gas equation:

$$(0.9583 \text{ atm})(0.542 \text{ L})/(0.0821 \text{ L} \cdot \text{atm/K} \cdot \text{mol})(300. \text{ K}) = n$$
$$0.02110 \text{ mol } O_2 = n$$

Now you can use the reaction stoichiometry to convert from moles O_2 to moles $KClO_3$ and then to grams $KClO_3$:

$$(0.02110 \text{ mol } O_2)\left(\frac{2 \text{ mol } KClO_3}{3 \text{ mol } O_2}\right)\left(\frac{122.55 \text{ g } KClO_3}{1 \text{ mol } KClO_3}\right) = 1.723 \text{ g } KClO_3$$

Non-Ideal Gases

We have been considering ideal gases, that is, gases that obey the postulates of the Kinetic Molecular Theory. But remember—a couple of those postulates were on shaky ground. The volume of the gas molecules was negligible, and there were no attractive forces between the gas particles. Many times approximations are fine and the ideal gas equation works well. But it would be nice to have a more accurate model for doing extremely precise work or when a gas exhibits a relatively large attractive force. In 1873, Johannes van der Waals introduced a modification of the ideal gas equation that attempted to take into account the volume and attractive forces of real gases by introducing two constants—*a* and *b*—into the ideal gas equation. Van der Waals realized that the actual volume of the gas is less than the ideal gas because gas molecules have a finite volume. He also realized that the more moles of gas present, the greater the real volume. He compensated for the volume of the gas particles mathematically with:

$$\text{corrected volume} = V - nb$$

where n is the number of moles of gas and b is a different constant for each gas. The larger the gas particles, the more volume they occupy and the larger the b value.

The attraction of the gas particles for each other tends to lessen the pressure of the gas, because the attraction slightly reduces the force of gas particle collisions with the container walls. The amount of attraction depends on the concentration of gas particles and the magnitude of the particles' intermolecular force. The greater the intermolecular forces of the gas, the higher the attraction is, and the less the real pressure. Van der Waals compensated for the attractive force with:

$$\text{corrected pressure} = P + an^2/V^2$$

where a is a constant for individual gases. The greater the attractive force between the molecules, the larger the value of a. The n^2/V^2 term corrects for the concentration. Substituting these corrections into the ideal gas equation gives **van der Waals equation**:

$$(P + an^2/V^2)(V - nb) = nRT$$

The larger, more concentrated, and stronger the intermolecular forces of the gas, the more deviation from the ideal gas equation one can expect and the more useful the van der Waals equation becomes.

Experiments

Gas law experiments generally involve pressure, volume, and temperature measurements. In a few cases, other measurements such as mass and time are necessary. You should remember that ΔP, for example, is NOT a measurement; the initial and final pressure measurements are the actual measurements made in the laboratory. Another common error is the application of gas law type information and calculations for non-gaseous materials.

A common consideration is the presence of water vapor, $H_2O(g)$. Water generates a vapor pressure, which varies with the temperature. Dalton's law is used in these cases to adjust the pressure of a gas sample for the presence of water vapor. The total pressure (normally atmospheric pressure) is the pressure of the gas or gases being collected and the water vapor. When the pressure of an individual gas is needed, the vapor pressure of water is subtracted from the total pressure. Finding the vapor pressure of water requires measuring the temperature and using a table showing vapor pressure of water versus temperature.

In experiments on Graham's law, time is measured. The amount of time required for a sample to effuse is the measurement. The amount of material effusing divided by the time elapsed is the rate of effusion.

Most gas law experiments use either the combined gas law or the ideal gas equation. Moles of gas are a major factor in many of these experiments. The combined gas law can generate the moles of a gas by adjusting the volume to STP and using Avogadro's relationship of 22.4 L/mol at STP. The ideal gas equation gives moles from the relationship $n = PV/RT$.

HINT: Make sure the conditions are STP before using 22.4 L/mol.

Two common gas law experiments are "Determination of Molar Mass by Vapor Density" and "Determination of the Molar Volume of a Gas." While it is possible to use the combined gas law (through 22.4 L/mol at STP) for either of these, the ideal gas equation is easier to use. The values for P, V, T, and n must be determined.

The temperature may be determined easily using a thermometer. The temperature measurement is normally in °C. The °C must then be converted to a Kelvin temperature (K = °C + 273).

Pressure is measured using a barometer. If water vapor is present, a correction is needed in the pressure to compensate for its presence. The vapor pressure of water is found in a table of vapor pressure versus temperature. Subtract the value found in this table from the measured pressure (Dalton's law). Values from tables are not considered to be measurements for an experiment. If you are going to use 0.0821 L atm/mol K for *R*, convert the pressure to atmospheres.

The value of *V* may be measured or calculated. A simple measurement of the volume of a container may be made, or a measurement of the volume of displaced water may be required. Calculating the volume requires knowing the number of moles of gas present. No matter how you get the volume, don't forget to convert it to liters when using $PV = nRT$ or STP.

The values of *P*, *T*, and *V* discussed above may be used, through the use of the ideal gas equation, to determine the number of moles present in a gaseous sample. Stoichiometry is the alternate method of determining the number of moles present. A quantity of a substance is converted to a gas. This conversion may be accomplished in a variety of ways. The most common stoichiometric methods are through volatilization or reaction. The volatilization method is the simplest. A weighed quantity (measure the mass) of a substance is converted to moles by using the molar mass (molecular weight). If a reaction is taking place, the quantity of one of the substances must be determined (normally with the mass and molar mass) and then, through the use of the mole-to-mole ratio, this value is converted to moles.

The values of *P*, *T*, and *n* may be used to determine the volume of a gas. If this volume is to be used with Avogadro's law of 22.4 L/mol, the combined gas law must be employed to adjust the volume to STP. This equation will use the measured values for *P* and *T* along with the calculated value of *V*. These values are combined with STP conditions (0°C [273.15 K] and 1.00 atm) to determine the molar volume of a gas.

Combining the value of *n* with the measured mass of a sample will allow you to calculate the molar mass of the gas.

Do not forget: Values found in tables and conversions from one unit to another are not experimental measurements.

Common Mistakes to Avoid

1. When using any of the gas laws, be sure you are dealing with gases, not liquids or solids. We've lost track of how many times we've seen people apply gas laws in situations in which no gases were involved.
2. In any of the gas laws, be sure to express the **temperature in kelvin**. Failure to do so is a quite common mistake.
3. Be sure, especially in stoichiometry problems involving gases, that you are calculating the volume, pressure, etc. of the correct gas. You can avoid this mistake by clearly labeling your quantities (*moles of* O_2 instead of just *moles*).
4. Make sure your **answer is reasonable**. Analyze the problem; don't just write a number down from your calculator. Be sure to check your number of significant figures.
5. If you have a gas at a certain set of volume/temperature/pressure conditions and the conditions change, you will probably use the combined gas equation. If moles of gas are involved, the ideal gas equation will probably be useful.
6. Make sure your **units cancel**.
7. In using the combined gas equation, make sure you group all initial-condition quantities on one side of the equals sign and all final-condition quantities on the other side.
8. Be sure to use the correct molecular mass for those gases that exist as diatomic molecules—H_2, N_2, O_2, F_2, Cl_2, and Br_2 and I_2 vapors.
9. If the value 22.4 L/mol is to be used, make absolutely sure that it is applied to a **gas** at **STP**.

› Review Questions

Use these questions to review the content of this chapter and practice for the AP Chemistry exam. First are 16 multiple-choice questions similar to what you will encounter in Section I of the AP Chemistry exam. Following those is a multipart free-response question like the ones in Section II of the exam. To make these questions an even more authentic practice for the actual exam, time yourself following the instructions provided.

Multiple-Choice Questions

Answer the following questions in 20 minutes. You may not use a calculator. You may use the periodic table and the equation sheet at the back of this book.

1. A sample of argon gas is sealed in a container. The volume of the container is doubled. If the pressure remains constant, what happens to the absolute temperature?

(A) It does not change.
(B) It is halved.
(C) It is doubled.
(D) It is squared.

2. A sealed, rigid container is filled with three ideal gases: A, B, and C. The partial pressure of each gas is known. The temperature and volume of the system are known. What additional information is necessary to determine the masses of the gases in the container?

(A) the average distance traveled between molecular collisions
(B) the intermolecular forces
(C) the molar masses of the gases
(D) the total pressure

3. Two balloons are at the same temperature and pressure. One contains 14 g of nitrogen and the other contains 20.0 g of argon. Which of the following is true?

(A) The density of the nitrogen sample is greater than the density of the argon sample.
(B) The average speed of the nitrogen molecules is greater than the average speed of the argon molecules.
(C) The average kinetic energy of the nitrogen molecules is greater than the average kinetic energy of the argon molecules.
(D) The volume of the nitrogen container is less than the volume of the argon container.

4. An experiment to determine the molar mass of a gas begins by heating a solid to produce a gaseous product. The gas passes through a tube and displaces water in an inverted, water-filled bottle. Which of the following items may be determined after the experiment is completed?

(A) vapor pressure of water
(B) temperature of the displaced water
(C) barometric pressure in the room
(D) mass of the solid used

5. The true volume of a real gas is larger than that calculated from the ideal gas equation. This occurs because the ideal gas equation does consider which of the following?

(A) the attraction between the molecules
(B) the shape of the molecules
(C) the volume of the molecules
(D) the mass of the molecules

6. Aluminum metal reacts with HCl to produce aluminum chloride and hydrogen gas. How many grams of aluminum metal must be added to an excess of HCl to produce 33.6 L of hydrogen gas, if the gas is at STP?

(A) 18.0 g
(B) 35.0 g
(C) 27.0 g
(D) 4.50 g

7. A reaction produces a gaseous mixture of carbon dioxide, carbon monoxide, and water vapor. After one reaction, the mixture was analyzed and found to contain 0.60 mol of carbon dioxide, 0.30 mol of carbon monoxide, and 0.10 mol of water vapor. If the total pressure of the mixture was 0.80 atm, what was the partial pressure of the carbon monoxide?

(A) 0.080 atm
(B) 0.34 atm
(C) 0.13 atm
(D) 0.24 atm

8. A sample of methane gas was collected over water at 35°C. The sample had a total pressure of 756 mm Hg. Determine the partial pressure of the methane gas in the sample. (The vapor pressure of water at 35°C is 41 mm Hg.)

(A) 760 mm Hg
(B) 41 mm Hg
(C) 715 mm Hg
(D) 797 mm Hg

9. A 1.15 mol sample of carbon monoxide gas has a temperature of 27°C and a pressure of 0.300 atm. If the temperature were lowered to 17°C, at constant volume, what would be the new pressure?

(A) 0.290 atm
(B) 0.519 atm
(C) 0.206 atm
(D) 0.338 atm

10. An ideal gas sample weighing 1.28 g at 127°C and 1.00 atm has a volume of 0.250 L. Determine the molar mass of the gas.

(A) 322 g/mol
(B) 168 g/mol
(C) 0.00621 g/mol
(D) 80.5 g/mol

11. Increasing the temperature of an ideal gas from 50°C to 75°C at constant volume will cause which of the following to increase for the gas?

(A) the average molecular mass of the gas
(B) the average distance between the molecules
(C) the average speed of the molecules
(D) the density of the gas

12. If a sample of CH_4 effuses at a rate of 9.0 mol per hour at 35°C, which of the gases below will effuse at approximately twice the rate under the same conditions?

(A) CO
(B) He
(C) O_2
(D) F_2

13. A steel tank containing argon gas has additional argon gas pumped into it at constant temperature. Which of the following is true for the gas in the tank?

(A) There is no change in the number of gas atoms.
(B) There is an increase in the volume of the gas.
(C) There is a decrease in the pressure exerted by the gas.
(D) The gas atoms travel with the same average speed.

14. Choose the gas that probably shows the greatest deviation from ideal gas behavior.

(A) He
(B) O_2
(C) SF_4
(D) SiH_4

15. Determine the formula for a gaseous silane (Si_nH_{2n+2}) if it has a density of 5.47 g per L at 0°C and 1.00 atm.

(A) SiH_4
(B) Si_2H_6
(C) Si_3H_8
(D) Si_4H_{10}

16. Which of the following best explains why a hot air balloon rises?

(A) The heating of the air causes the pressure inside the balloon to increase.
(B) The cool outside air pushes the balloon higher.
(C) The temperature difference between the inside and outside air causes convection currents.
(D) Hot air has a lower density than cold air.

Answers and Explanations for the Multiple-Choice Questions

1. C—This question relates to the combined gas law: $P_1V_1/T_1 = P_2V_2/T_2$. Since the pressure remains constant, the pressures may be removed from the combined gas law to produce Charles's law: $V_1/T_1 = V_2/T_2$. This equation may be rearranged to: $T_2 = V_2T_1/V_1$. The doubling of the volume means $V_2 = 2\ V_1$. On substituting: $T_2 = 2V_1T_1/V_1$; giving $T_2 = 2T_1$. The identity of the gas is irrelevant in this problem.

2. C—This problem depends on the ideal gas equation: $PV = nRT$. R, V, and T are known, and by using the partial pressure for a gas, the number of moles of that gas may be determined. To convert from moles to mass, the molar mass of the gas is necessary.

3. B—Since T and P are known, and since the moles (n) can be determined from the masses given, this question could use the ideal gas equation. The number of moles of each gas is 0.50. Equal moles of gases, at the same T and P, have equal volumes, which eliminates answer choice D. Equal volume also means that the greater mass has the greater density, eliminating choice A. The average kinetic energy of a gas depends on the temperature. If the temperatures are the same, then the average kinetic energy is the same, eliminating C. Finally, at the same temperature, heavier gases travel slower than lighter gases. Nitrogen is lighter than argon, so it travels at a faster average speed, making B the correct answer. You may find this type of reasoning process beneficial on any question in which you do not immediately know the answer.

4. A—This experiment requires the ideal gas equation. The mass of the solid is needed (to convert to moles); this eliminates answer choice D. The volume, temperature, and pressure must also be measured during the experiment, eliminating choices B and C. The measured pressure is the total pressure. Eventually the total pressure must be converted to the partial pressure of the gas using Dalton's law. The total pressure is the sum of the pressure of the gas plus the vapor pressure of water. The vapor pressure of water can be looked up in a table when the calculations are performed (only the temperature is needed to find the vapor pressure in a table). Answer A is correct.

5. C—Real gases are different from ideal gases because of two basic factors (see the van der Waals equation): molecules have a volume, and molecules attract each other. The molecules' volume is subtracted from the observed volume for a real gas (giving a smaller volume), and the pressure has a term added to compensate for the attraction of the molecules (correcting for a smaller pressure). Since these are the only two directly related factors, answers B and D are eliminated. The question is asking about volume; thus, the answer is C.

6. C—A balanced chemical equation is necessary:

$$2\ Al(s) + 6\ HCl(aq) \rightarrow 2\ AlCl_3(aq) + 3\ H_2(g)$$

The reaction produced 33.6 L/22.4 L or 1.50 mol, at STP. To produce this quantity of hydrogen, (2 mol Al/3 mol H_2) × 1.50 mol H_2 = 1.00 mol of Al is needed. The atomic weight of Al is 27.0 g/mol; thus, 27.0 g of Al are required.

7. D—The partial pressure of any gas is equal to its mole fraction times the total pressure. The mole fraction of carbon monoxide is [0.30/(0.60 + 0.30 + 0.10)] = 0.30, and the partial pressure of CO is 0.30 × 0.80 atm = 0.24 atm.

8. C—Using Dalton's law ($P_{Total} = P_A + P_B + \ldots$), the partial pressure may be found by:

$$756\text{ mm Hg} - 41\text{ mm Hg} = 715\text{ mm Hg}$$

9. A—You can begin by removing the volume (constant) from the combined gas law to produce Gay-Lussac's law = $P_1/T_1 = P_2/T_2$. This equation rearranges to $P_2 = P_1T_2/T_1$ = 0.300 atm × 290.0 K)/(300.0 K) = 0.290 atm. Estimation works well in the question as the "slight" temperature change should give a slight decrease in pressure. The moles are not important since they do not change. Some of the other answers result from common errors.

10. B—The molar mass may be obtained by dividing the grams by the number of moles (calculated from the ideal gas equation). Estimation works in this case as $n = PV/RT$ = (0.25)/(0.1 × 400). Do not forget to convert the temperature to kelvin.

11. C—Choice B requires an increase in volume, not allowed by the problem. Choice C requires an increase in temperature. Choice A requires a change in the composition of the gas. Choice D requires a decrease in the volume.

12. B—Lighter gases effuse faster. The only gas among the choices that is lighter than methane is helium. To calculate the molar mass, you would begin with the molar mass of methane and divide by the rate difference squared.

13. D—A steel tank will have a constant volume, and the problem states that the temperature is constant. Adding gas to the tank will increase the number of moles of the gas and the pressure (forcing the argon atoms closer together). A constant temperature means there will be a constant average speed.

14. C—Deviations from ideal behavior depend on the size and the intermolecular forces between the molecules. The greatest deviation would be for a large polar molecule. Sulfur tetrafluoride is the largest molecule, and it is the only polar molecule listed.

15. D—The molar mass of gas must be determined. The simplest method to find the molar mass is: (5.47 g/L) × (22.4 L/mol) = 123 g/mol (simple factor label). The molar mass may also be determined by dividing the mass of the gas by the moles (using 22.4 L/mol for a gas at STP and using 1 L). If you did not recognize the conditions as STP, you could find the moles from the ideal gas equation. The correct answer is the gas with the molar mass closest to 123 g/mol.

16. D—The hot air balloon rises because it has a lower density than air. Less dense objects will float on more dense objects. In other words, "lighter" objects will float on "heavy" objects.

Free-Response Question

You have 20 minutes to do the following question. You may use a calculator and the tables in the back of the book.

Question 1

A sample containing 2/3 mol of potassium chlorate, $KClO_3$, is heated until it decomposes to potassium chloride, KCl, and oxygen gas, O_2. The oxygen is collected in an inverted bottle through the displacement of water. Answer the following questions using this information.

(a) Write a balanced chemical equation for the reaction.

(b) Calculate the number of moles of oxygen gas produced.

(c) The temperature and pressure of the sample are adjusted to STP. The volume of the sample is slightly greater than 22.4 liters. Explain.

(d) An excess of sulfur, S, is burned in one mole of oxygen, in the presence of a catalyst, to form gaseous sulfur trioxide, SO_3. Write a balanced chemical equation and calculate the number of moles of gas formed.

(e) After the sulfur had completely reacted, a sample of the residual water was removed from the bottle and found to be acidic. Explain.

Answer and Explanation for the Free-Response Question

(a) $2\ KClO_3(s) \rightarrow 2\ KCl(s) + 3\ O_2(g)$

You get 1 point if you have the above equation.

(b) $(2/3 \text{ mol } KClO_3)(3 \text{ mol } O_2/2 \text{ moles } KClO_3) = 1 \text{ mol } O_2$

You get 1 point for the correct answer and 1 point for the work. You can get these points if you correctly use information from an incorrect equation in part **a**.

(c) At STP, the volume of 1 mol of O_2 should be 22.4 L. The volume is greater because oxygen was not the only gas in the sample. Water vapor was present. The presence of the additional gas leads to a larger volume.

You get 1 point for discussing STP and 22.4 L, and 1 point for discussing the presence of water vapor.

(d) The equation is:

$$2\ S(s) + 3\ O_2(g) \rightarrow 2\ SO_3(g)$$

According to this equation:

$$(1 \text{ mol } O_2)(2 \text{ mol } SO_3/3 \text{ mol } O_2) = 2/3 \text{ mol } SO_3$$

You get 1 point for each of these solutions. You can get the 1 point if you used an incorrect number of moles of O_2 from an incorrectly balanced equation.

(e) A nonmetal oxide, such as sulfur trioxide, will dissolve in water to produce an acid.

This will get you 1 point; you may wish to include the following equation:

$$SO_3(g) + H_2O(l) \rightarrow H_2SO_4(aq)$$

Total your points for the different parts. There are 8 points possible.

› Rapid Review

- Kinetic Molecular Theory—Gases are small particles of negligible volume moving in a random straight-line motion, colliding with the container walls (that is the gas pressure) and with each other. During these collisions no energy is lost, but energy may be transferred from one particle to another; the Kelvin temperature is proportional to the average kinetic energy. There is assumed to be no attraction between the particles.
- Pressure—Know how a barometer operates and the different units used in atmospheric pressure.
- Boyle's law—The volume and pressure of a gas are inversely proportional if the temperature and amount are constant.
- Charles's law—The volume and temperature of a gas are directly proportional if the amount and pressure are constant.
- Gay-Lussac's law—The pressure and temperature of a gas are directly proportional if the amount and volume are constant.
- Combined gas law—Know how to use the combined gas equation $P_1V_1/T_1 = P_2V_2/T_2$.

- Avogadro's law—The number of moles and volume of a gas are directly proportional if the pressure and temperature are constant. Remember that 1 mol of an ideal gas at STP (1 atm and 0°C) occupies a volume of 22.4 L. Remember that you should not use the 22.4 L unless the gas is at STP.
- Ideal gas equation—Know how to use the ideal gas equation $PV = nRT$.
- Dalton's law—The sum of the partial pressures of the individual gases in a gas mixture is equal to the total pressure: $P_{Total} = P_A + P_B + P_c + \ldots$
- Graham's law—The lower the molecular mass of a gas, the faster it will effuse/diffuse. Know how to use Graham's law: $\frac{r_1}{r_2} = \sqrt{\frac{M_2}{M_1}}$.
- Gas stoichiometry—Know how to apply the gas laws to reaction stoichiometry problems.
- Non-ideal gases—Know how the van der Waals equation accounts for the non-ideal behavior of real gases.
- Tips—Make sure the **temperature is in kelvin**; gas laws are being applied to gases only; the **units cancel**; and the **answer is reasonable.**
- Gas laws are very useful for gases, but not for liquids and solids. Before applying a gas law, be sure you are dealing with a gas.

CHAPTER 9

Thermodynamics

IN THIS CHAPTER

Summary: Thermodynamics is the study of heat and its transformations. **Thermochemistry** is the part of thermodynamics that deals with changes in heat that take place during chemical processes. We will be describing energy changes in this chapter. Energy can be of two types: kinetic or potential. **Kinetic energy** is energy of motion, while **potential energy** is stored energy. Energy can be converted from one form to another but, unless a nuclear reaction occurs, energy cannot be created or destroyed (Law of Conservation of Energy). We will discuss energy exchanges between a system and the surroundings. The **system** is that part of the universe that we are studying. It may be a beaker or it may be Earth. The **surroundings** are the rest of the universe.

The most common units of energy used in the study of thermodynamics are the joule and the calorie. The **joule (J)** is defined as:

$$1 \text{ J} = 1 \text{ kg m}^2/\text{s}^2$$

The **calorie** was originally defined as the amount of energy needed to raise the temperature of 1 g of water 1°C. Now it is defined in terms of its relationship to the joule:

$$1 \text{ cal} = 4.184 \text{ J}$$

It is important to realize that this is not the same calorie that is commonly associated with food and diets. That is the nutritional Calorie, Cal, which is really a kilocalorie (1 Cal = 1,000 cal).

Keywords and Equations

$S°$ = standard entropy
$H°$ = standard enthalpy
$G°$ = standard free energy
q = heat
c = specific heat capacity
C_p = molar heat capacity at constant pressure

$\Delta S° = \Sigma S°$ products $- \Sigma S°$ reactants

$\Delta H° = \Sigma\, \Delta H_f°$ products $- \Sigma\, \Delta H_f°$ reactants

$\Delta G° = \Sigma\, \Delta G_f°$ products $- \Sigma\, \Delta G_f°$ reactants

$$\Delta G° = \Delta H° - T\Delta S°$$

$$= -RT \ln K = -2.303\ RT \log K$$

$$= -n\,F\,E°$$

$$\Delta G = \Delta G° + RT \ln Q = \Delta G° + 2.303\ RT \log Q$$

$$q = mc\Delta T$$

$$C_p = \frac{\Delta H}{\Delta T}$$

Gas constant, $R = 8.31\ \text{J mol}^{-1}\ \text{K}^{-1}$

Calorimetry

Calorimetry is the laboratory technique used to measure the heat released or absorbed during a chemical or physical change. The quantity of heat absorbed or released during a chemical or physical change is represented as q and is proportional to the change in temperature of the system being studied. This system has what is called a **heat capacity** (C_p), which is the quantity of heat needed to change the temperature 1 K. It has the form:

$$C_p = \text{heat capacity} = q/\Delta T$$

Heat capacity most commonly has units of J/K. The **specific heat capacity (or specific heat) (c)** is the quantity of heat needed to raise the temperature of 1 g of a substance 1 K:

$$c = q/(m \times \Delta T) \text{ or } q = cm\Delta T,$$

where m is the mass of the substance.

The specific heat capacity commonly has units of J/g • K. Because of the original definition of the calorie, the specific heat capacity of water is 4.184 J/g • K. If the specific heat capacity, the mass, and the change of temperature are all known, the amount of energy absorbed can easily be calculated.

Another related quantity is the **molar heat capacity (*C*)**, the amount of heat needed to change the temperature of 1 mol of a substance by 1 K.

Calorimetry involves the use of a laboratory instrument called a calorimeter. Two types of calorimeter, a simple coffee-cup calorimeter and a more sophisticated bomb calorimeter, are shown in Figure 9.1, on the next page. In both, a process is carried out with known amounts of substances and the change in temperature is measured.

The coffee-cup calorimeter can be used to measure the heat changes in reactions or processes that are open to the atmosphere: q_p, constant-pressure reactions. These might be reactions that occur in open beakers and the like. This type of calorimeter is also commonly used to measure the specific heats of solids. A known mass of solid is heated to a certain temperature and then is added to the calorimeter containing a known mass of water at a known temperature. The final temperarure is then measured allowing us to calculate the ΔT. We know that the heat lost by the solid (the system) is equal to the heat gained by the surroundings (the water and calorimeter, although for simple coffee-cup calorimetry the heat gained by the calorimeter is small and is ignored):

$$-q_{\text{solid}} = q_{\text{water}}$$

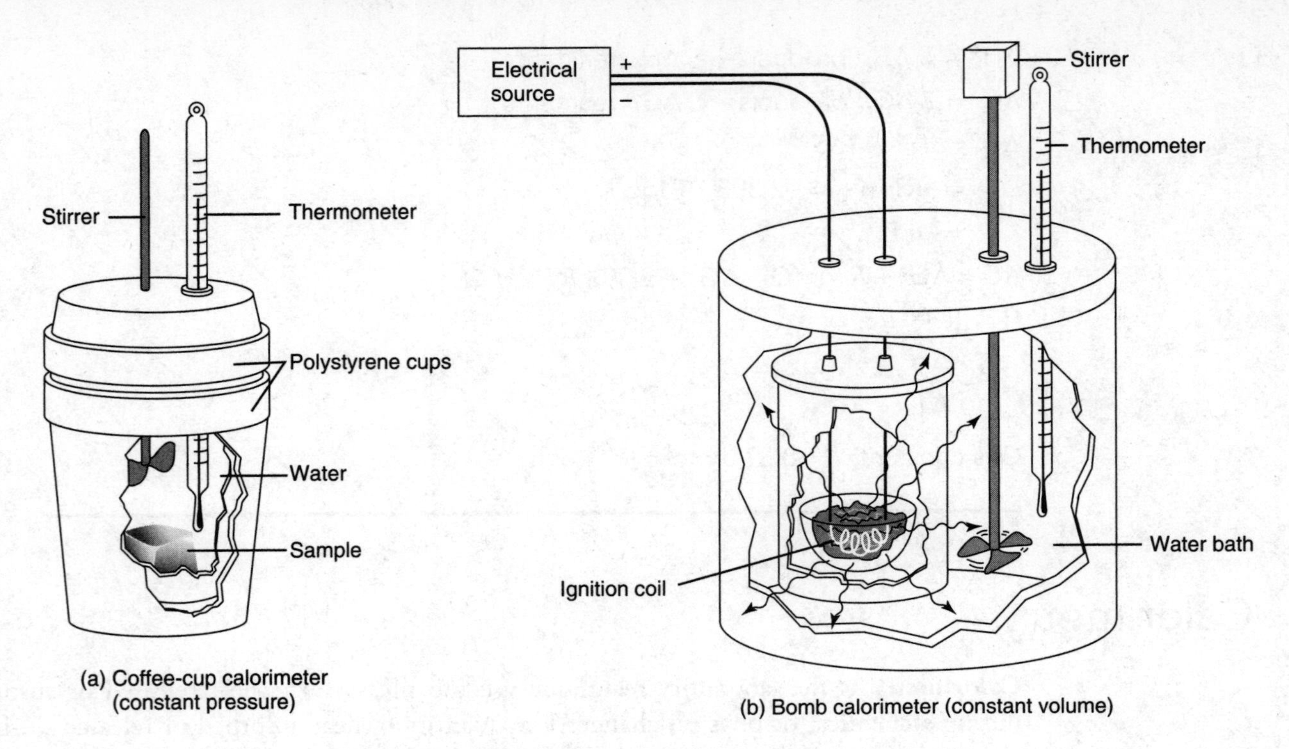

Figure 9.1 Two types of calorimeters.

Substituting the mathematical relationship for q gives:

$$-(c_{\text{solid}} \times m_{\text{solid}} \times \Delta T_{\text{solid}}) = c_{\text{water}} \times m_{\text{water}} \times \Delta T_{\text{water}}$$

This equation can then be solved for the specific heat capacity of the solid.

The constant-volume bomb calorimeter is used to measure the energy changes that occur during combustion reactions. A weighed sample of the substance being investigated is placed in the calorimeter, and compressed oxygen is added. The sample is ignited by a hot wire, and the temperature change of the calorimeter and a known mass of water is measured. The heat capacity of the calorimeter/water system is sometimes known.

For example, a 1.5886 g sample of glucose ($C_6H_{12}O_6$) was ignited in a bomb calorimeter. The temperature increased by 3.682°C. The heat capacity of the calorimeter was 3.562 kJ/°C, and the calorimeter contained 1.000 kg of water. Find the molar heat of reaction (i.e., kJ/mole) for:

$$C_6H_{12}O_6(s) + 6\ O_2(g) \rightarrow 6\ CO_2(g) + 6\ O(l)$$

Answer:

$$\frac{(3.562\,\text{kJ})}{(°\text{C})}(3.682°\text{C}) = 13.12\,\text{kJ}$$

$$(1.000\ \text{kg})\left(\frac{1{,}000\ \text{g}}{1\ \text{kg}}\right)\left(\frac{4.184\ \text{J}}{\text{g°C}}\right)\left(\frac{1\ \text{kJ}}{1{,}000\ \text{J}}\right)(3.682°\text{C}) = 15.40\ \text{kJ}$$

$$\text{total heat} = 13.12\ \text{kJ} + 15.40\ \text{kJ} = 28.52\ \text{kJ}$$

Note: The temperature increased so the reaction was exothermic (–)

$$\rightarrow -28.52\ \text{kJ}$$

This is not molar (yet)

$$(1.5886\,\text{g})\frac{(1\,\text{mol})}{(180.16\,\text{g})} = 8.8177 \times 10^{-3}\,\text{mol}$$

Thus:

$$\frac{-28.52\,\text{kJ}}{8.8177 \times 10^{-3}\,\text{mol}} = -3{,}234\ \text{kJ/mol}$$

Laws of Thermodynamics

The **First Law of Thermodynamics** states that the total energy of the universe is constant. This is simply the Law of Conservation of Energy. This can be mathematically stated as:

$$\Delta E_{\text{universe}} = \Delta E_{\text{system}} + \Delta E_{\text{surroundings}} = 0$$

The Second Law of Thermodynamics involves a term called entropy. **Entropy** (***S***) is related to the disorder of a system. The **Second Law of Thermodynamics** states that all processes that occur spontaneously move in the direction of an increase in entropy of the universe (system + surroundings). Mathematically, this can be stated as:

$$\Delta S_{\text{universe}} = \Delta S_{\text{system}} + \Delta S_{\text{surroundings}} > 0 \qquad \text{for a spontaneous process}$$

For a reversible process $\Delta S_{\text{universe}} = 0$. The qualitative entropy change (increase or decrease of entropy) for a system can sometimes be determined using a few simple rules:

1. Entropy increases when the number of molecules increases during a reaction.
2. Entropy increases with an increase in temperature.
3. Entropy increases when a gas is formed from a liquid or solid.
4. Entropy increases when a liquid is formed from a solid.

Let us now look at some applications of these first two laws of thermodynamics.

Products Minus Reactants

Enthalpies

Many of the reactions that chemists study are reactions that occur at constant pressure. During the discussion of the coffee-cup calorimeter, the heat change at constant temperature was defined as q_p. Because this constant-pressure situation is so common in chemistry, a special thermodynamic term is used to describe this energy: enthalpy. The **enthalpy change, ΔH,** is equal to the heat gained or lost by the system under constant-pressure conditions. The following sign conventions apply:

If $\Delta H > 0$ the reaction is endothermic.

If $\Delta H < 0$ the reaction is exothermic.

If a reaction is involved, ΔH is sometimes called $\Delta H_{\text{reaction}}$. ΔH is often given in association with a particular reaction. For example, the enthalpy change associated with the formation of water from hydrogen and oxygen gases can be shown in this fashion:

$$2\ H_2(g) + O_2(g) \rightarrow 2\ H_2O(g) \qquad \Delta H = -483.6\ \text{kJ}$$

The negative sign indicates that this reaction is exothermic. This value of ΔH is for the production of 2 mol of water. If 4 mol were produced, ΔH would be twice −483.6 kJ. The techniques developed in working reaction stoichiometry problems (see the Stoichiometry chapter) also apply here.

If the previous reaction for the formation of water were reversed, the sign of ΔH would be reversed. That would indicate that it would take 483.6 kJ of energy to decompose 2 mol of water. This would then become an endothermic process.

ΔH is dependent upon the state of matter. The enthalpy change would be different for the formation of liquid water instead of gaseous water.

ΔH can also indicate whether a reaction will be spontaneous. A negative (exothermic) value of ΔH is associated with a spontaneous reaction. However, in many reactions this is not the case. There is another factor to consider in predicting a reaction's spontaneity. We will cover this other factor a little later in this chapter.

Enthalpies of reaction can be measured using a calorimeter. However, they can also be calculated in other ways. **Hess's law** states that if a reaction occurs in a series of steps, then the enthalpy change for the overall reaction is simply the sum of the enthalpy changes of the individual steps. If, in adding the equations of the steps together, it is necessary to reverse one of the given reactions, then the sign of ΔH must also be reversed. Also, particular attention must be used if the reaction stoichiometry has to be adjusted. The value of an individual ΔH may need to be adjusted.

It doesn't matter whether the steps used are the actual steps in the mechanism of the reaction, because $\Delta H_{reaction}$ (ΔH_{rxn}) is a **state function**, a function that doesn't depend on the pathway, but only on the initial and final states.

Let's see how Hess's law can be applied, given the following information:

$C(s) + O_2(g) \rightarrow CO_2(g)$ $\Delta H = -393.5$ kJ

$H_2(g) + (1/2)O_2(g) \rightarrow H_2O(l)$ $\Delta H = -285.8$ kJ

$C_2H_2(g) + (5/2)O_2(g) \rightarrow 2\,CO_2(g) + H_2O(l)$ $\Delta H = -1{,}299.8$ kJ

find the enthalpy change for:

$$2C(s) + H_2(g) \rightarrow C_2H_2(g)$$

Answer:

$2[C(s) + O_2(g) \rightarrow CO_2(g)]$	2 (−393.5 kJ)
$H_2(g) + (1/2)\,O_2(g) \rightarrow H_2O(l)$	−285.8 kJ
$2\,CO_2(g) + H_2O(l) \rightarrow C_2H_2(g) + (5/2)\,O_2(g)$	−(−1,299.8 kJ)
$2C(s) + H_2(g) \rightarrow C_2H_2(g)$	227.0 kJ

Enthalpies of reaction can also be calculated from individual enthalpies of formation (or heats of formation), ΔH_f, for the reactants and products. Because the temperature, pressure, and state of the substance will cause these enthalpies to vary, it is common to use a standard state convention. For gases, the standard state is 1 atm pressure. For a substance in an aqueous solution, the standard state is 1 molar concentration. And for a pure substance (compound or element), the standard state is the most stable form at 1 atm pressure and 25°C. A degree symbol to the right of the H indicates a standard state, $\Delta H°$. The **standard enthalpy of formation** of a substance (ΔH_f°) is the change in enthalpy when 1 mol of the substance is formed from its elements when all substances are in their standard states. These values are then tabulated and can be used in determining $\Delta H°_{rxn}$.

ΔH_f° of an element in its standard state is zero.

$\Delta H_f^\circ{}_{rxn}$ can be determined from the tabulated ΔH_f° of the individual reactants and products. It is the sum of the ΔH_f° of the products minus the sum of the ΔH_f° of the reactants:

$$\Delta H°_{rxn} = \Sigma\,\Delta H_f^\circ \text{ products} - \Sigma\,\Delta H_f^\circ \text{ reactants}$$

In using this equation be sure to consider the number of moles of each, because ΔH_f° for the individual compounds refer to the formation of 1 mol.

For example, let's use standard enthalpies of formation to calculate ΔH_{rxn} for:

$$6\ H_2O(g) + 4\ NO(g) \rightarrow 5\ O_2(g) + 4\ NH_3(g)$$

Answer:

$$\Delta H_{rxn} = \{5[\Delta H_f^\circ O_2(g)] + 4[\Delta H_f^\circ NH_3(g)]\} - \{[6\Delta H_f^\circ H_2O(g)] + 4[\Delta H_f^\circ NO(g)]\}$$

Using tabulated standard enthalpies of formation gives:

$$\Delta H_{rxn} = [5(0.00\ kJ) + 4(-46.19\ kJ)] - [6(-241.85\ kJ) + 4(90.37)]$$
$$= 904.68\ kJ$$

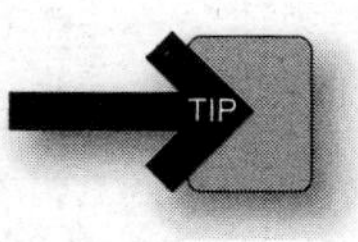

People commonly forget to subtract *all* the reactants from the products.

The values of ΔH_f° will be given to you on the AP exam, or you will be asked to stop before putting the numbers into the problem.

An alternative means of estimating the heat of reaction is to take the sum of the average bond energies of the reactant molecules and subtract the sum of the average bond energies of the product molecules.

Entropies

In much the same way as ΔH° was determined, the **standard molar entropies** (S°) of elements and compounds can be tabulated. The standard molar entropy is the entropy associated with 1 mol of a substance in its standard state. Entropies are also tabulated, but unlike enthalpies, the entropies of elements are not zero. For a reaction, the standard entropy change is calculated in the same way as the enthalpies of reaction:

$$\Delta S^\circ = \Sigma\ S^\circ \text{ products} - \Sigma\ S^\circ \text{ reactants}$$

Calculate ΔS° for the following. If you do not have a table of S° values, just set up the problems.

TIP

Note: These are thermochemical equations, so fractions are allowed.

a. $H_2(g) + \frac{1}{2}\ O_2(g) \rightarrow H_2O(g)$

b. $H_2(g) + \frac{1}{2}\ O_2(g) \rightarrow H_2O(l)$

c. $CaCO_3(s) + H_2SO_4(l) \rightarrow CaSO_4(s) + H_2O(g) + CO_2(g)$

Answers:

a. H_2O $\qquad H_2$ $\qquad O_2$

188.7 J/mol K − [131.0 + 1/2(205.0)]J/mol K

= −44.8 J/mol K

b. H_2O $\qquad H_2$ $\qquad O_2$

69.9 J/mol K − [131.0 + 1/2(205.0)]J/mol K

= −163.6 J/mol K

c. $CaSO_4$ H_2O CO_2 $CaCO_3$ H_2SO_4
[107 + 188.7 + 213.6] − [92.9 + 157] J/mol K
= 259 J/mol K

One of the goals of chemists is to be able to predict whether or not a reaction will be spontaneous. Some general guidelines for a spontaneous reaction have already been presented (negative ΔH and positive ΔS), but neither is a reliable predictor by itself. Temperature also plays a part. A thermodynamic factor that takes into account the entropy, enthalpy, and temperature of the reaction should be the best indicator of spontaneity. This factor is called the Gibbs free energy.

Gibbs Free Energy

The **Gibbs free energy** (***G***) is a thermodynamic function that combines the enthalpy, entropy, and temperature:

$$G = H - TS, \text{ where } T \text{ is the Kelvin temperature}$$

Like most thermodynamic functions, only the change in Gibbs free energy can be measured, so the relationship becomes:

$$\Delta G = \Delta H - T\Delta S$$

ΔG is the best indicator chemists have as to whether or not a reaction is spontaneous:

- If $\Delta G > 0$, the reaction is not spontaneous; energy must be supplied to cause the reaction to occur.
- If $\Delta G < 0$, the reaction is spontaneous.
- If $\Delta G = 0$, the reaction is at equilibrium.

If there is a ΔG associated with a reaction and that reaction is then reversed, the sign of ΔG changes.

Just like with the enthalpy and entropy, the standard Gibbs free energy change, ($\Delta G°$), is calculated:

$$\Delta G° = \Sigma\, \Delta G_f° \text{ products} - \Sigma\, \Delta G_f° \text{ reactants}$$

$\Delta G_f°$ of an element in its standard state is zero.

$\Delta G°$ for a reaction may also be calculated by using the standard enthalpy and standard entropy of reaction:

$$\Delta G° = \Delta H°_{rxn} - T\Delta S°_{rxn}$$

Calculate $\Delta G°$ for:

(If you do not have a table of $\Delta G°$ values, just set up the problems.)

a. $2\ NH_4Cl(s) + CaO(s) \rightarrow CaCl_2(s) + H_2O(l) + 2\ NH_3(g)$

b. $C_2H_4(g) + H_2O(g) \rightarrow C_2H_5OH(l)$

c. $Ca(s) + 2\ H_2SO_4(l) \rightarrow CaSO_4(s) + SO_2(g) + 2\ H_2O(l)$

Answers:

a. $CaCl_2(s)$ $H_2O(l)$ $NH_3(g)$ $NH_4Cl(s)$ $CaO(s)$
−750.2 −237.2 −16.6 −203.9 −604.2 kJ/mol

$[-750.2 + (-237.2) + 2(-16.6)] - [2(-203.9) + (-604.2)]$

$= -8.6$ kJ/mole

b. $C_2H_5OH(l)$ $C_2H_4(g)$ $H_2O(g)$

−174.18 68.12 −228.6

$-174.18 - [68.12 + (-228.6)] = -13.7$ kJ/mol

c. $CaSO_4(s)$ $SO_2(g)$ $H_2O(l)$ $Ca(s)$ $H_2SO_4(l)$

−1,320.3 −300.4 −237.2 0.0 −689.9 kJ/mol

$[(-1{,}320.3) + (-300.4) + 2(-237.2)] - [(0.0) + 2(-689.9)]$

$= -715.3$ kJ/mol

Thermodynamics and Equilibrium

Thus far, we have considered only situations under standard conditions. But how do we cope with nonstandard conditions? The change in Gibbs free energy under nonstandard conditions is:

$$\Delta G = \Delta G° + RT \ln Q = \Delta G° + 2.303 \log Q$$

Q is the activity quotient, products over reactants. This equation allows the calculation of ΔG in those situations in which the concentrations or pressures are not 1.

Using the previous concept, calculate ΔG for the following at 500 • K:

$$2\ NO(g) + O_2(g) \rightarrow 2\ NO_2(g)$$

2.00 M 0.500 M 1.00 M

(Assume $\Delta G_f° = \Delta G_f^{500}$)

$\Delta G_f°$ (86.71 0.000 51.84) kJ/mol

$$\Delta G_{rxn} = 2(51.84) - [2(86.71) + 0.000] = -69.74 \text{ kJ/mol}$$

$$\Delta G^{500} = \Delta G_{rxn} + RT \ln Q$$

$$Q = \frac{[NO_2]^2}{[NO]^2[O_2]}$$

$$= (-69.74\,\text{kJ})(1{,}000\,\text{J/kJ}) + \left(8.314\frac{\text{J}}{\text{mol K}}\right)(500.\,\text{K})\ln\frac{(1.00)^2}{(2.00)^2(0.500)}$$

$$= -7.262 \times 10^4 \text{ J/mol}$$

Note that Q, when at equilibrium, becomes K. This equation gives us a way to calculate the equilibrium constant, K, from a knowledge of the standard Gibbs free energy of the reaction and the temperature.

If the system is at equilibrium, then $\Delta G = 0$ and the equation above becomes:

$$\Delta G° = -RT \ln K = -2.303\ RT \log K$$

For example, calculate $\Delta G°$ for:

$$2O_3(g) \rightleftharpoons 3O_2(g) \quad K_p = 4.17 \times 10^{14}$$

Note: ° = 298 K
Answer:

$$\Delta G^\circ = -RT \ln K$$
$$= \frac{-(8.314\,\text{J})}{(\text{mol K})}(298\,\text{K})\ln 4.17\times 10^{14}$$
$$= -8.34\times 10^{4}\,\text{J/mol}$$

Experiments

The most common thermodynamic experiment is a calorimetry experiment. In this experiment the heat of transition or heat of reaction is determined.

The experiment will require a balance to determine the mass of a sample and possibly a pipet to measure a volume, from which a mass may be calculated using the density. A calorimeter, usually a polystyrene (Styrofoam) cup, is needed to contain the reaction. Finally, a thermometer is required. Tables of heat capacities or specific heats may be provided.

Mass and possible volume measurements, along with the initial and final temperatures, are needed. Remember: you *measure* the initial and final temperature so you can *calculate* the change in temperature.

After the temperature change is calculated, there are several ways to proceed. If the calorimeter contains water, the heat may be calculated by multiplying the specific heat of water by the mass of water by the temperature change. The heat capacity of the calorimeter may be calculated by dividing the heat by the temperature change. If a reaction is carried out in the same calorimeter, the heat from that reaction is the difference between the heat with and without a reaction.

KEY IDEA

Do not forget, if the temperature increases, the process is exothermic and the heat has a negative sign. The opposite is true if the temperature drops.

Common Mistakes to Avoid

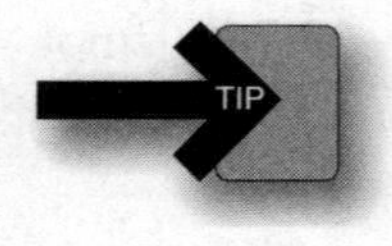

1. Be sure your units cancel giving you the unit desired in the final answer.
2. Check your significant figures.
3. Don't mix energy units, joules, and calories.
4. Watch your signs in all the thermodynamic calculations. They are extremely important.
5. Don't confuse enthalpy, ΔH, and entropy, ΔS.
6. Pay close attention to the state of matter for your reactants and products, and choose the corresponding value for use in your calculated entropies and enthalpies.
7. Remember: **products minus reactants**.
8. ΔH_f and ΔG_f are for 1 mol of substance. Use appropriate multipliers if needed.
9. ΔG_f and ΔH_f for an element in its standard state are zero.
10. All temperatures are in kelvin.
11. When using $\Delta G^\circ = \Delta H^\circ_{rxn} - T\Delta S^\circ_{rxn}$, pay particular attention to your enthalpy and entropy units. Commonly, enthalpies will use kJ and entropies J.

› Review Questions

Use these questions to review the content of this chapter and practice for the AP Chemistry exam. First are 16 multiple-choice questions similar to what you will encounter in Section I of the AP Chemistry exam. Following those is a multipart free-response question like the ones in Section II of the exam. To make these questions an even more authentic practice for the actual exam, time yourself following the instructions provided.

Multiple-Choice Questions

Answer the following questions in 20 minutes. You may not use a calculator. You may use the periodic table and the equation sheet at the back of this book.

1. Which of the following is the minimum energy required to initiate a reaction?

(A) free energy
(B) lattice energy
(C) kinetic energy
(D) activation energy

2. What is the minimum energy required to force a nonspontaneous reaction to occur?

(A) free energy
(B) lattice energy
(C) kinetic energy
(D) activation energy

3. The average ______________ is the same for any ideal gas at a given temperature.

(A) free energy
(B) lattice energy
(C) kinetic energy
(D) activation energy

4. What is the energy released when the gaseous ions combine to form an ionic solid?

(A) free energy
(B) lattice energy
(C) kinetic energy
(D) activation energy

5. Given the following information:

$C(s) + O_2(g) \rightarrow CO_2(g)$ $\quad \Delta H = -393.5$ kJ

$H_2(g) + (1/2)\ O_2(g) \rightarrow H_2O(l)$ $\quad \Delta H = -285.8$ kJ

$C_2H_2(g) + (5/2)\ O_2(g) \rightarrow 2\ CO_2(g) + H_2O(l)$ $\quad \Delta H = -1{,}299.8$ kJ

Find the enthalpy change for:

$2C(s) + H_2(g) \rightarrow C_2H_2(g)$

(A) 454.0 kJ
(B) −227.0 kJ
(C) 0.0 kJ
(D) 227.0 kJ

6. A sample of gallium metal is sealed inside a well-insulated, rigid container. The temperature inside the container is at the melting point of gallium metal. What can be said about the energy and the entropy of the system after equilibrium has been established? Assume the insulation prevents any energy change with the surroundings.

(A) The total energy increases. The total entropy will increase.
(B) The total energy is constant. The total entropy is constant.
(C) The total energy is constant. The total entropy will decrease.
(D) The total energy is constant. The total entropy will increase.

7. When ammonium chloride dissolves in water, the temperature drops. Which of the following conclusions may be related to this?

(A) Ammonium chloride is more soluble in hot water.
(B) Ammonium chloride produces an ideal solution in water.
(C) The heat of solution for ammonium chloride is exothermic.
(D) Ammonium chloride has a low lattice energy.

8. Choose the reaction expected to have the greatest increase in entropy.

(A) $H_2O(g) \rightarrow H_2O(l)$
(B) $2\ KClO_3(s) \rightarrow 2\ KCl(s) + 3\ O_2(g)$
(C) $Ca(s) + H_2(g) \rightarrow CaH_2(s)$
(D) $N_2(g) + 3H_2(g) \rightarrow 2\ NH_3(g)$

9. Under standard conditions, calcium metal reacts readily with chlorine gas. What conclusions may be drawn from the fact?

(A) $K_{eq} < 1$ and $\Delta G° > 0$
(B) $K_{eq} > 1$ and $\Delta G° = 0$
(C) $K_{eq} < 1$ and $\Delta G° < 0$
(D) $K_{eq} > 1$ and $\Delta G° < 0$

10. Which of the following combinations is true when sodium chloride melts?

(A) $\Delta H > 0$ and $\Delta S > 0$
(B) $\Delta H = 0$ and $\Delta S > 0$
(C) $\Delta H > 0$ and $\Delta S < 0$
(D) $\Delta H < 0$ and $\Delta S < 0$

11. Which of the following reactions have a negative entropy change?

(A) $2\ C_2H_6(g) + 7\ O_2(g) \rightarrow 4\ CO_2(g) + 6\ H_2O(g)$
(B) $2\ NH_3(g) \rightarrow N_2(g) + 3\ H_2(g)$
(C) $CaCl_2(s) \rightarrow Ca(s) + Cl_2(g)$
(D) $2\ H_2(g) + O_2(g) \rightarrow 2\ H_2O(l)$

12. A certain reaction is nonspontaneous under standard conditions, but becomes spontaneous at higher temperatures. What conclusions may be drawn under standard conditions?

(A) $\Delta H < 0$, $\Delta S > 0$, and $\Delta G > 0$
(B) $\Delta H > 0$, $\Delta S < 0$, and $\Delta G > 0$
(C) $\Delta H > 0$, $\Delta S > 0$, and $\Delta G > 0$
(D) $\Delta H < 0$, $\Delta S < 0$, and $\Delta G > 0$

13. $2\ H_2(g) + O_2(g) \rightarrow 2\ H_2O(g)$

From the table below, determine the enthalpy change for the above reaction.

BOND	AVERAGE BOND ENERGY (kJ/mol)
H–H	436
O=O	499
H–O	464

(A) 0 kJ
(B) 485 kJ
(C) –485 kJ
(D) 464 kJ

14. Which of the following reactions would be accompanied by the greatest decrease in entropy?

(A) $N_2(g) + 3\ H_2(g) \rightarrow 2\ NH_3(g)$
(B) $C(s) + O_2(g) \rightarrow CO_2(g)$
(C) $2\ H_2(g) + O_2(g) \rightarrow 2\ H_2O(g)$
(D) $2\ Na(s) + Cl_2(g) \rightarrow 2\ NaCl(s)$

15. $CO(g) + 2\ H_2(g) \rightarrow CH_3OH(g)\ \Delta H = -91 kJ$

Determine ΔH for the above reaction if $CH_3OH(l)$ were formed in the above reaction instead of $CH_3OH(g)$. The ΔH of vaporization for CH_3OH is 37 kJ/mol.

(A) –128 kJ
(B) –54 kJ
(C) +128 kJ
(D) +54 kJ

16. A solution is prepared by dissolving solid ammonium nitrate, NH_4NO_3, in water. The initial temperature of the water was 25°C, but after the solid had dissolved, the temperature had fallen to 20°C. What conclusions may be made about ΔH and ΔS?

(A) $\Delta H < 0$ $\quad \Delta S > 0$
(B) $\Delta H > 0$ $\quad \Delta S > 0$
(C) $\Delta H > 0$ $\quad \Delta S < 0$
(D) $\Delta H < 0$ $\quad \Delta S < 0$

Answers and Explanations for the Multiple-Choice Questions

1. D—You may wish to review the Kinetics chapter if you have forgotten what the activation energy is.

2. A—The free energy is the minimum energy required for a nonspontaneous reaction and the maximum energy available for a spontaneous reaction.

3. C—This is a basic postulate of kinetic molecular theory.

4. B—This is the reverse of the lattice energy definition.

5. D

$2[C(s) + O_2(g) \rightarrow CO_2(g)]$	2(−393.5 kJ)
$H_2(g) + (1/2)\ O_2(g) \rightarrow H_2O(l)$	−285.8 kJ
$2\ CO_2(g) + H_2O(l) \rightarrow C_2H_2(g) + (5/2)\ O_2(g)$	−(−1,299.8 kJ)
$2\ C(s) + H_2(g) \rightarrow C_2H_2(g)$	227.0 kJ

Simple rounding to the nearest 100 kJ gives 200 kJ.

6. D—The system is insulated and no work can be done on or by the system (rigid container); thus, the energy is constant. At the melting point, some of the gallium will spontaneously melt; changing a solid to a liquid increases the entropy.

7. A—The process is endothermic (the ammonium chloride is absorbing heat to cool the water). Endothermic processes are "helped" by higher temperatures. Answers A and C, and possibly D would give an increase in temperature. There is insufficient information about answer B.

8. B—The reaction showing the greatest increase in the number of moles of gas will show the greatest entropy increase. If no gases are present, then the greatest increase in the number of moles of liquid would yield the greatest increase.

9. D—If the reaction occurs readily, it must be spontaneous. Spontaneous reactions require $\Delta G° < 0$. A negative free energy leads to a large K (>1).

10. A—Heat is required to melt something ($\Delta H > 0$). A transformation from a solid to a liquid gives an increase in entropy ($\Delta S > 0$).

11. D—This equation has an overall decrease in the amount of gas (high entropy) present. The other answers produce more gas (increases entropy).

12. C—Nonspontaneous means that $\Delta G > 0$. Since the reaction becomes spontaneous, the sign must change. Recalling: $\Delta G = \Delta H - T\Delta S$. The sign change at higher temperature means that the entropy term (with $\Delta S > 0$) must become more negative than the enthalpy term ($\Delta H > 0$).

13. C—[2(436 kJ) + 499 kJ] − {2[2(464 kJ)]} = −485 kJ

14. A—The reaction that produces the most gas will have the greatest increase in entropy; the one losing the most gas would have the greatest decrease.

15. A—$CO(g) + 2\ H_2(g) \rightarrow CH_3OH(g)$ $\quad \Delta H = -91$ kJ

$CH_3OH(g) \rightarrow CH_3OH(l)$ $\quad \Delta H = -37$ kJ

Total $CO(g) + 2\ H_2(g) \rightarrow CH_3OH(l)$ $\quad \Delta H = -128$ kJ

16. B—Dissolving almost always has: $\Delta S > 0$. A decrease in temperature means the process has: $\Delta H > 0$ (the system is absorbing energy from the surroundings).

Free-Response Question

You have 10 minutes to answer the following question. You may use a calculator and the tables in the back of the book.

Question 1

$$Xe(g) + 3F_2\ (g) \rightleftharpoons XeF_6(g)$$

Under standard conditions, the enthalpy change for the reaction going from left to right (forward reaction) is $\Delta H^\circ = -294$ kJ.

(a) Is the value of ΔS°, for the above reaction, positive or negative? Justify your conclusion.

(b) The above reaction is spontaneous under standard conditions. Predict what will happen to ΔG for this reaction as the temperature is increased. Justify your prediction.

(c) Will the value of K remain the same, increase, or decrease as the temperature increases? Justify your prediction.

(d) Show how the temperature at which the reaction changes from spontaneous to non-spontaneous can be predicted. What additional information is necessary?

Answer and Explanation for the Free-Response Question

(a) The value is negative. The decrease in the number of moles of gas during the reaction means there is a decrease in entropy.

Give yourself 1 point if you predicted this. Give yourself 1 point for discussing the number of moles of gas. You may get this point even if you did not get the first point.

(b) Recalling $\Delta G = \Delta H - T\Delta S$ (this equation is given on the equation page of the AP exam).

The value of ΔG will increase (become less negative).

Give yourself 1 point for this answer if it is obvious that increasing means less negative.

In general, both ΔH and ΔS are relatively constant with respect to small temperature changes. As the temperature increases, the value of the entropy term, $T\Delta S$, becomes more negative. The negative sign in front of this term leads to a positive contribution. The value of ΔG will first become less negative (more positive), and eventually the value will be positive (no longer spontaneous).

Give yourself 1 point for the $\Delta G = \Delta H - T\Delta S$ argument even if you did not get the first point.

(c) Recalling $\Delta G = -RT \ln K$ (this equation is given on the equation page of the AP exam).

The value of K will decrease.

You get 1 point for this answer.

As the value of ΔG increases (see part **b**), the value of K will decrease.

You get 1 point for using $\Delta G = -RT \ln K$ in your discussion.

If you got the justification for part **b** wrong, and you used the same argument here, you will not be penalized twice. You still get your point.

(d) Recalling $\Delta G = \Delta H - T\Delta S$

Rearranging this equation to: $T = (\Delta G - \Delta H)/\Delta S$ will allow the temperature to be estimated.

This is worth 1 point. To do the calculation, the value of ΔS is necessary. Give yourself 1 point for this.

There are a total of 8 points possible. All of the mathematical relations presented in the answers are in the exam booklet.

Note that many students lose points on a question like this because their answers are too long. Keep your answers short and to the point, even if it appears that you have multiple pages available.

❯ Rapid Review

- Thermodynamics is the study of heat and its transformations.
- Kinetic energy is energy of motion, while potential energy is stored energy.
- The common units of energy are the joule, J, and the calorie, cal.
- A calorimeter is used to measure the heat released or absorbed during a chemical or physical change. Know how a calorimeter works.
- The specific heat capacity is the amount of heat needed to change the temperature of 1 gram of a substance by 1 K, while the molar heat capacity is the heat capacity per mole.
- The heat lost by the system in calorimetry is equal to the heat gained by the surroundings.
- The specific heat (c) of a solid can be calculated by: $-(c_{solid} \times m_{solid} \times \Delta T_{solid}) = c_{water} \times m_{water} \times \Delta T_{water}$ or by $g = cm\Delta T$.
- The First Law of Thermodynamics states that the total energy of the universe is constant. (Energy is neither created nor destroyed.)
- The Second Law of Thermodynamics states that all spontaneous processes move in a way that increases the entropy (disorder) of the universe.
- The enthalpy change, ΔH, is equal to the heat lost or gained by the system under constant pressure conditions.
- ΔH values are associated with a specific reaction. If that reaction is reversed, the sign of ΔH changes. If one has to use a multiplier on the reaction, it must also be applied to the ΔH value.
- The standard enthalpy of formation of a compound, ΔH_f°, is the enthalpy change when 1 mol of the substance is formed from its elements and all substances are in their standard states.
- The standard enthalpy of formation of an element in its standard state is zero.
- $\Delta H^\circ_{rxn} = \Sigma\, \Delta H_f^\circ$ products $- \Sigma\, \Delta H_f^\circ$ reactants. Know how to apply this equation.
- ΔH°_{rxn} is usually negative for a spontaneous reaction.
- $\Delta S^\circ = \Sigma\, S^\circ$ products $- \Sigma\, S^\circ$ reactants. Know how to apply this equation.
- ΔS° is usually positive for a spontaneous reaction.
- The Gibbs free energy is a thermodynamic quantity that relates the enthalpy and entropy, and is the best indicator for whether or not a reaction is spontaneous.
- If $\Delta G^\circ > 0$ the reaction is not spontaneous; if $\Delta G^\circ < 0$, the reaction is spontaneous; and if $\Delta G^\circ = 0$, the reaction is at equilibrium.
- $\Delta G^\circ = \Sigma\, \Delta G_f^\circ$ products $- \Sigma\, \Delta G_f^\circ$ reactants. Know how to apply this equation.
- $\Delta G^\circ = \Delta H^\circ_{rxn} - T\Delta S^\circ_{rxn}$. Know how to apply this equation.
- For a system not at equilibrium: $\Delta G = \Delta G^\circ + RT \ln Q = \Delta G^\circ + 2.303\, RT \log Q$. Know how to apply this equation.
- For a system at equilibrium: $\Delta G^\circ = -RT \ln K = -2.303\, RT \log K$. Know how to apply this equation to calculate equilibrium constants.

CHAPTER 10

Spectroscopy, Light, and Electrons

IN THIS CHAPTER

Summary: In developing the model of the atom it was thought initially that all subatomic particles obeyed the laws of classical physics—that is, they were tiny bits of matter behaving like macroscopic pieces of matter. Later, however, it was discovered that this particle view of the atom could not explain many of the observations being made. About this time the dual particle/wave model of matter began to gain favor. It was discovered that in many cases, especially when dealing with the behavior of electrons, describing some of their behavior in terms of waves explained the observations much better. Thus, the quantum mechanical model of the atom was born.

Keywords and Equations

Speed of light, $c = 3.0 \times 10^8$ ms^{-1}

E = energy $\quad$ ν = frequency $\quad$ λ = wavelength

p = momentum $\quad$ v = velocity $\quad$ n = principal quantum number

m = mass $\quad$ $E = h\nu$ $\quad$ $c = \lambda\nu$ $\quad$ $\lambda = \dfrac{h}{mv}$ $\quad$ $p = mv$

$$E_n = \frac{-2.178 \times 10^{-18}}{n^2} \text{ joule}$$

Planck's constant, $h = 6.63 \times 10^{-34}$ Js

The Nature of Light

Light is a part of the **electromagnetic spectrum**—radiant energy composed of gamma rays, X-rays, ultraviolet light, visible light, etc. Figure 10.1 shows the electromagnetic spectrum.

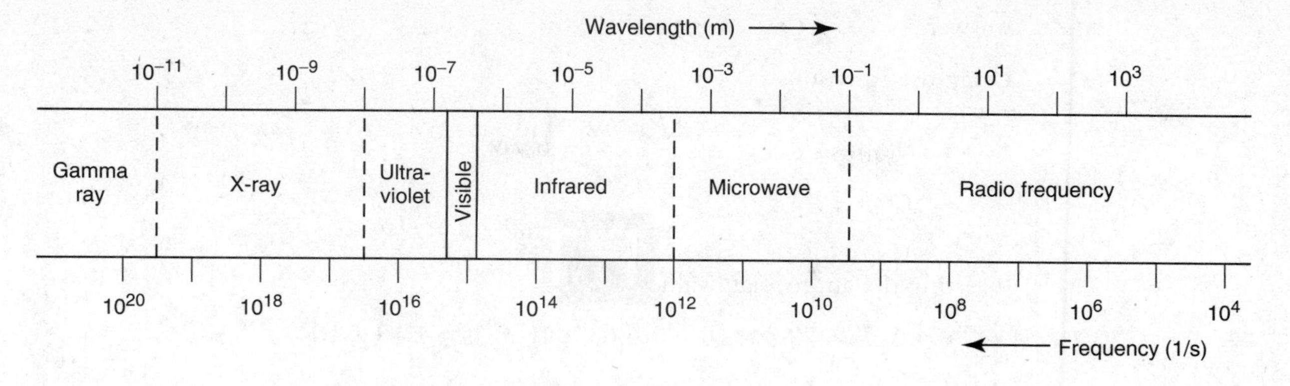

Figure 10.1 The electromagnetic spectrum.

The energy of the electromagnetic spectrum moves through space as waves that have three associated variables—frequency, wavelength, and amplitude. The **frequency, ν,** is the number of waves that pass a point per second. **Wavelength** (λ) is the distance between two identical points on a wave. **Amplitude** is the height of the wave and is related to the intensity (or brightness, for visible light) of the wave. Figure 10.2 shows the wavelength and amplitude of a wave.

The energy associated with a certain frequency of light is related by the equation:

$$E = h\nu \text{ where } h \text{ is Planck's constant} = 6.63 \times 10^{-34} \text{ Js}$$

In developing the quantum mechanical model of the atom, it was found that the electrons can have only certain distinct quantities of energy associated with them, and that in order for the atom to change its energy it has to absorb or emit a certain amount of energy. The energy that is emitted or absorbed is really the difference in the two energy states and can be calculated by:

$$\Delta E = h\nu$$

All electromagnetic radiation travels at about the same speed in a vacuum, 3.0×10^8 m/s. This constant is called the **speed of light (*c*)**. The product of the frequency and the wavelength is the speed of light:

$$c = \nu\lambda$$

Let's apply some of the relationships. What wavelength of radiation has photons of energy 7.83×10^{-19} J?

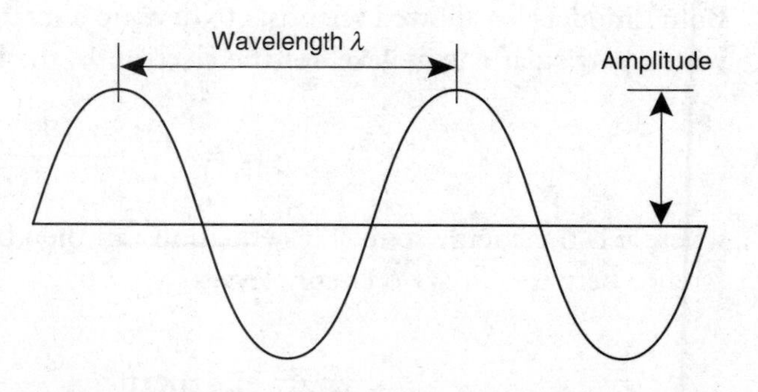

Figure 10.2 Wavelength and amplitude of a wave.

Answer:

Using the equations

$$\Delta E = h\nu \quad \text{and} \quad c = \nu\lambda$$

we get

$$\nu = \Delta E / h \quad \text{and} \quad \lambda = c/\nu$$

Inserting the appropriate values:

$$\nu = \Delta E/h = 7.83 \times 10^{-19}\ \text{J}/6.63 \times 10^{-34}\ \text{Js} = 1.18 \times 10^{15}\ \text{s}^{-1}$$

Then:

$$\lambda = c/\nu = (3.0 \times 10^{8}\ \text{m/s})/(1.18 \times 10^{15}\text{s}^{-1}) = 2.5 \times 10^{-7}\text{m}$$

This answer could have been calculated more quickly by combining the original two equations to give:

$$\lambda = hc/\Delta E$$

Wave Properties of Matter

The concept that matter possesses both particle and wave properties was first postulated by de Broglie in 1925. He introduced the equation $\lambda = h/mv$, which indicates a mass (m) moving with a certain velocity (v) would have a specific wavelength (λ) associated with it. (Note that this v is the velocity not ν the frequency.) If the mass is very large (a locomotive), the associated wavelength is insignificant. However, if the mass is very small (an electron), the wavelength is measurable. The denominator may be replaced with the momentum of the particle ($p = mv$).

Atomic Spectra

Late in the 19th century scientists discovered that when the vapor of an element was heated it gave off a **line spectrum**, a series of fine lines of colors instead of a continuous spectrum like a rainbow. This was used in the developing quantum mechanical model as evidence that the energy of the electrons in an atom was **quantized**, that is, there could only be certain distinct energies (lines) associated with the atom. Niels Bohr developed the first modern atomic model for hydrogen using the concepts of quantized energies. The Bohr model postulated a **ground state** for the electrons in the atom, an energy state of lowest energy, and an **excited state**, an energy state of higher energy. In order for an electron to go from its ground state to an excited state, it must absorb a certain amount of energy. If the electron dropped back from that excited state to its ground state, that same amount of energy would be emitted. Bohr's model also allowed scientists to develop a method of calculating the energy associated with a particular energy level for the electron in the hydrogen atom:

$$E_n = \frac{-2.18 \times 10^{-18}}{n^2}\ \text{joule}$$

where n is the energy state. This equation can then be modified to calculate the energy difference between any two energy levels:

$$\Delta E = -2.18 \times 10^{-18}\,\text{J}\left(\frac{1}{n^2_{\text{final}}} - \frac{1}{n^2_{\text{initial}}}\right)$$

Atomic Orbitals

Bohr's model worked well for hydrogen, the simplest atom, but didn't work very well for any others. In the early 1900s Schrödinger developed a more involved model and set of equations that better described atoms by using quantum mechanical concepts. His model introduced a mathematical description of the electron's motion called a **wave function** or **atomic orbital**. Squaring the wave function (orbital) gives the volume of space in which the probability of finding the electron is high. This is commonly referred to as the **electron cloud**.

Schrödinger's equation required the use of three **quantum numbers** to describe each electron within an atom, corresponding to the orbital size, shape, and orientation in space. It was also found that a quantum number concerning the spin of the electron was needed.

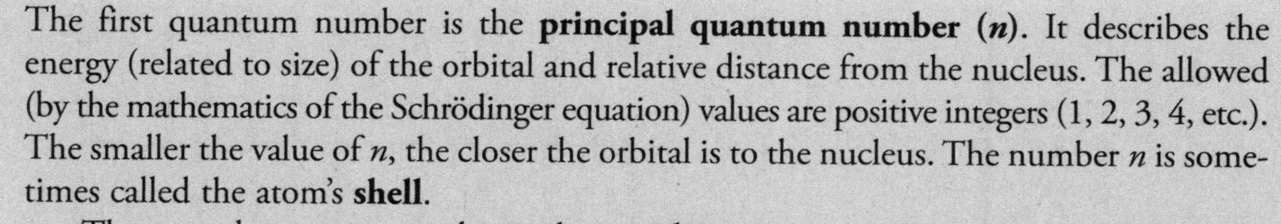

The first quantum number is the **principal quantum number (*n*)**. It describes the energy (related to size) of the orbital and relative distance from the nucleus. The allowed (by the mathematics of the Schrödinger equation) values are positive integers (1, 2, 3, 4, etc.). The smaller the value of n, the closer the orbital is to the nucleus. The number n is sometimes called the atom's **shell**.

The second quantum number is the **angular momentum quantum number (*l*)**. Its value is related to the principal quantum number and has allowed values of 0 up to $(n - 1)$. For example, if $n = 3$, then the possible values of l would be 0, 1, and 2 (3 − 1). This value of l defines the shape of the orbital:

- If $l = 0$, the orbital is called an s orbital and has a spherical shape with the nucleus at the center of the sphere. The greater the value of n, the larger the sphere.
- If $l = 1$, the orbital is called a p orbital and has two lobes of high electron density on either side of the nucleus. This makes for an hourglass or dumbbell shape.
- If $l = 2$, the orbital is a d orbital and can have variety of shapes.
- If $l = 3$, the orbital is an f orbital, with more complex shapes.

Figure 10.3, on the next page, shows the shapes of the s, p, and d orbitals. These are sometimes called **sublevels** or **subshells**.

The third quantum number is the **magnetic quantum number (m_l)**. It describes the orientation of the orbital around the nucleus. The possible values of m_l depend on the value of the angular momentum quantum number, l. The allowed values for m_l are $-l$ through zero to $+l$. For example, for $l = 2$ the possible values of m_l would be −2, −1, 0, +1, +2. This is why, for example, if $l = 1$ (a p orbital), then there are three p orbitals corresponding to m_l values of −1, 0, +1. This is also shown in Figure 10.3.

The fourth quantum number, the **spin quantum number (m_s)**, indicates the direction the electron is spinning. There are only two possible values for m_s, +½ and −½.

The quantum numbers for the six electrons in carbon would be:

QUANTUM NUMBER	FIRST ELECTRON	SECOND ELECTRON	THIRD ELECTRON	FOURTH ELECTRON	FIFTH ELECTRON	SIXTH ELECTRON
n	1	1	2	2	2	2
l	0	0	0	0	1	1
m_l	0	0	0	0	1	0
m_s	+½	−½	+½	−½	+½	+½

Therefore, the electron configuration of carbon is $1s^2 2s^2 2p^2$.

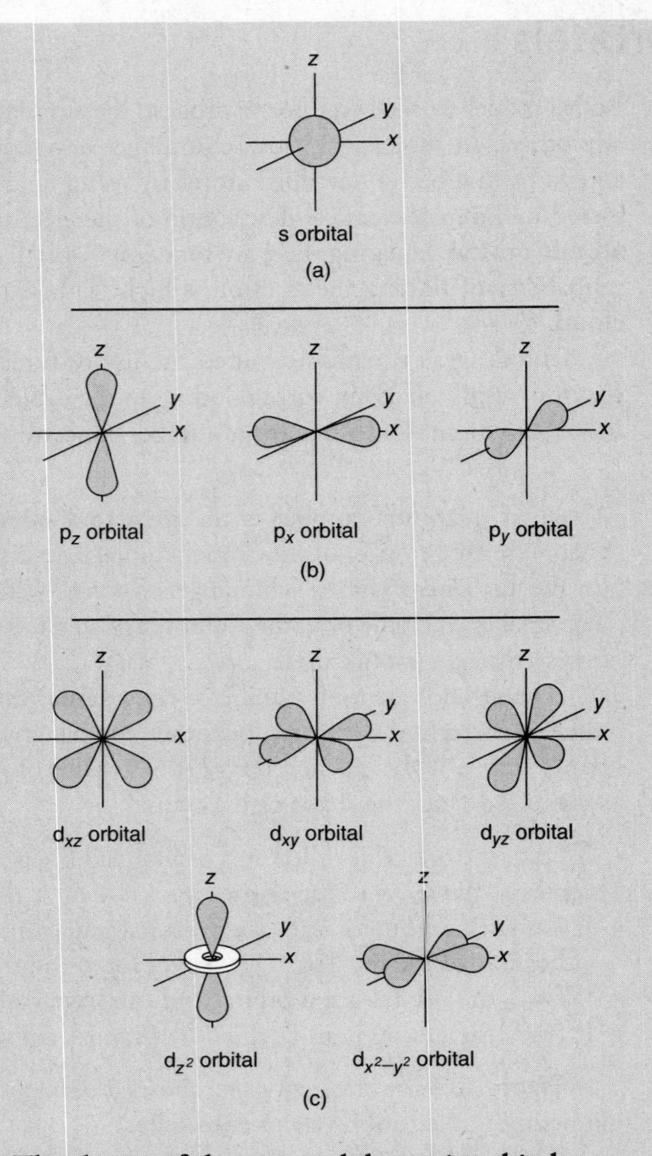

Figure 10.3 The shapes of the s, p, and d atomic orbitals.

Photoelectron (Photoemission) Spectroscopy (PES)

Photoelectron spectroscopy is one of a group of related techniques where high-energy photons remove an electron from an atom in a photoelectric effect process. The method relies on a measurement of the kinetic energy of the emitted electron. The kinetic energy is equal to the energy of the photon minus the binding energy of the electron. The binding energy is the energy holding the electron in the atom and can be rather difficult to measure.

X-ray photons can excite core electrons. For example, it is possible to focus on the 1s electrons of an oxygen atom. The binding energy is in part related to the effective nuclear charge experienced by the electron. In compounds, other atoms bonded to the atom of interest can influence the effective nuclear charge. Atoms donating electron density to the atom of interest decrease the effective nuclear charge, while electron-withdrawing atoms lead to an increase in the effective nuclear charge. An important factor in whether an atom donates or withdraws electron density is the relative electronegativity of the two atoms. This experimental method can be used to give information on which atoms are bonded to each other.

Experiments

No experimental questions related to this chapter have appeared on the AP exam in recent years.

Common Mistakes to Avoid

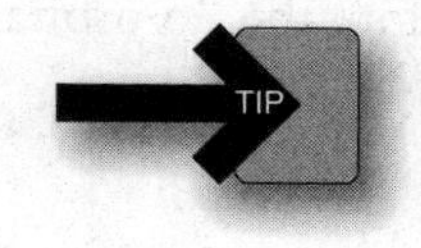

1. Be sure not to confuse wavelength and frequency.
2. The speed of light is 3.0×10^8 m/s. The exponent is positive.
3. The value of n is never zero.
4. The values of l and m_l include zero.
5. Do not confuse velocity (v) and frequency (ν).

❯ Review Questions

Use these questions to review the content of this chapter and practice for the AP Chemistry exam. First are 16 multiple-choice questions similar to what you will encounter in Section I of the AP Chemistry exam. Following those is a multipart free-response question like the ones in Section II of the exam. To make these questions an even more authentic practice for the actual exam, time yourself following the instructions provided.

Multiple-Choice Questions

Answer the following questions in 20 minutes. You may not use a calculator. You may use the periodic table and the equation sheet at the back of this book.

1. Which of the following represents the electron arrangement for the least reactive element?

(A) 1s ↑ 2s ↑↓
(B) 1s ↑↓ 2s ↑
(C) [Kr] 5s ↑↓ 4d ↑↑_ _ _
(D) 1s ↑↓ 2s ↑↓ 2p ↑↓↑↓↑↓

2. Which of the following might refer to a transition element?

(A) 1s ↑ 2s ↑↓
(B) 1s ↑↓ 2s ↑
(C) [Kr] 5s ↑↓ 4d ↑↑_ _ _
(D) 1s ↑↓ 2s ↑↓ 2p ↑↓↑↓↑↓

3. Which of the following electron arrangements refers to the most chemically reactive element?

(A) 1s ↑ 2s ↑↓
(B) 1s ↑↓ 2s ↑
(C) [Kr] 5s ↑↓ 4d ↑↑_ _ _
(D) 1s ↑↓ 2s ↑↓ 2p ↑↓↑↓↑↓

4. Which of the following electron arrangements represents an atom in an excited state?

(A) 1s ↑ 2s ↑↓
(B) 1s ↑↓ 2s ↑
(C) [Kr] 5s ↑↓ 4d ↑↑_ _ _
(D) 1s ↑↓ 2s ↑↓2p ↑↓↑↓↑↓

5. The ground-state configuration of Fe^{2+} is which of the following?

(A) $1s^22s^22p^63s^23p^63d^54s^1$
(B) $1s^22s^22p^63s^23p^63d^6$
(C) $1s^22s^22p^63s^23p^63d^64s^2$
(D) $1s^22s^22p^63s^23p^63d^84s^2$

6. Which of the following contains only atoms that are diamagnetic in their ground state?

(A) Kr, Ca, and P
(B) Ne, Be, and Zn
(C) Ar, K, and Ba
(D) He, Sr, and C

7. Which of the following is the electron configuration of a halogen?

(A) $1s^21p^62s^22p^3$
(B) $1s^22s^22p^63s^23p^64s^23d^{10}4p^65s^24d^1$
(C) $1s^22s^22p^63s^23p^63d^3$
(D) $1s^22s^22p^5$

8. Which of the following is a possible configuration for a transition metal atom?

(A) $1s^21p^62s^22p^3$
(B) $1s^22s^22p^63s^23p^64s^23d^{10}4p^65s^24d^1$
(C) $1s^22s^22p^63s^23p^63d^3$
(D) $1s^22s^22p^5$

9. Which of the following electron configurations is not possible?

(A) $1s^21p^62s^22p^3$
(B) $1s^22s^22p^63s^23p^64s^23d^{10}4p^65s^24d^1$
(C) $1s^22s^22p^63s^23p^63d^3$
(D) $1s^22s^22p^5$

10. This is a possible configuration of a transition metal ion.

(A) $1s^21p^62s^22p^3$
(B) $1s^22s^22p^63s^23p^64s^23d^{10}4p^65s^24d^1$
(C) $1s^22s^22p^63s^23p^63d^3$
(D) $1s^22s^22p^5$

11. Which idea relates to the fact that the exact position of an electron is not known?

(A) Pauli exclusion principle
(B) electron shielding
(C) Hund's rule
(D) Heisenberg uncertainty principle

12. Which of the following explains why oxygen atoms, in their ground state, are paramagnetic?

(A) Pauli exclusion principle
(B) electron shielding
(C) Hund's rule
(D) Heisenberg uncertainty principle

13. An atomic orbital can hold no more than two electrons; this is a consequence of which of the following?

(A) Pauli exclusion principle
(B) electron shielding
(C) Hund's rule
(D) Heisenberg uncertainty principle

14. Why does the 4s orbital fill before the 3d orbital starts to fill?

(A) Pauli exclusion principle
(B) electron shielding
(C) Hund's rule
(D) Heisenberg uncertainty principle

15. Calcium reacts with element *X* to form an ionic compound. If the ground-state electron configuration of *X* is $1s^22s^22p^4$, what is the simplest formula for this compound?

(A) CaX
(B) CaX_2
(C) Ca_4X_2
(D) Ca_2X_2

16. Which of the following best explains the diffraction of electrons?

(A) Pauli exclusion principle
(B) Hund's rule
(C) the wave properties of matter
(D) Heisenberg uncertainty principle

Answers and Explanations for the Multiple-Choice Questions

1. D—This configuration represents a noble gas (neon). The outer s and p orbitals are filled.

2. C—Transition elements have partially filled d orbitals. This configuration is for the metal zirconium, Zr.

3. B—The single electron in the s orbital indicates that this is the very reactive alkali metal lithium.

4. A—The 1s orbital is not filled. One indication of excited states is to have one or more inner orbitals unfilled.

5. B—The electron configuration for iron is $1s^2 2s^2 2p^6 3s^2 3p^6 3d^6 4s^2$. To produce an iron(II) ion, the two 4s electrons are removed first.

6. B—The elements that are normally diamagnetic are those in the same columns of the periodic table as Be, Zn, and He because all of the electrons are paired. All other columns are normally paramagnetic.

7. D—Halogens have a valence shell with s^2p^5.

8. B—Transition metals have partially filled d orbitals (d^{1-10}), along with an s^1 or s^2.

9. A—The 1p orbital does not exist.

10. C—The outer s electrons are not present in most transition metal ions; however, d electrons may be present. C could be V^{2+}, Cr^{3+}, or Mn^{4+} (among other choices).

11. D—This is part of the Heisenberg uncertainty principle.

12. C—The four electrons in the oxygen 2p orbitals are arranged with one pair and two unpaired electrons with spins parallel. This makes the oxygen atom paramagnetic. This arrangement is due to Hund's rule.

13. A—The Pauli exclusion principle restricts the number of electrons that can occupy a single orbital.

14. B—The d orbitals are shielded more efficiently than the s orbitals. Thus, the less shielded d orbitals do not fill as readily as s orbitals with similar energy.

15. A—Calcium will form a +2 ion (Ca^{2+}), and *X* will need to gain two electrons to fill its outer shell and become a −2 ion (X^{2-}). The simplest formula for a compound containing a +2 ion and a −2 ion would be Ca*X*. The other answers involve different charges or a formula that has not been simplified.

16. C—Diffraction is a wave phenomenon.

Free-Response Question

You have 15 minutes to answer the following question. You may use a calculator and the tables in the back of the book.

Question 1

(a) The bond energy of fluorine is 159 kJ mol^{-1}.

i. Determine the energy, in J, of a photon of light needed to break an F–F bond.

ii. Determine the frequency of this photon in s^{-1}.

iii. Determine the wavelength of this photon in nanometers.

(b) Barium imparts a characteristic green color to a flame. The wavelength of this light is 551 nm. Determine the energy involved in kJ/mol.

Answer and Explanation for the Free-Response Question

(a) If you do not remember them, several of the equations given at the beginning of the exam are necessary. In addition, the values of Planck's constant, Avogadro's number, and the speed of light are necessary. These constants are also given on the exam.

i. This is a simple conversion problem:

$$\left(\frac{159\ \text{kJ}}{\text{mol}}\right)\left(\frac{1\ \text{mol}}{6.022\times10^{23}}\right)\left(\frac{1{,}000\ \text{J}}{1\ \text{kJ}}\right) = 2.64\times10^{-19}\ \text{J}$$

Give yourself 1 point if you got this answer.

ii. This part requires the equation $\Delta E = h\nu$ (this equation is given on the equation page of the AP exam).

$$\nu = \frac{\Delta E}{h} = \frac{2.64\times10^{-19}\,\text{J}}{6.63\times10^{-34}\,\text{Js}} = 3.98\times10^{14}\,\text{s}^{-1}$$

Give yourself 1 point for this answer. If you got the wrong answer in the preceding part, but used it correctly here (in place of the 2.64×10^{-19} J), you still get 1 point.

iii. (a) The equation $c = \lambda\nu$ is needed. (This equation is given on the equation page of the AP exam.)

$$\lambda = \frac{c}{\nu} = \left(\frac{3.0\times10^{8}\,\text{m/s}}{3.98\times10^{14}\,\text{s}^{-1}}\right)\left(\frac{1\,\text{nm}}{10^{-9}\,\text{m}}\right) = 7.5\times10^{2}\,\text{nm}$$

Again, give yourself 1 point for the correct answer. If you correctly used a wrong answer from the preceding part, you still get 1 point.

(b) This can be done as a one-step or a two-step problem. The AP test booklet gives you the equations to solve this directly as a two-step problem. This method will be followed here. The two equations may be combined to produce an equation that will allow you to do the problem in one step.

Using $c = \lambda\nu$:

$$\nu = c/\lambda = \left(\frac{3.0\times10^{8}\,\text{m/s}}{551\,\text{nm}}\right)\left(\frac{1\,\text{nm}}{10^{-9}\,\text{m}}\right) = 5.4\times10^{14}\,\text{s}^{-1}$$

Using $\Delta E = h\nu$:

$$\Delta E = (6.63 \times 10^{-34}\text{ Js})(5.4 \times 10^{14}\text{ s}^{-1})(1\text{ kJ}/1{,}000\text{ J})(6.022 \times 10^{23}/\text{mol})$$

$$= 2.2 \times 10^{2}\text{ kJ/mol}$$

Give yourself 1 point for each of these answers. If you did the problem as a one-step problem, give yourself 2 points if you got the final answer correct or 1 point if you left out any of the conversions.

The total for this question is 5 points, minus 1 point if any answer does not have the correct number of significant figures.

› Rapid Review

- Know the regions of the electromagnetic spectrum.
- The frequency, ν, is defined as the number of waves that pass a point per second.
- The wavelength, λ, is the distance between two identical points on a wave.
- The energy of light is related to the frequency by $E = h\nu$.
- The product of the frequency and wavelength of light is the speed of light: $c = \nu\lambda$.

- An orbital or wave function is a quantum mechanical, mathematical description of the electron.
- If all electrons in an atom are in their lowest possible energy level, then the atom is said to be in its ground state.
- If any electrons in an atom are in a higher energy state, then the atom is said to be in an excited state.
- The energy of an atom is quantized, existing in only certain distinct energy states.
- Quantum numbers are numbers used in Schrödinger's equation to describe the orbital size, shape, and orientation in space, and the spin of an electron.
- The principal quantum number, n, describes the size of the orbital. It must be a positive integer. It is sometimes referred to as the atom's shell.
- The angular momentum quantum number, l, defines the shape of the electron cloud. If $l = 0$, it is an s orbital; if $l = 1$, it is a p orbital; if $l = 2$, it is a d orbital; if $l = 3$, it is an f orbital, etc.
- The magnetic quantum number, m_l, describes the orientation of the orbital around the nucleus. It can be integer values ranging from $-l$ through 0 to $+l$.
- The spin quantum number, m_s, describes the spin of the electron and can only have values of $+\frac{1}{2}$ and $-\frac{1}{2}$.
- Be able to write the quantum numbers associated with the first 20 electrons.

CHAPTER 11

Bonding

IN THIS CHAPTER

Summary: The difference between elements and compounds was discussed in the Basics chapter, and chemical reactions were discussed in the Reactions and Periodicity chapter. But what are the forces holding together a compound? What is the difference in bonding between table salt and sugar? What do these compounds look like in three-dimensional space?

Compounds have a certain fixed proportion of elements. The periodic table often can be used to predict the type of bonding that might exist between elements. The following general guidelines apply:

metals + nonmetals → ionic bonds
nonmetal + nonmetal → covalent bonds
metal + metal → metallic bonding

We will discuss the first two types of bonding, ionic and covalent, in some depth. Metallic bonding is a topic that is very rarely encountered on the AP exam. Suffice it to say that metallic bonding is a bonding situation between metals in which the valence electrons are donated to a vast electron pool (sometimes called a "sea of electrons"), so that the valence electrons are free to move throughout the entire metallic solid.

The basic concept that drives bonding is related to the stability of the noble gas family (the group VIIIA or group 18 elements). Their extreme stability (lower energy state) is due to the fact that they have a filled valence shell, a full complement of eight valence electrons. (Helium is an exception. Its valence shell, the 1s, is filled with two electrons.) This is called the octet rule. During chemical reactions, atoms lose, gain, or share electrons in order to achieve a filled valence shell, to complete their octet. By completing their valence shell in this fashion they become isoelectronic, having the same number and arrangement of electrons, as the closest noble gas. There are numerous exceptions to the octet rule; for example, some atoms may have more than an octet.

Keywords and Equations

There are no keywords or equations on the AP exam specific to this chapter.

Lewis Electron-Dot Structures

The Lewis electron-dot symbol is a way of representing the element and its valence electrons. The chemical symbol is written, which represents the atom's nucleus and all inner-shell electrons. The valence, or outer-shell, electrons are represented as dots surrounding the atom's symbol. Take the valence electrons, distribute them as dots one at a time around the four sides of the symbol, and then pair them up until all the valence electrons are distributed. Figure 11.1 shows the Lewis symbol for several different elements.

The Lewis symbols will be used in the discussion of bonding, especially covalent bonding, and will form the basis of the discussion of molecular geometry.

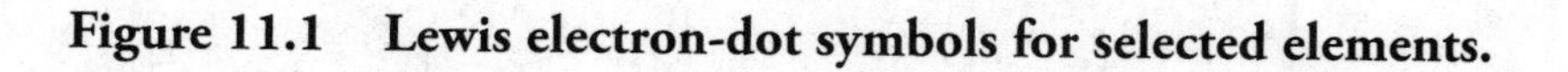

Figure 11.1 Lewis electron-dot symbols for selected elements.

Ionic and Covalent Bonding

Ionic Bonding

Ionic bonding results from the transfer of electrons from a metal to a nonmetal with the formation of cations (positively charged ions) and anions (negatively charged ions). The attraction of the opposite charges forms the **ionic bond**. The metal loses electrons to form a cation (the positive charge results from having more protons than electrons), and the nonmetal becomes an anion by gaining electrons (it now has more electrons than protons). This is shown in Figure 11.2 for the reaction of sodium and chlorine to form sodium chloride.

The number of electrons to be lost by the metal and gained by the nonmetal is determined by the number of electrons lost or gained by the atom in order to achieve a full octet. There is a rule of thumb that an atom can gain or lose one or two and, on rare occasions, three electrons, but not more than that. Sodium has one valence electron in energy level 3.

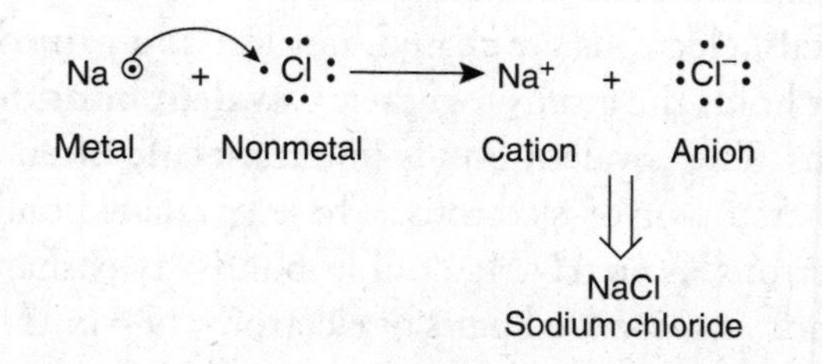

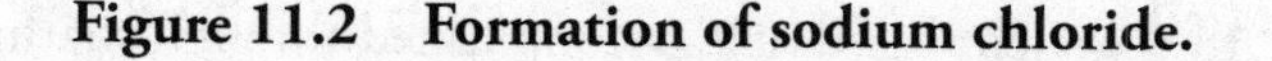

Figure 11.2 Formation of sodium chloride.

If it lost that one, the valence shell, now energy level 2, would be full (a more common way of showing this is with zero electrons). Chlorine, having seven valence electrons, needs to gain one more in order to complete its octet. So an electron is transferred from sodium to chlorine, completing the octet for both.

If magnesium, with two valence electrons to be lost, reacts with chlorine (which needs one additional electron), then magnesium will donate one valence electron to each of *two* chlorine atoms, forming the ionic compound $MgCl_2$. Make sure the formula has the lowest whole-number ratio of elements.

If aluminum, with three valence electrons to be lost, reacts with oxygen, which needs two additional electrons to complete its octet, then the lowest common factor between 3 and 2 must be found—6. Two aluminum atoms would each lose 3 electrons (total of 6 electrons lost) to three oxygen atoms, which would each gain 2 electrons (total 6 electrons gained). The total number of electrons lost must equal the total number of electrons gained.

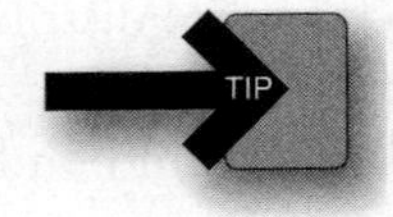

Another way of deriving the formula of the ionic compound is the **crisscross rule**. In this technique the cation and anion are written side by side. The numerical value of the superscript charge on the cation (without the sign) becomes the subscript on the nonmetal in the compound, and the superscript charge on the anion becomes the subscript on the metal in the compound. Figure 11.3 illustrates the crisscross rule for the reaction between aluminum and oxygen.

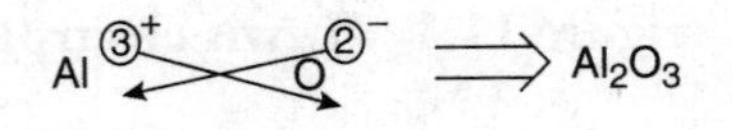

Figure 11.3 Using the crisscross rule.

If magnesium reacts with oxygen, then automatic application of the crisscross rule would lead to the formula Mg_2O_2, which is incorrect because the subscripts are not in the lowest whole-number ratio. For the same reason, lead(IV) oxide would have the formula PbO_2 and not Pb_2O_4. Make sure the formula has the lowest whole-number ratio of elements.

Ionic bonding may also involve polyatomic ions. The polyatomic ion(s) simply replace(s) one or both of the monoatomic ions.

Covalent Bonding

Consider two hydrogen atoms approaching each other. Both have only one electron, and each requires an additional electron to become isoelectronic with the nearest noble gas, He. One hydrogen atom could lose an electron; the other could gain that electron. One atom would have achieved its noble gas arrangement; but the other, the atom that lost its electron, has moved farther away from stability. The formation of the very stable H_2 cannot be explained by the loss and gain of electrons. In this situation, like that between any two nonmetals, electrons are shared, not lost and gained. No ions are formed. It is a covalent bond that holds the atoms together. **Covalent bonding** is the sharing of one or more *pairs* of electrons. The covalent bonds in a **molecule** often are represented by a dash, which represents a shared *pair* of electrons. These covalent bonds may be single bonds, one pair of shared electrons as in H–H; double bonds, two shared pairs of electrons $H_2C{=}CH_2$; or triple bonds, three shared pairs of electrons, N≡N. The same driving force forms a covalent bond as an ionic bond—establishing a stable (lower energy) electron arrangement. In the case of the covalent bond, it is accomplished through sharing electrons.

In the hydrogen molecule the electrons are shared equally. Each hydrogen nucleus has one proton equally attracting the bonding pair of electrons. A bond like this is called a **non-polar covalent bond**. In cases where the two atoms involved in the covalent bond are not the same, the attraction is not equal, and the bonding electrons are pulled toward the atom with the greater attraction. The bond becomes a **polar covalent bond**, with the atom that has the greater attraction taking on a partial negative charge and the other atom a partial positive charge. Consider for example, HF(g). The fluorine has a greater attraction for the bonding pair of electrons (greater electronegativity) and so takes on a partial negative charge. Many times, instead of using a single line to indicate the covalent bond, an arrow is used with the arrow head pointing toward the atom that has the greater attraction for the electron pair:

$$^{\delta+}H-F^{\delta-}$$

$$\longleftrightarrow$$

The **electronegativity (EN)** is a measure of the attractive force that an atom exerts on a bonding pair of electrons. Electronegativity values are tabulated. In general, electronegativities increase from left to right on the periodic table, except for the noble gases, and decrease going from top to bottom. This means that fluorine has the highest electronegativity of any element. If the difference in the electronegativities of the two elements involved in the bond is great (>1.7), the bond is considered to be mostly ionic in nature. If the difference is slight (<0.4), it is mostly nonpolar covalent. Anything in between is polar covalent.

Many times the Lewis structure will be used to indicate the bonding pattern in a covalent compound. In Lewis formulas the valence electrons that are not involved in bonding are shown as dots surrounding the element symbols, while a bonding pair of electrons is represented as a dash. There are several ways of deriving the Lewis structure, but here is one that works well for those compounds that obey the octet rule.

Draw the Lewis structural formula for CH_4O.

First, write a general framework for the molecule. In this case the carbon must be bonded to the oxygen, because hydrogen can form only one bond. Hydrogen is *never* central. Remember: **Carbon forms four bonds.**

```
     H
   H C O H
     H
```

To determine where all the electrons are to be placed, apply the N – A = S rule where:

N = sum of valence electrons needed for each atom. The two allowed values are two for hydrogen and eight for all other elements.
A = sum of all available valence electrons
S = # of electrons shared and S/2 = # bonds

For CH_4O, we would have:

	1 C		4 H		1 O	
N	8	+	4 (2) = 8	+	8	= 24
A	4	+	4(1) = 4	+	6	= 14

S = N – A = 24 – 14 = 10 bonds = S/2 = 10/2 = 5

Place the electron pairs, as dashes, between the adjacent atoms in the framework and then distribute the remaining available electrons so that each atom has its full octet, eight electrons—bonding or nonbonding, shared or not—for every atom except hydrogen, which gets two. Figure 11.4 shows the Lewis structural formula of CH_4O.

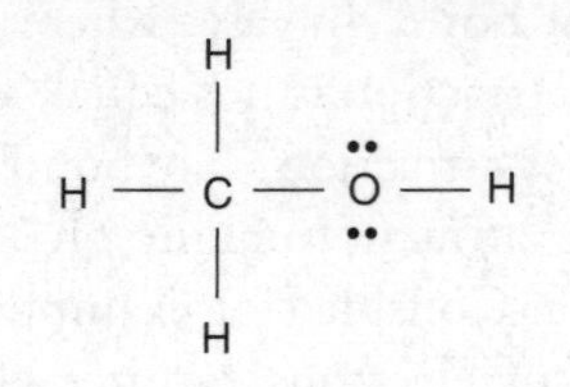

Figure 11.4 Lewis structure of CH_4O.

Lewis structures may also be written for polyatomic anions or cations. The N – A = S rule can be used, but if the ion is an anion, extra electrons equal to the magnitude of the negative charge must be added to the electrons available. If the ion is a cation, electrons must be subtracted.

As we have mentioned previously, there are many exceptions to the octet rule. In these cases, the N – A = S rule does not apply, as illustrated by the following example.

Draw the Lewis structure for XeF_4.
Answer:

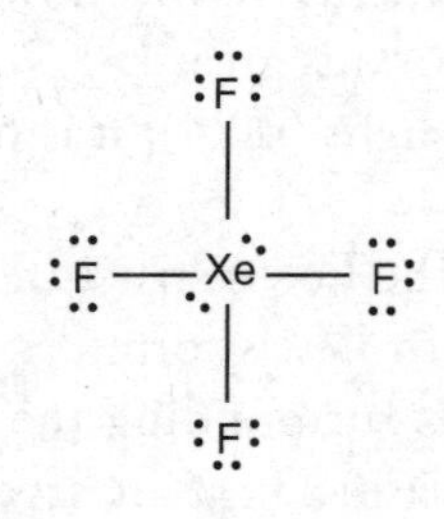

Each of the fluorines will have an additional three pairs of electrons. Only the four fluorine atoms have their octets.

This process will usually result in the correct Lewis structure. However, there will be cases when more than one structure may seem to be reasonable. One way to eliminate inappropriate structures is by using the formal charge.

There is a formal charge associated with each atom in a Lewis structure. To determine the formal charge for an atom, enter the number of electrons for each atom into the following relationship:

Formal Charge = (number of valence electrons) – (number of nonbonding electrons + 1/2 number of bonding electrons)

A formal charge of zero for each atom in a molecule is a very common result for a favorable Lewis structure. In other cases, a favorable Lewis structure will follow these rules: The formal charges are:

1. Small numbers, preferably 0.
2. No like charges are adjacent to each other, but unlike charges are close together.
3. The more electronegative element(s), the lower the formal charge(s) will be.
4. The total of the formal charges equals the charge on the ion.

Now we will apply this formal-charge concept to the cyanate ion OCN^-. We chose this example because many students incorrectly write the formula as CNO^-, and then try to use this as the atomic arrangement in the Lewis structure. Based on the number of electrons needed, the carbon should be the central atom. We will work this example using both the incorrect atom arrangement and the correct atom arrangement. Notice that in both structures all atoms have a complete octet.

	$[\ddot{\underset{..}{O}}::N::\ddot{C}:]^-$			$[\ddot{\underset{..}{O}}::C::\ddot{N}:]^-$		
Number of valence electrons	6	5	4	6	4	5
– Number of nonbonding electrons	–4	–0	–4	–4	–0	–4
– 1/2 Number of bonding electrons	–2	–4	–2	–2	–4	–2
Formal Charges	0	+1	–2	0	0	–1

The formal charges make the OCN arrangement the better choice.

Molecular Geometry—VSEPR

The shape of a molecule has quite a bit to do with its reactivity. This is especially true in biochemical processes, where slight changes in shape in three-dimensional space might make a certain molecule inactive or cause an adverse side effect. One way to predict the shape of molecules is the **VSEPR** (valence-shell electron-pair repulsion) **theory**. The basic idea behind this theory is that the valence electron pairs surrounding a central atom, whether involved in bonding or not, will try to move as far away from each other as possible to minimize the repulsion between the like charges. Two geometries can be determined; the *electron-group geometry,* in which all electron pairs surrounding a nucleus are considered, and *molecular geometry,* in which the nonbonding electrons become "invisible" and only the geometry of the atomic nuclei are considered. For the purposes of geometry, double and triple bonds count the same as single bonds. To determine the geometry:

KEY IDEA

1. Write the Lewis electron-dot formula of the compound.
2. Determine the number of electron-pair groups surrounding the central atom(s). Remember that double and triple bonds are treated as a single group.
3. Determine the geometric shape that maximizes the distance between the electron groups. This is the geometry of the electron groups.
4. Mentally allow the nonbonding electrons to become invisible. They are still there and are still repelling the other electron pairs, but we don't "see" them. The molecular geometry is determined by the remaining arrangement of atoms (as determined by the bonding electron groups) around the central atom.

Figure 11.5, on the next page, shows the electron-group and molecular geometry for two to six electron pairs.

For example, let's determine the electron-group and molecular geometry of carbon dioxide, CO_2, and water, H_2O. At first glance, one might imagine that the geometry of these two compounds would be similar, since both have a central atom with two groups attached. Let's see if that is true.

First, write the Lewis structure of each. Figure 11.6 shows the Lewis structures of these compounds.

Next, determine the electron-group geometry of each. For carbon dioxide, there are two electron groups around the carbon, so it would be linear. For water, there are four electron pairs around the oxygen—two bonding and two nonbonding electron pairs—so the electron-group geometry would be tetrahedral.

Finally, mentally allow the nonbonding electron pairs to become invisible and describe what is left in terms of the molecular geometry. For carbon dioxide, all groups are involved in bonding so the molecular geometry is also linear. However, water has two nonbonding

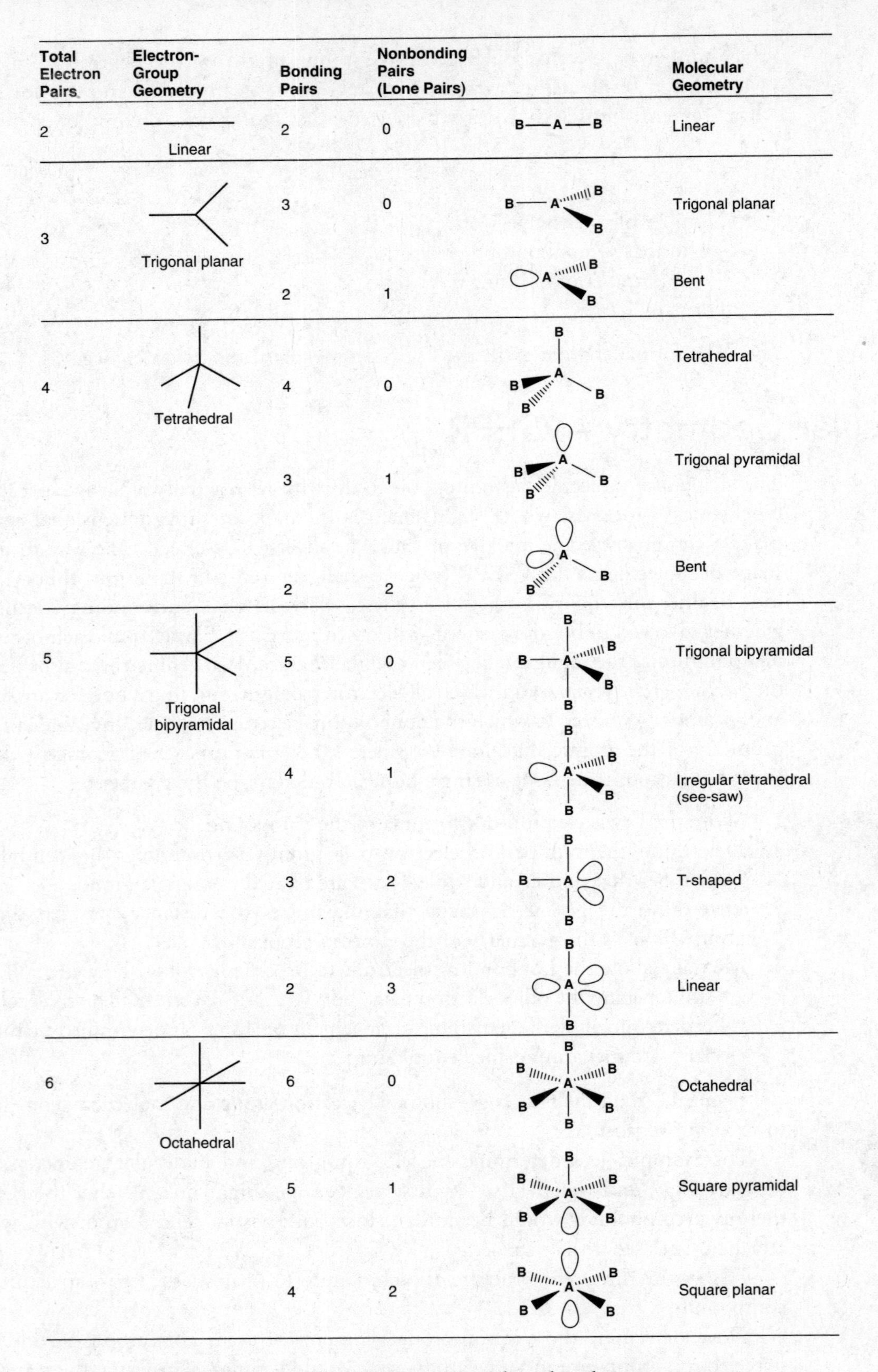

Total Electron Pairs	Electron-Group Geometry	Bonding Pairs	Nonbonding Pairs (Lone Pairs)	Molecular Geometry
2	Linear	2	0	Linear
3	Trigonal planar	3	0	Trigonal planar
		2	1	Bent
4	Tetrahedral	4	0	Tetrahedral
		3	1	Trigonal pyramidal
		2	2	Bent
5	Trigonal bipyramidal	5	0	Trigonal bipyramidal
		4	1	Irregular tetrahedral (see-saw)
		3	2	T-shaped
		2	3	Linear
6	Octahedral	6	0	Octahedral
		5	1	Square pyramidal
		4	2	Square planar

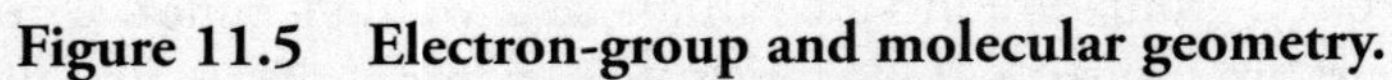

Figure 11.5 Electron-group and molecular geometry.

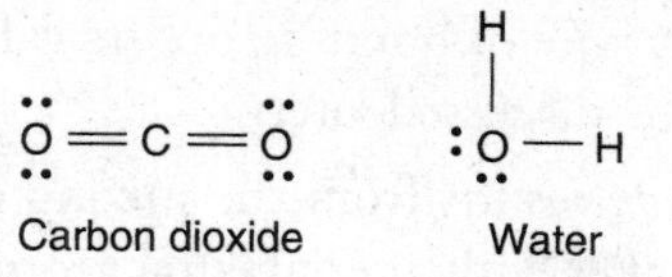

Figure 11.6 Lewis structures of carbon dioxide and water.

pairs of electrons so the remaining bonding electron pairs (and hydrogen nuclei) are in a bent arrangement.

This determination of the molecular geometry of carbon dioxide and water also accounts for the fact that carbon dioxide does not possess a dipole and water has one, even though both are composed of polar covalent bonds. Carbon dioxide, because of its linear shape, has partial negative charges at both ends and a partial charge in the middle. To possess a dipole, one end of the molecule must have a positive charge and the other a negative end. Water, because of its bent shape, satisfies this requirement. Carbon dioxide does not.

Valence Bond Theory

The VSEPR theory is only one way in which the molecular geometry of molecules may be determined. Another way involves the valence bond theory. The **valence bond theory** describes covalent bonding as the mixing of atomic orbitals to form a new kind of orbital, a hybrid orbital. **Hybrid orbitals** are atomic orbitals formed as a result of mixing the atomic orbitals of the atoms involved in the covalent bond. The number of hybrid orbitals formed is the same as the number of atomic orbitals mixed, and the type of hybrid orbital formed depends on the types of atomic orbital mixed. Figure 11.7 shows the hybrid orbitals resulting from the mixing of s, p, and d orbitals.

	Linear	Trigonal planar	Tetrahedral	Trigonal bipyramidal	Octahedral
Atomic orbitals mixed	one s one p	one s two p	one s three p	one s three p one d	one s three p two d
Hybrid orbitals formed	two sp	three sp^2	four sp^3	five sp^3d	six sp^3d^2
Unhybridized orbitals remaining	two p	one p	none	four d	three d
Orientation					

Figure 11.7 Hybridization of s, p, and d orbitals.

sp hybridization results from the overlap of an s orbital with one p orbital. Two sp hybrid orbitals are formed with a bond angle of 180°. This is a linear orientation.

sp^2 hybridization results from the overlap of an s orbital with two p orbitals. Three sp^2 hybrid orbitals are formed with a trigonal planar orientation and a bond angle of 120°.

One place this type of bonding occurs is in the formation of the carbon-to-carbon double bond, as will be discussed later.

sp^3 hybridization results from the mixing of one s orbital and three p orbitals, giving four sp^3 hybrid orbitals with a tetrahedral geometric orientation. This sp^3 hybridization is found in carbon when it forms four single bonds.

ENRICHMENT

sp^3d hybridization results from the blending of an s orbital, three p orbitals, and one d orbital. The result is five sp^3d orbitals with a trigonal bipyramidal orientation. This type of bonding occurs in compounds like PCl_5. Note that this hybridization is an exception to the octet rule.

sp^3d^2 hybridization occurs when one s, three p, and two d orbitals are mixed, giving an octahedral arrangement. SF_6 is an example. Again, this hybridization is an exception to the octet rule. If one starts with this structure and one of the bonding pairs becomes a lone pair, then a square pyramidal shape results, while two lone pairs gives a square planar shape.

Figure 11.8 shows the hybridization that occurs in ethylene, $H_2C{=}CH_2$. Each carbon has undergone sp^2 hybridization. On each carbon, two of the hybrid orbitals have overlapped with an s orbital on a hydrogen atom, to form a carbon-to-hydrogen covalent bond. The third sp^2 hybrid orbital has overlapped with the sp^2 hybrid on the other carbon to form a carbon-to-carbon covalent bond. Note that the remaining p orbital on each carbon that has not undergone hybridization is also overlapping above and below a line joining the carbons. In ethylene there are two types of bond. In **sigma** (σ) **bonds,** the overlap of the orbitals occurs on a line between the two atoms involved in the covalent bond. In ethylene, the C–H bonds and one of the C–C bonds are sigma bonds. In **pi** (π) **bonds,** the overlap of orbitals occurs above and below a line through the two nuclei of the atoms involved in the bond. A double bond always is composed of one sigma and one pi bond. A carbon-to-carbon triple bond results from the overlap of an sp hybrid orbital and two p orbitals on one carbon, with the same on the other carbon. In this situation there will be one sigma bond (overlap of the sp hybrid orbitals) and two pi bonds (overlap of two sets of p orbitals).

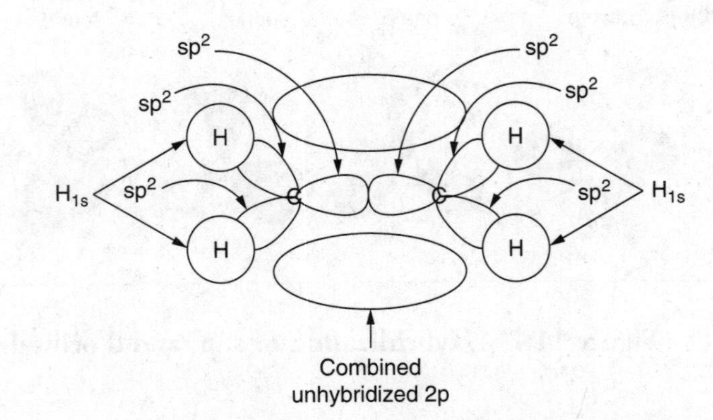

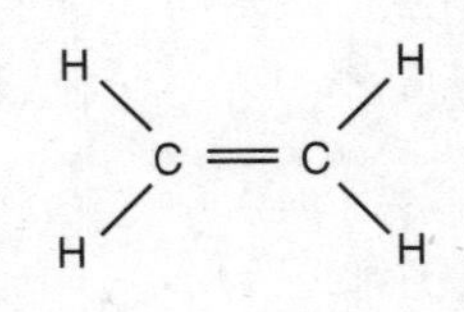

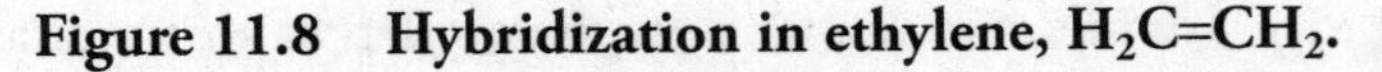
Figure 11.8 Hybridization in ethylene, $H_2C{=}CH_2$.

Molecular Orbital Theory

Still another model to represent the bonding that takes place in covalent compounds is the molecular orbital theory. In the **molecular orbital (MO) theory** of covalent bonding, atomic orbitals (AOs) on the individual atoms combine to form orbitals that encompass the entire molecule. These are called molecular orbitals (MOs). These molecular orbitals have definite shapes and energies associated with them. When two atomic orbitals are added, two molecular orbitals are formed, one bonding and one antibonding. The bonding MO is of lower energy than the antibonding MO. In the molecular orbital model the atomic orbitals are added together to form the molecular orbitals. Then the electrons are added to the molecular orbitals, following the rules used previously when filling orbitals: lowest-energy orbitals get filled first, maximum of two electrons per orbital, and half fill orbitals of equal energy before pairing electrons (see Chapter 5). When s atomic orbitals are added, one sigma bonding (σ) and one sigma antibonding (σ^*) molecular orbital are formed. Figure 11.9 shows the molecular orbital diagram for H_2.

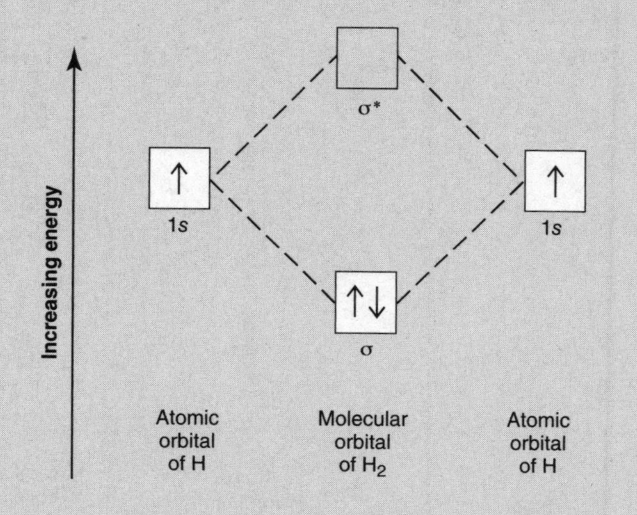

Figure 11.9 Molecular orbital diagram of H_2.

Note that the two electrons (one from each hydrogen) have both gone into the sigma bonding MO. The bonding situation can be calculated in the molecular orbital theory by calculating the MO bond order. The MO bond order is the number of electrons in bonding MOs minus the number of electrons in antibonding MOs, divided by 2. For H_2 in Figure 11.9 the bond order would be $(2 - 0)/2 = 1$. A stable bonding situation exists between two atoms when the bond order is greater than zero. The larger the bond order, the stronger the bond.

When two sets of p orbitals combine, one sigma bonding and one sigma antibonding MO are formed, along with two bonding pi MOs and two pi antibonding (π^*) MOs. Figure 11.10, on the next page, shows the MO diagram for O_2. For the sake of simplicity, the 1s orbitals of each oxygen and the MOs for these elections are not shown, just the valence-electron orbitals.

The bond order for O_2 would be $(10 - 6)/2 = 2$. (Don't forget to count the bonding and antibonding electrons at energy level 1.)

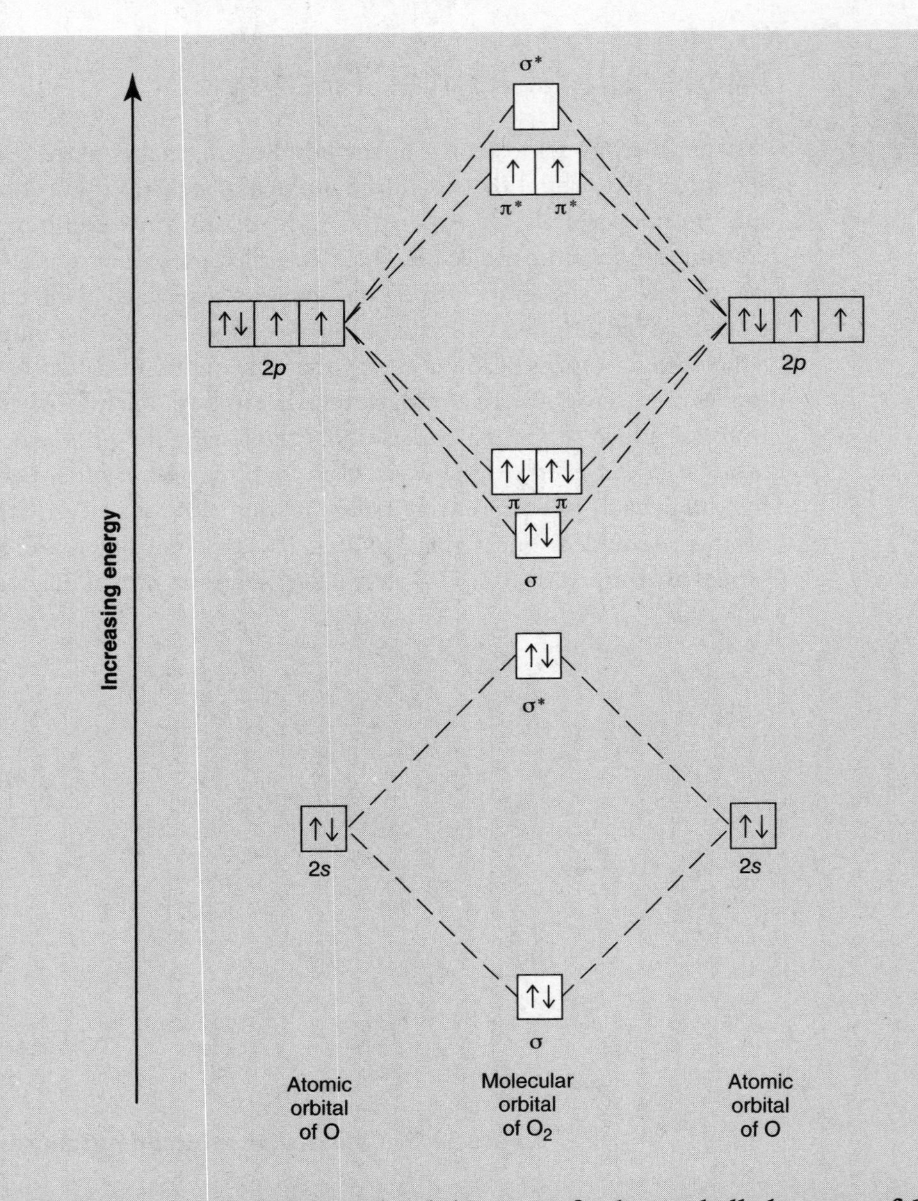

Figure 11.10 Molecular orbital diagram of valence-shell electrons of O_2.

Resonance

Sometimes when writing the Lewis structure of a compound, more than one possible structure is generated for a given molecule. The nitrate ion, NO_3^-, is a good example. Three possible Lewis structures can be written for this polyatomic anion, differing in which oxygen is double bonded to the nitrogen. None truly represents the actual structure of the nitrate ion; that would require an average of all three Lewis structures. **Resonance** theory is used to describe this situation. Resonance occurs when more than one Lewis structure can be written for a molecule. The individual structures are called resonance structures (or forms) and are written with a two-headed arrow (↔) between them. Figure 11.11 shows the three resonance forms of the nitrate ion.

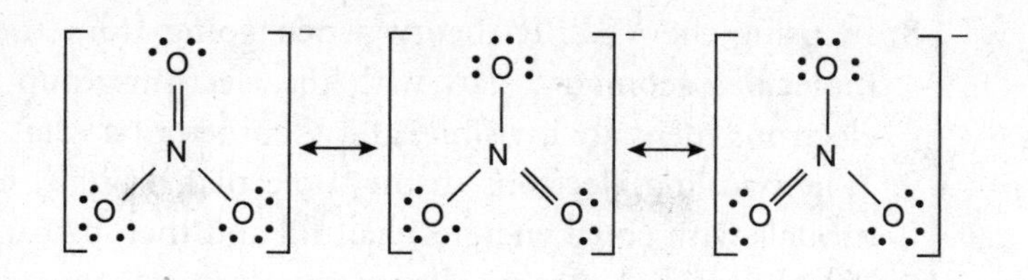

Figure 11.11 Resonance structures of the nitrate ion, NO_3^-.

Again, let us emphasize that the actual structure of the nitrate is not any of the three shown. Neither is it flipping back and forth among the three. It is an average of all three. All the bonds are the same, intermediate between single bonds and double bonds in strength and length.

Bond Length, Strength, and Magnetic Properties

The length and strength of a covalent bond is related to its bond order. The greater the bond order, the shorter and stronger the bond. Diatomic nitrogen, for example, has a short, extremely strong bond due to its nitrogen-to-nitrogen triple bond.

One of the advantages of the molecular orbital model is that it can predict some of the magnetic properties of molecules. If molecules are placed in a strong magnetic field, they exhibit one of two magnetic behaviors—attraction or repulsion. **Paramagnetism**, the attraction to a magnetic field, is due to the presence of unpaired electrons; **diamagnetism**, the slight repulsion from a magnetic field, is due to the presence of only paired electrons. Look at Figure 11.10, the MO diagram for diatomic oxygen. Note that it does have two unpaired electrons in the π_{2p}^* antibonding orbitals. Thus one would predict, based on the MO model, that oxygen should be paramagnetic, and that is exactly what is observed in the laboratory.

Experiments

There have been no experimental questions concerning this material on recent AP Chemistry exams.

Common Mistakes to Avoid

1. Remember that metals + nonmetals form ionic bonds, while the reaction of two nonmetals forms a covalent bond.
2. The octet rule does not always work, but for the representative elements it works the majority of the time.
3. Atoms that lose electrons form cations; atoms that gain electrons form anions.
4. In writing the formulas of ionic compounds, make sure the subscripts are in the lowest ratio of whole numbers.
5. When using the crisscross rule be sure the subscripts are reduced to the lowest whole-number ratio.
6. When using the $N - A = S$ rule in writing Lewis structures, be sure you add electrons to the A term for a polyatomic anion, and subtract electrons for a polyatomic cation.
7. In the $N - A = S$ rule, only the *valence* electrons are counted.

8. In using the VSEPR theory, when going from the electron-group geometry to the molecular geometry, start with the electron-group geometry; make the nonbonding electrons mentally invisible; and then describe what is left.
9. When adding electrons to the molecular orbitals, remember: lowest energy first. On orbitals with equal energies, half fill and then pair up.
10. When writing Lewis structures of polyatomic ions, don't forget to show the charge.
11. When you draw resonance structures, you can move only electrons (bonds). Never move the atoms.
12. When answering questions, the stability of the noble-gas configurations is a result, not an explanation. Your answers will require an explanation, i.e., lower energy state.

› Review Questions

Use these questions to review the content of this chapter and practice for the AP Chemistry exam. First are 16 multiple-choice questions similar to what you will encounter in Section I of the AP Chemistry exam. Following those is a multipart free-response question like the ones in Section II of the exam. To make these questions an even more authentic practice for the actual exam, time yourself following the instructions provided.

Multiple-Choice Questions

Answer the following questions in 20 minutes. You may not use a calculator. You may use the periodic table and the equation sheet at the back of this book.

1. VSEPR predicts an SbF_5 molecule will be which of the following shapes?

(A) tetrahedral
(B) trigonal bipyramidal
(C) square pyramid
(D) trigonal planar

2. The shortest bond would be present in which of the following substances?

(A) I_2
(B) CO
(C) CCl_4
(D) O_2^{2-}

3. Which of the following does not have one or more π bonds?

(A) H_2O
(B) HNO_3
(C) O_2
(D) N_2

4. Which of the following is polar?

(A) SF_4
(B) XeF_4
(C) CF_4
(D) SbF_5

5. Resonance structures are necessary to describe the bonding in which of the following?

(A) H_2O
(B) ClF_3
(C) HNO_3
(D) CH_4

For questions 6 and 7, pick the best choice from the following:

(A) ionic bonds
(B) hybrid orbitals
(C) resonance structures
(D) van der Waals attractions

6. An explanation of the equivalent bond lengths of the nitrite ion is:

7. Most organic substances have low melting points. This may be because, in most cases, the intermolecular forces are:

8. Which of the following has more than one unshared pair of valence electrons on the central atom?

(A) BrF_5
(B) NF_3
(C) IF_7
(D) ClF_3

9. What is the expected hybridization of the central atom in a molecule of $TiCl_4$? This molecule is tetrahedral.

(A) sp^3
(B) sp^3d
(C) sp
(D) sp^2

10. The only substance listed below that contains ionic, σ, and π bonds is:

(A) Na_2CO_3
(B) $HClO_2$
(C) H_2O
(D) NaCl

11. The electron pairs point toward the corners of which geometrical shape for a molecule with sp^2 hybrid orbitals?

(A) trigonal planar
(B) octahedron
(C) trigonal bipyramid
(D) trigonal pyramid

12. Regular tetrahedral molecules or ions include which of the following?

(A) SF_4
(B) NH_4^+
(C) XeF_4
(D) ICl_4^-

13. Which molecule or ion in the following list has the greatest number of unshared electrons around the central atom?

(A) CF_4
(B) ClF_3
(C) BF_3
(D) NH_4^+

14. Which of the following molecules is the least polar?

(A) PH_3
(B) CH_4
(C) H_2O
(D) NO_2

15. Which of the following molecules is the most polar?

(A) NH_3
(B) N_2
(C) CH_3I
(D) BF_3

16. Which of the following processes involves breaking an ionic bond?

(A) $H_2(g) + Cl_2(g) \rightarrow 2\ HCl(g)$
(B) $2\ KBr(s) \rightarrow 2\ K(g) + Br_2(g)$
(C) $Na(s) \rightarrow Na(g)$
(D) $2\ C_2H_6(g) + 7\ O_2(g) \rightarrow 4\ CO_2(g) + 6\ H_2O(g)$

Answers and Explanations for the Multiple-Choice Questions

1. B—The Lewis (electron-dot) structure has five bonding pairs around the central Sb and no lone pairs. VSEPR predicts this number of pairs to give a trigonal bipyramidal structure.

2. B—All the bonds except in CO are single bonds. The CO bond is a triple bond. Triple bonds are shorter than double bonds, which are shorter than single bonds. Drawing Lewis structures might help you answer this question.

3. A—Answers B through D contain molecules or ions with double or triple bonds. Double and triple bonds contain π bonds. Water has only single (σ) bonds. If any of these are not obvious to you, draw a Lewis structure.

4. A—The VSEPR model predicts all the other molecules to be nonpolar.

5. C—All the other answers involve species containing only single bonds. Substances without double or triple bonds seldom need resonance structures.

6. C—Resonance causes bonds to have the same average length.

7. D—Many organic molecules are nonpolar. Nonpolar substances are held together by weak van der Waals attractions.

8. D—Lewis structures are required. You do not need to draw all of them. A and B have one unshared pair, while C does not have an unshared pair. D has two unshared pairs of electrons.

9. A—Tetrahedral molecules are normally sp^3 hybridized.

10. A—Only A and D are ionic. The chloride ion has no internal bonds, so σ and π bonds are not possible.

11. A—This hybridization requires a geometrical shape with three corners.

12. B—One or more Lewis structures may help you. A is an irregular tetrahedron (seesaw); C and D are square planar.

13. B—A has 0. B has 2. C and D have 0. You may need to draw one or more Lewis structures.

14. B—All the molecules are polar except B.

15. A—Drawing one or more Lewis structures may help you. Only A and C are polar. Only the ammonia has hydrogen bonding, which is very, very polar.

16. B—C is breaking metallic bonding. All the others involve covalently bonded molecules.

Free-Response Question

You have 15 minutes to answer the following question. You may use a calculator and the tables in the back of the book.

Question 1

Answer the following questions about structure and bonding.

(a) Which of the following tetrafluoride compounds is nonpolar? Use Lewis electron-dot structures to explain your conclusions.

SiF_4 SF_4 XeF_4

(b) Rank the following compounds in order of increasing melting point. Explain your answer. Lewis electron-dot structures may aid you.

SnF_2 SeF_2 KrF_2

(c) Use Lewis electron-dot structures to show why the carbon–oxygen bonds in the oxalate ion ($C_2O_4^{2-}$) are all equal.

(d) When PCl_5 is dissolved in a polar solvent, the solution conducts electricity. Explain why. Use an appropriate chemical equation to illustrate your answer.

Answer and Explanation for the Free-Response Question

(a) Silicon tetrafluoride is the only one of the three compounds that is not polar.

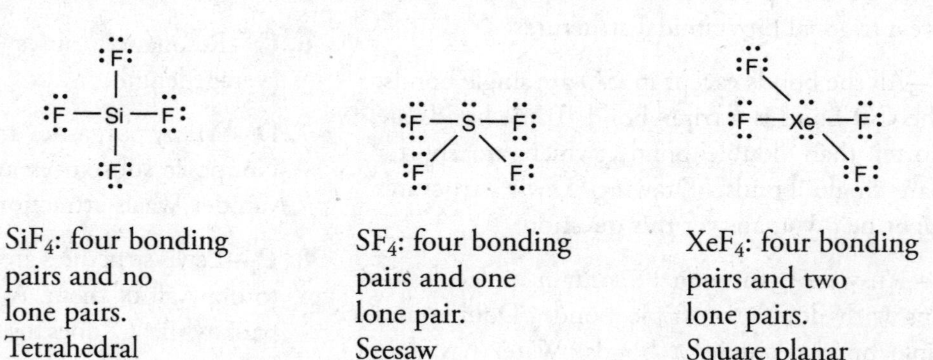

You get 1 point if you correctly predict only SiF_4 to be nonpolar. You get 1 additional point for each correct Lewis structure.

(b) The order is $KrF_2 < SeF_2 < SnF_2$.

:F—Kr—F:	:F—Se: | :F:	$[:Sn]^{2+}$ $2[:F:]^-$
KrF_2: two bonding pairs and three nonbonding pairs	SeF_2: two bonding pairs and two nonbonding pairs	SnF_2: ionic

The Lewis structure indicates that KrF_2 is nonpolar. Thus, it only has very weak London dispersion forces between the molecules. SeF_2 is polar, and the molecules are attracted by dipole-dipole attractions, which are stronger than London. SnF_2 has the highest melting point, because of the presence of strong ionic bonds.

You get 1 point for the order and 1 point for the discussion.

(c) It is possible to draw the following resonance structures for the oxalate ion. The presence of resonance equalizes the bonds.

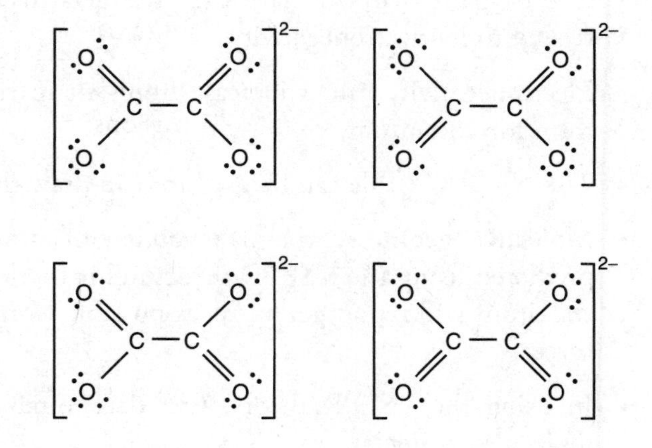

You get 1 point for any correct Lewis structure for $C_2O_4^{2-}$ and 1 point for showing or discussing resonance.

(d) PCl_5 must ionize. There are several acceptable equations, all of which must indicate the formation of ions. Here are two choices:

$$2\,PCl_5 \rightleftarrows PCl_4^+ + PCl_6^- \quad \text{or} \quad PCl_5 \rightleftarrows PCl_4^+ + Cl^-$$

You get 1 point for the explanation, and you get 1 point for either of the equations.

Total your points for the different parts. There are 10 points possible.

› Rapid Review

- Compounds are pure substances that have a fixed proportion of elements.
- Metals react with nonmetals to form ionic bonds, and nonmetals react with other nonmetals to form covalent bonds.

- The Lewis electron-dot structure is a way of representing an element and its valence electrons.
- Atoms tend to lose, gain, or share electrons to achieve the same electronic configuration (become isoelectronic to) as the nearest noble gas.
- Atoms are generally most stable when they have a complete octet (eight electrons).
- Ionic bonds result when a metal loses electrons to form cations and a nonmetal gains those electrons to form an anion.
- Ionic bonds can also result from the interaction of polyatomic ions.
- The attraction of the opposite charges (anions and cations) forms the ionic bond.
- The crisscross rule can help determine the formula of an ionic compound.
- In covalent bonding two atoms share one or more electron pairs.
- If the electrons are shared equally, the bond is a nonpolar covalent bond, but unequal sharing results in a polar covalent bond.
- The element that will have the greatest attraction for a bonding pair of electrons is related to its electronegativity.
- Electronegativity values increase from left to right on the periodic table and decrease from top to bottom.
- The $N - A = S$ rule can be used to help draw the Lewis structure of a molecule.
- Molecular geometry, the arrangement of atoms in three-dimensional space, can be predicted using the VSEPR theory. This theory says the electron pairs around a central atom will try to get as far as possible from each other to minimize the repulsive forces.
- In using the VSEPR theory, first determine the electron-group geometry, then the molecular geometry.
- The valence bond theory describes covalent bonding as the overlap of atomic orbitals to form a new kind of orbital, a hybrid orbital.
- The number of hybrid orbitals is the same as the number of atomic orbitals that were mixed together.
- There are a number of different types of hybrid orbital, such as sp, sp^2, and sp^3.
- In the valence bond theory, sigma bonds overlap on a line drawn between the two nuclei, while pi bonds result from the overlap of atomic orbitals above and below a line connecting the two atomic nuclei.
- A double or triple bond is always composed of one sigma bond and the rest pi bonds.
- In the molecular orbital (MO) theory of covalent bonding, atomic orbitals form molecular orbitals that encompass the entire molecule.
- The MO theory uses bonding and antibonding molecular orbitals.

- The bond order is (# electrons in bonding MOs – # electrons in anti-bonding MOs)/2.
- Resonance occurs when more than one Lewis structure can be written for a molecule. The actual structure of the molecule is an average of the Lewis resonance structures.
- The higher the bond order, the shorter and stronger the bond.
- Paramagnetism, the attraction of a molecule to a magnetic field, is due to the presence of unpaired electrons. Diamagnetism, the repulsion of a molecule from a magnetic field, is due to the presence of paired electrons.

CHAPTER 12

Solids, Liquids, and Intermolecular Forces

IN THIS CHAPTER

Summary: In the chapter on Gases we discussed the gaseous state. In this chapter, we will discuss the liquid and solid states and the forces that exist between the particles—the intermolecular forces. A substance's state of matter depends on two factors: the average kinetic energy of the particles, and the intermolecular forces between the particles. The kinetic energy tends to move the particles away from each other. The temperature of the substance is a measure of the average kinetic energy of the molecules. As the temperature increases, the average kinetic energy increases and the particles tend to move farther apart. This is consistent with our experience of heating ice, for example, and watching it move from the solid state to the liquid state and finally to the gaseous state. For this to happen, the kinetic energy overcomes the forces between the particles, the intermolecular forces.

In the solid state, the kinetic energy of the particles cannot overcome the intermolecular forces; the particles are held close together by the intermolecular forces. As the temperature increases, the kinetic energy increases and begins to overcome the attractive intermolecular forces. The substance will eventually melt, going from the solid to the liquid state. As this melting takes place, the temperature remains constant even though energy is being added. The temperature at which the solid converts into the liquid state is called the **melting point (m.p.)** of the solid.

After all the solid has been converted into a liquid, the temperature again starts to rise as energy is added. The particles are still relatively close together, but possess enough kinetic energy to move with respect to each other. Finally, if enough energy is added, the particles start to break free of the intermolecular forces keeping them relatively close together and they escape the liquid as essentially independent gas particles. This process of going from the liquid state to the gaseous state is called boiling, and the temperature at which this occurs is called the **boiling point (b.p.)** of the liquid. Sometimes, however, a solid can go directly from the solid state to the gaseous state without ever having become a liquid. This process is called **sublimation.** Dry ice, solid carbon dioxide, readily sublimes.

These changes of state, called **phase changes**, are related to temperature, but sometimes pressure can influence the changes. We will see how these relationships can be diagrammed later in this chapter.

KEY IDEA

Keywords and Equations

No specific keywords or equations are listed on the AP exam for this topic.

Structures and Intermolecular Forces

Intermolecular forces are attractive or repulsive forces between molecules, caused by partial charges. The attractive forces are the ones that work to overcome the randomizing forces of kinetic energy. The structure and type of bonding of a particular substance have quite a bit to do with the type of interaction and the strength of that interaction. Before we start examining the different types of intermolecular forces, recall from the Bonding chapter that those molecules that have polar covalent bonding (unequal sharing of the bonding electron pair) may possess dipoles (having positive and negative ends due to charge separation within the molecule). Dipoles are often involved in intermolecular forces.

Ion–Dipole Intermolecular Forces

These forces are due to the attraction of an ion and one end of a polar molecule (dipole). This type of attraction is especially important in aqueous salt solutions, where the ion attracts water molecules and may form a hydrated ion, such as $Al(H_2O)_6^{3+}$. This is one of the strongest of the intermolecular forces.

It is also important to realize that this intermolecular force requires two different species—an ion and a polar molecule.

Dipole–Dipole Intermolecular Forces

These forces result from the attraction of the positive end of one dipole to the negative end of another dipole. For example, in gaseous hydrogen chloride, HCl(g), the hydrogen end has a partial positive charge and the chlorine end has a partial negative charge, due to chlorine's higher electronegativity. Dipole–dipole attractions are especially important in polar liquids. They tend to be a rather strong force, although not as strong as ion–dipole attractions.

Hydrogen Bond Intermolecular Forces

Hydrogen bonding is a special type of dipole–dipole attraction in which a hydrogen atom is polar-covalently bonded to one of the following extremely electronegative elements: N, O, or F. These hydrogen bonds are extremely polar bonds by nature, so there is a great degree of charge separation within the molecule. Therefore, the attraction of the positively charged hydrogen of one molecule and the negatively charged N, O, or F of another molecule is extremely strong. These hydrogen bonds are in general, stronger than the typical dipole–dipole interaction.

Hydrogen bonding explains why HF(aq) is a weak acid, while HCl(aq), HBr(aq), etc. are strong acids. The hydrogen bond between the hydrogen of one HF molecule and the fluorine of another "traps" the hydrogen, so it is much harder to break its bonds and free the hydrogen to be donated as an H^+. Hydrogen bonding also explains why water has such unusual properties— for example, its unusually high boiling point and the fact that its solid phase is less dense than its liquid phase. The hydrogen bonds tend to stabilize the water molecules and keep them from readily escaping into the gas phase. When water freezes, the hydrogen bonds are stabilized and lock the water molecules into a framework with a lot of open space. Therefore, ice floats in liquid water. Hydrogen bonding also holds the strands of DNA together.

Ion-Induced Dipole and Dipole-Induced Dipole Intermolecular Forces

These types of attraction occur when the charge on an ion or a dipole distorts the electron cloud of a nonpolar molecule and induces a temporary dipole in the nonpolar molecule. Like ion–dipole intermolecular forces, these also require two different species. They are fairly weak interactions.

London (Dispersion) Intermolecular Force

This intermolecular attraction occurs in all substances, but is significant only when the other types of intermolecular forces are absent. It arises from a momentary distortion of the electron cloud, with the creation of a very weak dipole. The weak dipole induces a dipole in another nonpolar molecule. This is an extremely weak interaction, but it is strong enough to allow us to liquefy nonpolar gases such as hydrogen, H_2, and nitrogen, N_2. If there were no intermolecular forces attracting these molecules, it would be impossible to liquefy them.

The Liquid State

At the microscopic level, liquid particles are in constant flux. They may exhibit short-range areas of order, but these do not last very long. Clumps of particles may form and then break apart. At the macroscopic level, a liquid has a specific volume but no fixed shape. Three other macroscopic properties deserve discussion: surface tension, viscosity, and capillary action. In the body of a liquid the molecules are pulled in all different ways by the intermolecular forces between them. At the surface of the liquid, the molecules are only being pulled into the body of the liquid from the sides and below, not from above. The effect of this unequal attraction is that the liquid tries to minimize its surface area by forming a sphere. In a large pool of liquid, where this is not possible, the surface behaves as if it had a thin "skin" over it. It requires force to break the attractive forces at the surface. The amount of force required to break through this molecular layer at the surface is called the liquid's **surface tension**. The greater the intermolecular forces, the greater the surface tension. Polar liquids, especially those that undergo hydrogen bonding, have a much higher surface tension than nonpolar liquids.

Viscosity, the resistance of liquids to flow, is affected by intermolecular forces, temperature, and molecular shape. Liquids with strong intermolecular forces tend to have a higher viscosity than those with weak intermolecular forces. Again, polar liquids tend to have a higher viscosity than nonpolar liquids. As the temperature increases, the kinetic energy of the particles becomes greater, overcoming the intermolecular attractive forces. This causes a lower viscosity. Finally, the longer and more complex the molecules, the more contact the particles will have as they slip by each other, increasing the viscosity.

Capillary action is the spontaneous rising of a liquid through a narrow tube, against the force of gravity. It is caused by competition between the intermolecular forces in the liquid and those attractive forces between the liquid and the tube wall. The stronger the attraction between the liquid and the tube, the higher the level will be. Liquids that have weak attractions to the walls, like mercury in a glass tube, have a low capillary action. Liquids like water in a glass tube have strong attractions to the walls and will have a high capillary action.

As we have noted before, water, because of its stronger intermolecular forces (hydrogen bonding) has some very unusual properties. It will dissolve a great number of substances, both ionic and polar covalent, because of its polarity and ability to form hydrogen bonds. It is sometimes called the "universal solvent." It has a high **heat capacity**, the heat absorbed to cause the temperature to rise, and a high **heat of vaporization**, the heat needed to transform the liquid into a gas. Both of these thermal properties are due to the strong hydrogen bonding between the water molecules. Water has a high surface tension for the same reason. The fact that the solid form of water (ice) is less dense than liquid water is because water molecules in ice are held in a rigid, open, crystalline framework by the hydrogen bonds. As the ice starts melting, the crystal structure breaks and water molecules fill the holes in the structure, increasing the density. The density reaches a maximum at around 4°C; then the increasing kinetic energy of the particles causes the density to begin to decrease.

The Solid State

At the macroscopic level a **solid** is defined as a substance that has both a definite volume and a definite shape. At the microscopic level, solids may be one of two types—amorphous or crystalline. **Amorphous solids** lack extensive ordering of the particles. There is a lack of regularity of the structure. There may be small regions of order separated by large areas of disordered particles. They resemble liquids more than solids in this characteristic. Amorphous solids have no distinct, melting point. They simply get softer and softer as the temperature rises, leading to a decrease in viscosity. Glass, rubber, and charcoal are examples of amorphous solids.

Crystalline solids display a very regular ordering of the particles in a three-dimensional structure called the **crystal lattice**. In this crystal lattice there are repeating units called **unit cells**. Figure 12.1 shows the relationship of the unit cells to the crystal lattice.

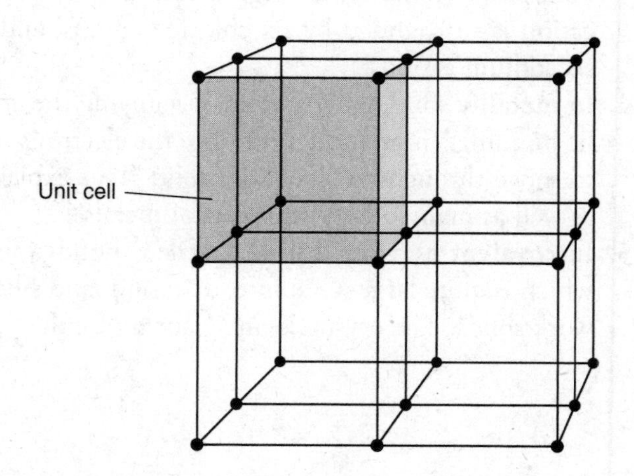

Figure 12.1 The crystal lattice for a simple cubic unit cell.

Several types of unit cells are found in solids. The cubic system is the most common type. Three types of unit cells are found in the cubic system:

1. The **simple cubic unit cell** has particles located at the corners of a simple cube.
2. The **body-centered unit cell** has particles located at the corners of the cube and in the center of the cube.
3. The **face-centered unit cell** has particles at the corners and one in the center of each face of the cube, but not in the center of the cube itself.

Figure 12.2 shows three types of cubic unit cells.

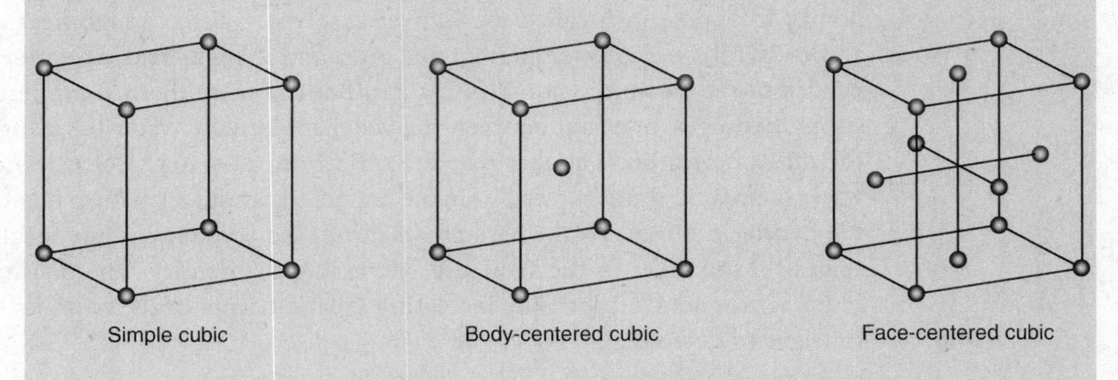

Figure 12.2 The three types of unit cell of the cubic lattice.

Five types of crystalline solid are known:

1. In **atomic solids,** individual atoms are held in place by London forces. The noble gases are the only atomic solids known to form.
2. In **molecular solids,** lattices composed of molecules are held in place by London forces, dipole–dipole forces, and hydrogen bonding. Solid methane and water are examples of molecular solids.
3. In **ionic solids,** lattices composed of ions are held together by the attraction of the opposite charges of the ions. These crystalline solids tend to be strong, with high melting points because of the strength of the intermolecular forces. NaCl and other salts are examples of ionic solids. Figure 12.3 shows the lattice structure of NaCl. Each sodium cation is surrounded by six chloride anions, and each chloride anion is surrounded by six sodium cations.
4. In **metallic solids**, metal atoms occupying the crystal lattice are held together by metallic bonding. In **metallic bonding**, the electrons of the atoms are delocalized and are free to move throughout the entire solid. This explains electrical and thermal conductivity, as well as many other properties of metals.
5. In **covalent network solids**, covalent bonds join atoms together in the crystal lattice, which is quite large. Graphite, diamond, and silicon dioxide (SiO_2) are examples of network solids. The crystal is one giant molecule.

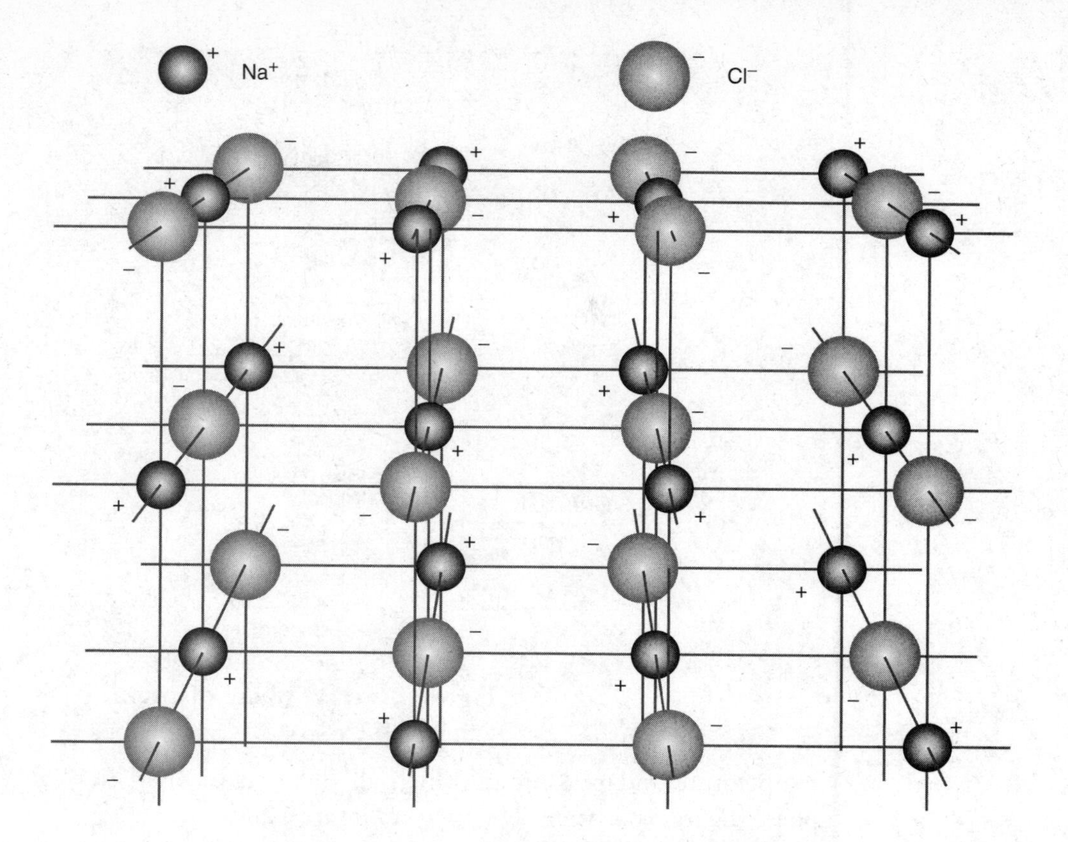

Figure 12.3 Sodium chloride crystal lattice.

Phase Diagrams

The equilibrium that exists between a liquid and its vapor is just one of several that can exist between states of matter. A **phase diagram** is a graph representing the relationship of a substance's states of matter to temperature and pressure. The diagram allows us to predict which state of matter a substance will assume at a certain combination of temperature and pressure. Figure 12.4, on the next page, shows a general form of the phase diagram.

Note that the diagram has three general areas corresponding to the three states of matter—solid, liquid, and gas. The line from A to C represents the solid's change in vapor pressure with changing temperature, for the sublimation equilibrium. The A-to-D line represents the variation in the melting point with varying pressure. The A-to-B line represents the variation of a liquid's vapor pressure with varying pressure. The B point shown on this phase diagram is called the **critical point** of the substance, the point beyond which the gas and liquid phases are indistinguishable from each other. At or beyond this critical point, no matter how much pressure is applied, the gas cannot be condensed into a liquid. Point A is the substance's **triple point**, the combination of

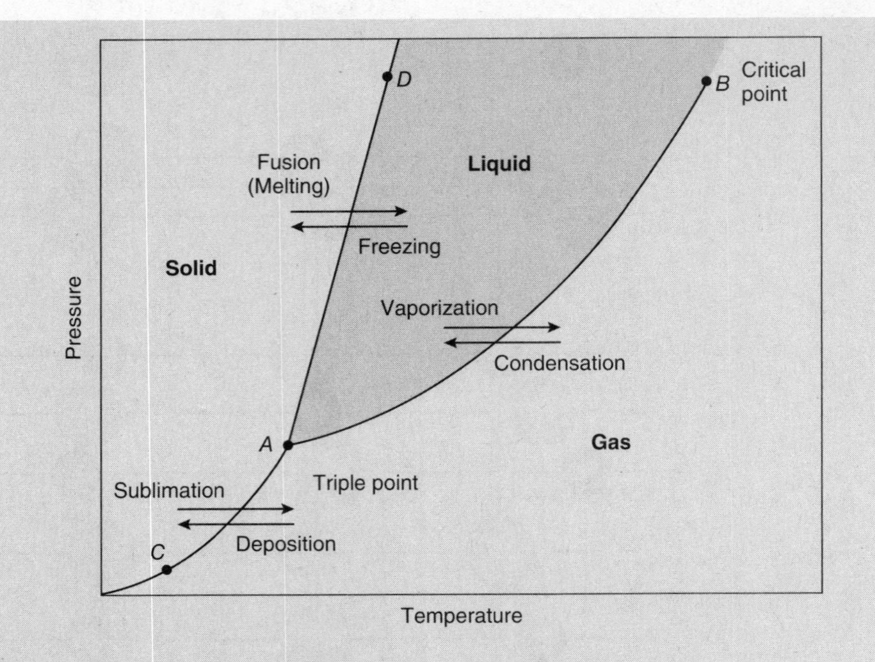

Figure 12.4 A phase diagram.

temperature and pressure at which all three states of matter can exist together. The phase diagram for water is shown in Figure 12.5.

For each of the phase transitions, there is an associated enthalpy change or heat of transition. For example, there are heats of vaporization, fusion, sublimation, and so on.

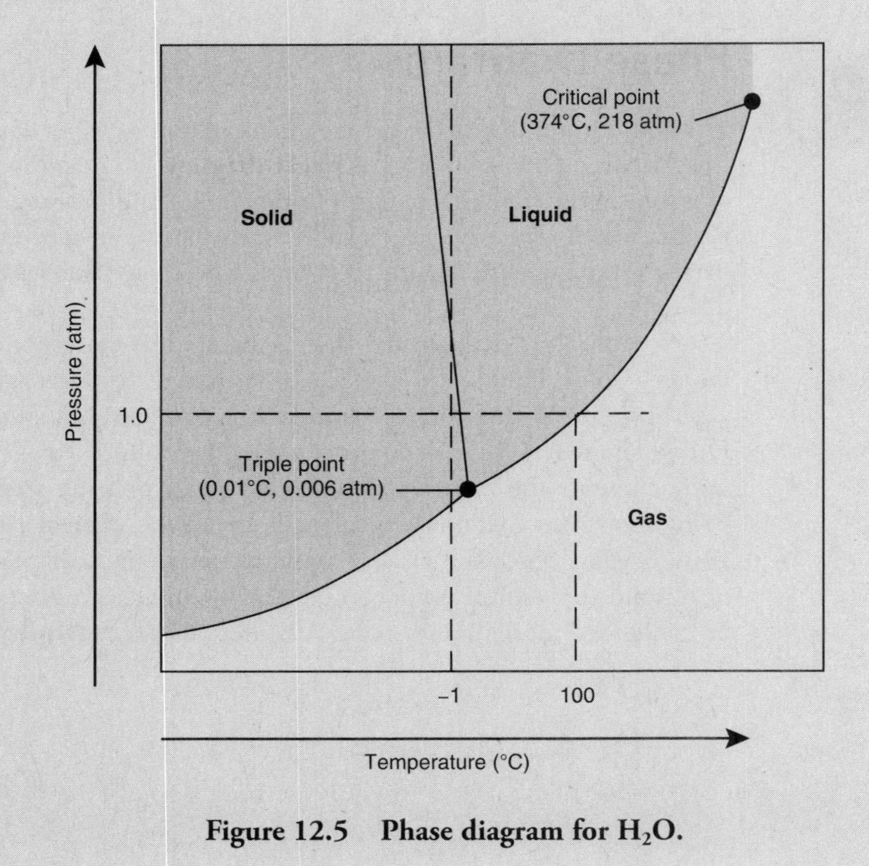

Figure 12.5 Phase diagram for H_2O.

Relationship of Intermolecular Forces to Phase Changes

The intermolecular forces can affect phase changes to a great degree. The stronger the intermolecular forces present in a liquid, the more kinetic energy must be added to convert it into a gas. Conversely, the stronger the intermolecular forces between the gas particles, the easier it will be to condense the gas into a liquid. In general, the weaker the intermolecular forces, the higher the vapor pressure. The same type of reasoning can be used about the other phase equilibria—in general, the stronger the intermolecular forces, the higher the heats of transition.

Example: Based on intermolecular forces, predict which will have the higher vapor pressure and higher boiling point, water or dimethyl ether, CH_3–O–CH_3.

Answer: Dimethyl ether will have the higher vapor pressure and the lower boiling point.

Explanation: Water is a polar substance with strong intermolecular hydrogen bonds. Dimethyl ether is a polar material with weaker intermolecular forces (dipole–dipole). It will take much more energy to vaporize water; thus, water has a lower vapor pressure and higher boiling point.

Experiments

The concept of intermolecular forces is important in the separation of the components of a mixture.

Common Mistakes to Avoid

1. Don't confuse the various types of intermolecular forces.
2. The melting point and the freezing point are identical.
3. Hydrogen bonding can occur only when a hydrogen atom is directly bonded to an N, O, or F atom.
4. When moving from point to point in a phase diagram, pay attention to which phase transitions the substance exhibits.
5. In looking at crystal lattice diagrams, be sure to count all the particles, in all three dimensions, that surround another particle.

❯ Review Questions

Use these questions to review the content of this chapter and practice for the AP Chemistry exam. Below are 14 multiple-choice questions similar to what you will encounter in Section I of the AP Chemistry exam. To make these questions an even more authentic practice for the actual exam, time yourself following the instructions provided.

Multiple-Choice Questions

Answer the following questions in 20 minutes. You may not use a calculator. You may use the periodic table and the equation sheet at the back of this book.

1. Which of the following best describes Fe(s)?

(A) composed of macromolecules held together by strong bonds
(B) composed of atoms held together by delocalized electrons
(C) composed of positive and negative ions held together by electrostatic attractions
(D) composed of molecules held together by intermolecular dipole–dipole interactions

2. The best description of the interactions in KNO_3(s) is which of the following?

(A) composed of macromolecules held together by strong bonds
(B) composed of atoms held together by delocalized electrons
(C) composed of positive and negative ions held together by electrostatic attractions
(D) composed of molecules held together by intermolecular dipole–dipole interactions

3. Sand is primarily SiO_2(s). Which of the following best describes the interactions inside a grain of sand?

(A) composed of macromolecules held together by strong bonds
(B) composed of atoms held together by delocalized electrons
(C) composed of positive and negative ions held together by electrostatic attractions
(D) composed of molecules held together by intermolecular dipole–dipole interactions

4. At sufficiently low temperatures, it is possible to form HCl(s). What best describes the interactions in this solid?

(A) composed of macromolecules held together by strong bonds
(B) composed of atoms held together by delocalized electrons
(C) composed of positive and negative ions held together by electrostatic attractions
(D) composed of molecules held together by intermolecular dipole–dipole interactions

5. Which of the following best describes diamond, C(s)?

(A) an ionic solid
(B) a metallic solid
(C) a molecular solid containing polar molecules
(D) a covalent network solid

6. What type of solid is solid sulfur dioxide, SO_2(s)?

(A) an ionic solid
(B) a metallic solid
(C) a molecular solid containing polar molecules
(D) a covalent network solid

7. The approximate boiling points for hydrogen compounds of some elements in the nitrogen family are: (SbH_3 – 15°C), (AsH_3 – 62°C), (PH_3 – 87°C), and (NH_3 – 33°C). The best explanation for the fact that NH_3 does not follow the trend of the other hydrogen compounds is:

(A) NH_3 is the only one to exhibit hydrogen bonding.
(B) NH_3 is the only one that is water-soluble.
(C) NH_3 is the only one that is nearly ideal in the gas phase.
(D) NH_3 is the only one that is a base.

8. Why is it possible to solidify argon at a sufficiently low temperature?

(A) London dispersion forces
(B) covalent bonding
(C) hydrogen bonding
(D) metallic bonding

9. Which of the following best describes why diamond is so hard?

(A) London dispersion forces
(B) covalent bonding
(C) hydrogen bonding
(D) metallic bonding

10. A sample of a pure liquid is placed in an open container and heated to the boiling point. Which of the following may increase the boiling point of the liquid?

(A) The moles of liquid are increased.
(B) The size of the container is increased.
(C) A vacuum is created over the liquid.
(D) The container is sealed.

11. Which of the following best explains why 1-butanol, $CH_3CH_2CH_2CH_2OH$, has a higher surface tension than its isomer, diethyl ether, $CH_3CH_2OCH_2CH_3$?

(A) the higher density of 1-butanol
(B) the lower specific heat of 1-butanol
(C) the lack of hydrogen bonding in 1-butanol
(D) the presence of hydrogen bonding in 1-butanol

12. Pick the answer that most likely represents the substances' relative solubilities in water.

(A) $CH_3CH_2CH_2CH_3 < CH_3CH_2CH_2OH < HOCH_2CH_2OH$
(B) $CH_3CH_2CH_2OH < CH_3CH_2CH_2CH_3 < HOCH_2CH_2OH$
(C) $CH_3CH_2CH_2CH_3 < HOCH_2CH_2OH < CH_3CH_2CH_2OH$
(D) $HOCH_2CH_2OH < CH_3CH_2CH_2OH < CH_3CH_2CH_2CH_3$

13. What is the energy change that accompanies the conversion of molecules in the gas phase to a liquid?

(A) heat of condensation
(B) heat of deposition
(C) heat of sublimation
(D) heat of fusion

14. Which of the following explains why the melting point of sodium chloride (NaCl 801°C) is lower than the melting point of calcium fluoride (CaF_2 1423°C)?

(A) Sodium is more reactive than calcium is.
(B) The chloride ion is smaller than the fluoride ion.
(C) The charge on a sodium ion is less than the charge on a calcium ion.
(D) The ratio of anions to cations is lower in sodium chloride.

Answers and Explanations for the Multiple-Choice Questions

1. B—This answer describes a metallic solid.

2. C—This answer describes an ionic solid.

3. A—This answer describes a covalent network solid.

4. D—This answers describes a solid consisting of discrete polar molecules. Even though HCl(aq) is a strong acid with ions in solution, there is no water here to lead to ionization.

5. D—Each of the carbon atoms is covalently bonded to four other carbon atoms.

6. C—Sulfur dioxide molecules are polar.

7. A—Hydrogen bonding occurs when hydrogen is directly bonded to F, O, and in this case N.

8. A—Argon is a noble gas; none of the other bonding choices is an option.

9. B—Diamond is a covalent network solid with a large number of strong covalent bonds between the carbon atoms.

10. D—The size of the container or the number of moles is irrelevant. Sealing the container will cause an increase in pressure that will increase the boiling point. A decrease in pressure will lower the boiling point.

11. D—The compound with the higher surface tension is the one with the stronger intermolecular force. The hydrogen bonding in 1-butanol is stronger than the dipole-dipole attractions in diethyl ether.

12. A—The sequence for these similar molecules is nonpolar, then one hydrogen bond, then two hydrogen bonds.

13. A—This change is condensation, so the energy is the heat of condensation.

14. C—The only applicable factor listed is the charge difference. The chloride ion is larger than the fluoride ion. The ion ratio is not important nor is the reactivity of the elements.

› Rapid Review

- The state of matter in which a substance exists depends on the competition between the kinetic energy of the particles (proportional to temperature) and the strength of the intermolecular forces between the particles.
- The melting point is the temperature at which a substance goes from the solid to the liquid state and is the same as the freezing point.
- The boiling point is the temperature at which a substance goes from the liquid to the gaseous state. This takes place within the body of the liquid, unlike evaporation which takes place only at the surface of the liquid.
- Sublimation is the conversion of a solid to a gas without ever having become a liquid. Deposition is the reverse process.
- Phase changes are changes of state.
- Intermolecular forces are the attractive or repulsive forces between atoms, molecules, or ions due to full or partial charges. Be careful not to confuse intermolecular forces with intramolecular forces, the forces within the molecule.
- Ion–dipole intermolecular forces occur between ions and polar molecules.
- Dipole–dipole intermolecular forces occur between polar molecules.
- Hydrogen bonds are intermolecular forces between dipoles in which there is a hydrogen atom attached to an N, O, or F atom.
- Ion-induced dipole intermolecular forces occur between an ion and a nonpolar molecule.
- London (dispersion) forces are intermolecular forces between nonpolar molecules.
- Liquids possess surface tension (liquids behaving as if they had a thin "skin" on their surface, due to unequal attraction of molecules at the surface of the liquid), viscosity (resistance to flow), and capillary action (flow up a small tube).
- Amorphous solids have very little structure in the solid state.
- Crystalline solids have a great deal of structure in the solid state.
- The crystal lattice of a crystalline solid is the regular ordering of the unit cells.
- Know the five types of crystalline solid: atomic, molecular, ionic, metallic, and network.
- Phase changes can be related to the strength of intermolecular forces.

CHAPTER 13

Solutions and Colligative Properties

IN THIS CHAPTER

Summary: A **solution** is a homogeneous mixture composed of a solvent and one or more solutes. The **solvent** is the substance that acts as the dissolving medium and is normally present in the greatest amount. Commonly the solvent is a liquid, but it doesn't have to be. Our atmosphere is a solution with nitrogen as the solvent; it is the gas present in the largest amount (79%). Many times you will be dealing with a solution in which water is the solvent, an **aqueous solution**. The **solute** is the substance that the solvent dissolves and is normally present in the smaller amount. You may have more than one solute in a solution. For example, if you dissolved table salt (sodium chloride) and table sugar (sucrose) in water, you would have one solvent (water) and two solutes (sodium chloride and sucrose).

Some substances will dissolve in a particular solvent and others will not. There is a general rule in chemistry that states that "*like dissolves like*." This general statement may serve as an answer in the multiple-choice questions, but does not serve as an explanation in the free-response questions. This simply means that polar substances (salts, alcohols, etc.) will dissolve in polar solvents such as water, and nonpolar solutes, such as iodine, will dissolve in nonpolar solvents such as carbon tetrachloride. The solubility of a particular solute is normally expressed in terms of grams solute per 100 mL of solvent (g/mL) at a specified temperature. The temperature must be specified because the solubility of a particular substance will vary with the temperature. Normally, the solubility of solids dissolving in liquids increases with increasing temperature, while the reverse is true for gases dissolving in liquids.

A solution in which one has dissolved the maximum amount of solute per given amount of solvent at a given temperature is called a **saturated solution**.

An **unsaturated solution** has less than the maximum amount of solute dissolved. Sometimes, if the temperature, purity of the solute and solvent, and other factors are just right, you might be able to dissolve more than the maximum amount of solute, resulting in a **supersaturated solution**. Supersaturated solutions are unstable, and sooner or later separation of the excess solute will occur, until a saturated solution and separated solute remain.

The formation of a solution depends on many factors, such as the nature of the solvent, the nature of the solute, the temperature, and the pressure. Some of these factors were addressed in the Reactions and Periodicity chapter. In general, the solubility of a solid or liquid will increase with temperature and be unaffected by pressure changes. The solubility of a gas will decrease with increasing temperature and will increase with increasing partial pressure of the gas **(Henry's law)**.

Keywords and Equations

π = osmotic pressure
i = van't Hoff factor
K_f = molal freezing-point depression constant
K_b = molal boiling-point elevation constant
K_f for H_2O = 1.86 K kg mol^{-1}
K_b for H_2O = 0.512 K kg mol^{-1}
$\Delta T_f = iK_f$ molality
$\Delta T_b = iK_b$ molality
$\pi = iMRT$
molarity, M = moles solute per liter solution
molality, m = moles solute per kilogram solvent

Concentration Units

There are many ways of expressing the relative amounts of solute(s) and solvent in a solution. The terms *saturated, unsaturated,* and *supersaturated* give a qualitative measure, as do the terms *dilute* and *concentrated*. The term **dilute** refers to a solution that has a relatively small amount of solute in comparison to the amount of solvent. **Concentrated**, on the other hand, refers to a solution that has a relatively large amount of solute in comparison to the solvent. However, these terms are very subjective. If you dissolve 0.1 g of sucrose per liter of water, that solution would probably be considered dilute; 100 g of sucrose per liter would probably be considered concentrated. But what about 25 g per liter—dilute or concentrated? In order to communicate effectively, chemists use quantitative ways of expressing the concentration of solutions. Several concentration units are useful, including percentage, molarity, and molality.

Percentage

One common way of expressing the relative amount of solute and solvent is through percentage, amount-per-hundred. Percentage can be expressed in three ways:

mass percent
mass/volume percent
volume/volume percent

Mass (Sometimes Called Weight) Percentage

The mass percentage of a solution is the mass of the solute divided by the mass of the solution, multiplied by 100% to get percentage. The mass is commonly measured in grams.

$$\text{mass \%} = (\text{grams of solute/grams solution}) \times 100\%$$

For example, a solution is prepared by dissolving 25.2 g of sodium chloride in 250.0 g of water. Calculate the mass percent of the solution.

Answer:

$$\text{mass \%} = \frac{(25.2 \text{ g solute})}{(25.2 + 250.0) \text{ g solution}} \times 100\% = 9.16\%$$

A common error is forgetting to add the solute and solvent masses together in the denominator.

When solutions of this type are prepared, the solute and solvent are weighed out separately and then mixed together to form a solution. The final volume of the solution is unknown.

Mass/Volume Percentage

The mass/volume percent of a solution is the mass of the solute divided by the volume of the solution, multiplied by 100% to yield percentage. The volume of the solution is generally expressed in milliliters.

$$\text{mass/volume \%} = (\text{grams solute/volume of solution}) \times 100\%$$

When mass/volume solutions are prepared, the grams of the solute are weighed out and dissolved and diluted to the required volume.

For example, a solution is prepared by mixing 125.0 g of benzene with 250.0 g of toluene. The density of benzene is 0.8765 g/mL, and the density of toluene is 0.8669 g/mL. Determine the mass/volume percentage of the solution. Assume that the volumes are additive.

Answer:

First, determine the volume of the solution.

$$\begin{aligned}\text{solution volume} &= (125.0 \text{ g benzene})(\text{mL}/0.8765 \text{ g benzene}) \\ &\quad + (250.0 \text{ g toluene})(\text{mL}/0.8669 \text{ g toluene}) \\ &= 431.0 \text{ mL}\end{aligned}$$

Then

$$\text{mass \%} = \frac{(125.0 \text{ g benzene})}{431.0 \text{ mL solution}} \times 100\% = 29.00\%$$

Notice that it is not necessary to know the chemical formula of either constituent. A common error is forgetting to add the solute and solvent volumes together.

Volume/Volume Percentage

The third case is one in which both the solute and solvent are liquids. The volume percent of the solution is the volume of the solute divided by the volume of the solution, multiplied by 100% to generate the percentage.

$$\text{volume \%} = (\text{volume solute/volume solution}) \times 100\%$$

When volume percent solutions are prepared, the mL of the solute are diluted with solvent to the required volume.

For example, determine the volume percentage of carbon tetrachloride in a solution prepared by dissolving 100.0 mL of carbon tetrachloride and 100.0 mL of methylene chloride in 750.0 mL of chloroform. Assume the volumes are additive.

Answer:

$$\text{volume }\% = \frac{(100.0\text{ mL carbon tetrachloride})}{(100.0 + 100.0 + 750.0)\text{ mL solution}} \times 100\% = 10.53\%$$

A common error is not to add all the volumes together to get the volume of the solution.

If the solute is ethyl alcohol and the solvent is water, then another concentration term is used, proof. The **proof** of an aqueous ethyl alcohol solution is twice the volume percent. A 45.0 volume % ethyl alcohol solution would be 90.0 proof.

Molarity

Percentage concentration is common in everyday life (3% hydrogen peroxide, 5% acetic acid, commonly called vinegar, etc.). The concentration unit most commonly used by chemists is molarity. **Molarity (M)** is the number of moles of solute per liter of solution.

$$\text{M} = \text{moles solute/liter solution}$$

In preparing a molar solution, the correct number of moles of solute (commonly converted to grams using the molar mass) is dissolved and diluted to the required volume.

Determine the molarity of sodium sulfate in a solution produced by dissolving 15.2 g of Na_2SO_4 in sufficient water to produce 750.0 mL of solution.

$$\text{molarity} = \frac{15.2\text{ g }Na_2SO_4}{750.0\text{ mL}} \times \frac{1\text{ mol g }Na_2SO_4}{142\text{ g }Na_2SO_4} \times \frac{1{,}000\text{ mL}}{1\text{L}} = 0.143\text{M}$$

The most common error is not being careful with the units. Grams must be converted to moles, and milliliters must be converted to liters.

Another way to prepare a molar solution is by dilution of a more concentrated solution to a more dilute one by adding solvent. The following equation can be used:

$$(\text{M}_{before})(V_{before}) = (\text{M}_{after})(V_{after})$$

In the preceding equation, *before* refers to before dilution and *after* refers to after dilution.

Let's see how to apply this relationship. Determine the final concentration when 500.0 mL of water is added to 400.0 mL of a 0.1111 M solution of HC1. Assume the volumes are additive.

$\text{M}_{before} = 0.1111\text{ M}$ $\text{M}_{after} = ?$

$V_{before} = 400.0\text{ mL}$ $V_{before} = (400.0 + 500.0)\text{ mL}$

$$\text{M}_{after} = (\text{M}_{before})(V_{before})/(V_{after}) = (0.1111\text{ M})(400.0\text{ mL})/(900.0\text{ mL})$$

$$= 0.04938\text{ M}$$

The most common error is forgetting to add the two volumes.

ENRICHMENT

Molality

Sometimes the varying volumes of a solution's liquid component(s) due to changes in temperature present a problem. Many times volumes are not additive, but mass is additive. The chemist then resorts to defining concentration in terms of the molality. **Molality (m)** is defined as the moles of solute per kilogram of solvent.

$$\text{m} = \text{moles solute/kilograms solvent}$$

Notice that this equation uses kilograms of solvent, not solution. The other concentration units use mass or volume of the entire solution. Molal solutions use only the mass of the *solvent*. For dilute aqueous solutions, the molarity and the molality will be close to the same numerical value.

For example, ethylene glycol ($C_2H_6O_2$) is used in antifreeze. Determine the molality of ethylene glycol in a solution prepared by adding 62.1 g of ethylene glycol to 100.0 g of water.

$$\text{molality} = \frac{62.1\,\text{g}\ C_2H_6O_2}{100.0\,\text{g}\ H_2O} \times \frac{1{,}000\,\text{g}}{1\,\text{kg}} \times \frac{1\,\text{mol}\ C_2H_6O_2}{62.1\,\text{g}\ C_2H_6O_2} = 10.0\ \text{m}\ C_2H_6O_2$$

The most common error is to use the total grams in the denominator instead of just the grams of solvent.

Electrolytes and Nonelectrolytes

An **electrolyte** is a substance that, when dissolved in a solvent or melted conducts an electrical current. A **nonelectrolyte** does not conduct a current when dissolved. The conduction of the electrical current is usually determined using a light bulb connected to a power source and two electrodes. The electrodes are placed in the aqueous solution or melt, and if a conducting medium is present, such as ions, the light bulb will light, indicating the substance is an electrolyte.

The ions that conduct the electrical current can result from a couple of sources. They may result from the dissociation of an ionically bonded substance (a salt). If sodium chloride (NaCl) is dissolved in water, it dissociates into the sodium cation (Na^+) and the chloride anion (Cl^-). But certain covalently bonded substances may also produce ions if dissolved in water, a process called ionization. For example, acids, both inorganic and organic, will produce ions when dissolved in water. Some acids, such as hydrochloric acid (HCl), will essentially completely ionize. Others, such as acetic acid (CH_3COOH), will only partially ionize. They establish an equilibrium with the ions and the unionized species (see Chapter 15 for more on chemical equilibrium).

$$HCl(aq) \rightarrow H^+(aq) + Cl^-(aq) \qquad \text{100\% ionization}$$

$$CH_3COOH(aq) \rightleftarrows H^+(aq) + CH_3COO^-(aq) \qquad \text{partial ionization}$$

Species such as HCl that completely ionize in water are called **strong electrolytes**, and those that only partially ionize are called **weak electrolytes**. Most soluble salts also fall into the strong electrolyte category.

Colligative Properties

Some of the properties of solutions depend on the chemical and physical nature of the individual solute. The blue color of a copper(II) sulfate solution and the sweetness of a sucrose solution are related to the properties of those solutes. However, some solution properties simply depend on the *number* of solute particles, not the type of solute. These properties are called **colligative properties** and include:

- vapor pressure lowering
- freezing-point depression
- boiling-point elevation
- osmotic pressure

Vapor Pressure Lowering

If a liquid is placed in a sealed container, molecules will evaporate from the surface of the liquid and eventually establish a gas phase over the liquid that is in equilibrium with the liquid phase. The pressure generated by this gas is the **vapor pressure** of the liquid. Vapor pressure is temperature-dependent; the higher the temperature, the higher the vapor pressure. If the liquid is made a solvent by adding a nonvolatile solute, the vapor pressure of the resulting solution is always less than that of the pure liquid. The vapor pressure has been lowered by the addition of the solute; the amount of lowering is proportional to the number of solute particles added and is thus a colligative property.

Solute particles are evenly distributed throughout a solution, even at the surface. Thus, there are fewer solvent particles at the gas–liquid interface where evaporation takes place. Fewer solvent particles escape into the gas phase, and so the vapor pressure is lower. The higher the concentration of solute particles, the less solvent is at the interface and the lower the vapor pressure. This relationship is referred to as **Raoult's law**.

Freezing-Point Depression

The freezing point of a solution of a nonvolatile solute is always lower than the freezing point of the pure solvent. It is the number of solute particles that determines the amount of the lowering of the freezing point. The amount of lowering of the freezing point is proportional to the molality of the solute and is given by the equation

$$\Delta T_f = iK_f \text{ molality}$$

where ΔT_f is the number of degrees that the freezing point has been lowered (the difference in the freezing point of the pure solvent and the solution); K_f is the freezing-point depression constant (a constant of the individual solvent); the molality is the molality of the solute; and i is the van't Hoff factor—the ratio of the number of moles of particles released into solution per mole of solute dissolved. For a nonelectrolyte, such as sucrose, the van't Hoff factor would be 1. For an electrolyte, such as sodium chloride, you must take into consideration that if 1 mol of NaCl dissolves, 2 mol of particles would result (1 mol Na^+, 1 mol Cl^-). Therefore, the van't Hoff factor should be 2. However, because sometimes there is a pairing of ions in solution, the observed van't Hoff factor is slightly less (for example, it is 1.9 for a 0.05 m NaCl solution). The more dilute the solution, the closer the observed van't Hoff factor should be to the expected factor. If you can calculate the molality of the solution, you can also calculate the freezing point of the solution.

Let's learn to apply the preceding equation. Determine the freezing point of an aqueous solution containing 10.50 g of magnesium bromide in 200.0 g of water.

$$\Delta T = iK_f m = 3(1.86\,\text{K kg mol}^{-1})\left[\frac{(10.50\,\text{g MgBr}_2)\left(\frac{1\,\text{mole MgBr}_2}{184.113\,\text{g MgBr}_2}\right)}{(200.0\,\text{g})\left(\frac{1\,\text{kg}}{1{,}000\,\text{g}}\right)}\right]$$

$$= 1.59\,\text{K}$$

$$T_{fp} = (273.15 - 1.59)\,\text{K} = 271.56\,\text{K}\ (= -1.59^\circ\text{C})$$

The most common mistake is to forget to subtract the ΔT value from the normal freezing point.

The freezing-point depression technique is also commonly used to calculate the molar mass of a solute.

For example, a solution is prepared by dissolving 0.490 g of an unknown compound in 50.00 mL of water. The freezing point of the solution is –0.201°C. Assuming the compound is a nonelectrolyte, what is the molecular mass of the compound? Use 1.00 g/mL as the density of water.

$$m = \Delta T/K_f = 0.201\,\text{K}/(1.86\,\text{K kg mol}^{-1}) = 0.108\,\text{mol/kg}$$

$$50.00\,\text{mL}\ (1.00\,\text{g/mL})\ (1\,\text{kg}/1{,}000\,\text{g}) = 0.0500\,\text{kg}$$

$$(0.108\,\text{mol/kg})\ (0.0500\,\text{kg}) = 0.00540\,\text{mol}$$

$$0.490\,\text{g}/0.00540\,\text{mol} = 90.7\,\text{g/mol}$$

Many students make the mistake of stopping before they complete this problem.

Boiling-Point Elevation

Just as the freezing point of a solution of a nonvolatile solute is always lower than that of the pure solvent, the boiling point of a solution is always higher than the solvent's. Again, only the number of solute particles affects the boiling point. The mathematical relationship is similar to the one for the freezing-point depression above and is

$$\Delta T_b = iK_b\ \text{molality}$$

where ΔT_b is the number of degrees the boiling point has been elevated (the difference between the boiling point of the pure solvent and the solution); K_b is the boiling-point elevation constant; the molality is the molality of the solute; and *i* is the van't Hoff factor. You can calculate a solution's boiling point if you know the molality of the solution. If you know the amount of the boiling-point elevation and the molality of the solution, you can calculate the value of the van't Hoff factor, *i*.

For example, determine the boiling point of a solution prepared by adding 15.00 g of NaCl to 250.0 g water. ($K_b = 0.512$ K kg mol^{-1})

$$\Delta T = iK_bm = 2(0.512\,\text{K kg mol}^{-1})\left[\frac{(15.00\text{ g NaCl})\left(\dfrac{1\text{ mole NaCl}}{58.44\text{ g NaCl}}\right)}{(250.0\text{ g})\left(\dfrac{1\text{ kg}}{1{,}000\text{ g}}\right)}\right]$$

$$= 1.05\,\text{K}$$

$$T_{bp} = (373.15 + 1.05)\,\text{K} = 374.20\,\text{K}\ (= 101.05°\text{C})$$

A 1.00 molal aqueous solution of trichloroacetic acid (CCl_3COOH) is heated to the boiling point. The solution has a boiling point of 100.18°C.

Determine the van't Hoff factor for trichloroacetic acid (K_b for water = 0.512 K kg mol^{-1}).

$$\Delta T = (101.18 - 100.00) = 0.18°\text{C} = 0.18\text{ K}$$

$$i = \Delta T/K_bm = 0.18\text{ K}/(0.512\text{ K kg mol}^{-1})(1.00\text{ mol kg}^{-1}) = 0.35$$

A common mistake is the assumption that the van't Hoff factor must be a whole number. This is true only for strong electrolytes at very low concentrations.

Osmotic Pressure

If you were to place a solution and a pure solvent in the same container but separate them by a **semipermeable membrane** (which allows the passage of some molecules, but not all particles) you would observe that the level of the solvent side would decrease while the solution side would increase. This indicates that the solvent molecules are passing through the semipermeable membrane, a process called **osmosis**. Eventually the system would reach equilibrium, and the difference in levels would remain constant. The difference in the two levels is related to the **osmotic pressure**. In fact, one could exert a pressure on the solution side exceeding the osmotic pressure, and solvent molecules could be forced back through the semipermeable membrane into the solvent side. This process is

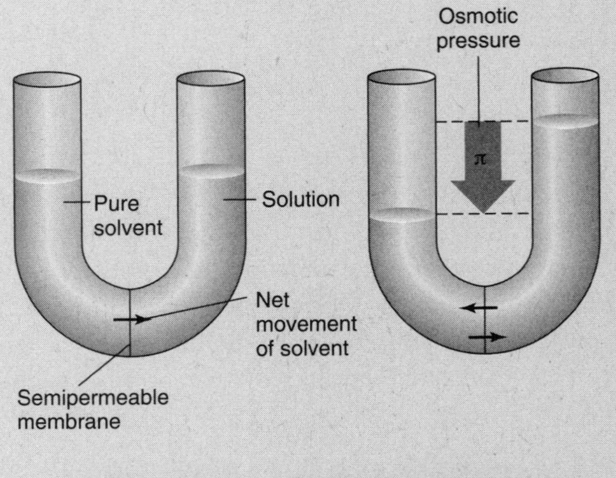

Figure 13.1 Osmotic pressure.

called **reverse osmosis** and is the basis of the desalination of seawater for drinking purposes. These processes are shown in Figure 13.1.

The osmotic pressure is a colligative property and mathematically can be represented as $\pi = (nRT/V)\ i$, where π is the osmotic pressure in atmospheres; n is the number of moles of solute; R is the ideal gas constant 0.0821 L · atm/K · mol; T is the Kelvin temperature; V is the volume of the solution; and i is the van't Hoff factor. Measurements of the osmotic pressure can be used to calculate the molar mass of a solute. This is especially useful in determining the molar mass of large molecules such as proteins.

For example, a solution prepared by dissolving 8.95 mg of a gene fragment in 35.0 mL of water has an osmotic pressure of 0.335 torr at 25.0°C. Assuming the fragment is a nonelectrolyte, determine the molar mass of the gene fragment.

Rearrange $\pi = (nRT/V)\ i$ to $n = \pi\ V/RT$ ($i = 1$ for a nonelectrolyte)

$$\frac{(0.335 \text{ torr})(35.0 \text{ mL})}{\left(0.0821\frac{\text{L atm}}{\text{mol K}}\right)(298.2 \text{ K})}\left(\frac{1 \text{ atm}}{760 \text{ torr}}\right)\left(\frac{1 \text{ L}}{1{,}000 \text{ mL}}\right) = 6.30 \times 10^{-7}$$

$$\frac{(8.95 \text{ mg})(0.001 \text{ g/mg})}{6.30 \times 10^{-7} \text{ mol}} = 1.42 \times 10^{4} \text{ g/mol}$$

Colloids

If you watch a glass of muddy water, you will see particles in the water settling out. This is a heterogeneous mixture where the particles are large (in excess of 1,000 nm), and it is called a **suspension**. In contrast, dissolving sodium chloride in water results in a true homogeneous **solution**, with solute particles less than 1 nm in diameter. True solutions do not settle out because of the very small particle size. But there are mixtures whose solute diameters fall in between solutions and suspensions. These are called **colloids** and have solute particles in the range of 1 to 1,000 nm diameter. Table 13.1 shows some representative colloids.

Many times it is difficult to distinguish a colloid from a true solution. The most common method is to shine a light through the mixture under investigation. A light shone through a true solution is invisible, but a light shown through a colloid is visible because the light reflects off the larger colloid particles. This is called the **Tyndall effect**.

Table 13.1 Common Colloid Types

COLLOID TYPE	SUBSTANCE DISPERSED	DISPERSING MEDIUM	EXAMPLES
aerosol	solid	gas	smoke
aerosol	liquid	gas	fog
solid foam	gas	solid	marshmallow
foam	gas	liquid	whipped cream
emulsion	liquid	liquid	milk, mayonnaise
solid emulsion	liquid	solid	cheese, butter
sol	solid	liquid	paint, gelatin

Experiments

Experimental procedures for solutions involve concentration units. Keeping close track of the units may simplify the problem.

Concentration problems are concerned with the definitions of the various units. It is possible to calculate the mass and/or volume of the solvent and solute by taking the difference between the final and initial measurements. The density, if not given, is calculated, not measured. It is important to recognize the difference between the values that must be measured and those that can be calculated. Moles are also calculated, not measured.

Do not forget that nearly all the concentration units use the total for the solution in the denominator. For these units it is important to remember to combine the quantities for the solvent and all solutes present.

Common Mistakes to Avoid

1. In molarity problems, be sure to use liters of **solution**.
2. Make sure your units cancel, leaving you with the units desired in your final answer.
3. Round off your final numerical answers to the correct number of significant figures.
4. Remember, most molecular compounds—compounds containing only nonmetals—do not ionize in solution. Acids are the most common exceptions.

› Review Questions

Use these questions to review the content of this chapter and practice for the AP Chemistry exam. First are 14 multiple-choice questions similar to what you will encounter in Section I of the AP Chemistry exam. Following those are two multipart free-response questions like the ones in Section II of the exam. To make these questions an even more authentic practice for the actual exam, time yourself following the instructions provided.

Multiple-Choice Questions

Answer the following questions in 20 minutes. You may not use a calculator. You may use the periodic table and the equation sheet at the back of this book.

1. A solution is prepared by dissolving 1.25 g of an unknown substance in 100.0 mL of water. Which procedure from the following list could be used to determine whether the solute is an electrolyte?

(A) Measure the specific heat of the solution.
(B) Measure the volume of the solution.
(C) Measure the freezing point of the solution.
(D) Determine the specific heat of the solution.

2. What is the final K^+ concentration in a solution made by mixing 300.0 mL of 1.0 M KNO_3 and 700.0 mL of 2.0 M K_3PO_4?

(A) 4.5 M
(B) 5.0 M
(C) 3.0 M
(D) 2.0 M

3. Strontium sulfate, $SrSO_4$, will precipitate when a solution of sodium sulfate is added to a strontium nitrate solution. What will be the strontium ion, Sr^{2+}, concentration remaining after 30.0 mL of 0.10 M Na_2SO_4 solution are added to 70.0 mL of 0.20 M $Sr(NO_3)_2$ solution?

(A) 0.14 M
(B) 0.15 M
(C) 0.11 M
(D) 0.20 M

4. Which of the following is a strong electrolyte when it is mixed with water?

(A) HNO_2
(B) KNO_3
(C) C_2H_5OH
(D) CH_3COOH

5. A solution with a total chloride ion, Cl^-, concentration of 1.0 M is needed. Initially, the solution is 0.30 M in $MgCl_2$. How many moles of solid $CaCl_2$ must be added to 400 mL of the $MgCl_2$ solution to achieve the desired concentration of chloride ion?

(A) 0.10
(B) 0.080
(C) 0.20
(D) 0.15

6. Assuming the volumes are additive, what is the final $H^+(aq)$ concentration produced by adding 30.0 mL of 0.50 M HNO_3 to 70.0 mL of 1.00 M HCl?

(A) 0.75 M
(B) 1.50 M
(C) 1.25 M
(D) 0.85 M

7. To prepare 3.0 L of a 0.20 molar K_3PO_4 solution (molecular weight 212 g/mol), a student should follow which of the following procedures?

(A) The student should weigh 42 g of solute and add sufficient water to obtain a final volume of 3.0 L.
(B) The student should weigh 42 g of solute and add 3.0 Kg of water.
(C) The student should weigh 130 g of solute and add sufficient water to obtain a final volume of 3.0 L.
(D) The student should weigh 42 g of solute and add 3.0 L of water.

8. How many grams of $MgSO_4$ (molecular weight 120.4 g/mol) are in 100.0 mL of a 5.0 molar solution?

(A) 600 g
(B) 5.0 g
(C) 12 g
(D) 60.0 g

9. How many milliliters of concentrated nitric acid (16.0 molar HNO_3) are needed to prepare 0.500 L of 6.0 molar HNO_3?

(A) 0.19 mL
(B) 250 mL
(C) 375 mL
(D) 190 mL

10. A solution has 10 grams of urea in 100 grams of solution. Which item(s) from the following list are needed to calculate the molarity of this solution?

(A) the density of the solution and the molecular weight of urea
(B) the density of the solution and the molecular weight of urea
(C) the density of the solvent and the density of the solution
(D) the molecular weight of urea and the density of the solvent

11. Which of the following aqueous solutions would have the greatest conductivity?

(A) 0.2 M NaOH
(B) 0.2 M RbCl
(C) 0.2 M K_3PO_4
(D) 0.2 M HNO_2

12. How many milliliters of water must be added to 50.0 mL of 10.0 M HNO_3 to prepare 4.00 M HNO_3, assuming that the volumes of nitric acid and water are additive?

(A) 50.0 mL
(B) 125 mL
(C) 500 mL
(D) 75.0 mL

13. The best method to isolate pure $MgSO_4$ from an aqueous solution of $MgSO_4$ is:

(A) evaporate the solution to dryness
(B) titrate the solution
(C) electrolyze the solution
(D) use paper chromatography

14. Pick the conditions that would yield the highest concentration of $N_2(g)$ in water.

(A) partial pressure of gas = 1.0 atm; temperature of water = 25°C
(B) partial pressure of gas = 0.50 atm; temperature of water = 55°C
(C) partial pressure of gas = 2.0 atm; temperature of water = 25°C
(D) partial pressure of gas = 2.0 atm; temperature of water = 85°C

Answers and Explanations for the Multiple-Choice Questions

1. C—If the solute is an electrolyte, the solution will conduct electricity and the van't Hoff factor, *i*, will be greater than 1. The choices do not include any conductivity measurements; therefore, the van't Hoff factor would need to be determined. This determination is done by measuring the osmotic pressure, the boiling-point elevation, or the freezing-point depression. The freezing point depression may be found by measuring the freezing point of the solution.

2. A—The potassium ion contribution from the KNO_3 is:

$(300.0\text{ mL})(1.0\text{ mol }KNO_3/1{,}000\text{ mL})(1\text{ mol }K^+/1\text{ mol }KNO_3) = 0.300\text{ mol }K^+$

The potassium ion contribution from K_3PO_4 is:

$(700.0\text{ mL})(2.0\text{ mol }K_3PO_4/1{,}000\text{ mL})(3\text{ mol }K^+/1\text{ mol }K_3PO_4) = 4.20\text{ mol }K^+$

The total potassium is 4.50 mol in a total volume of 1.000 L. Thus, the potassium concentration is 4.50 M.

3. C—The reaction is:

$$Sr^{2+}(aq) + SO_4^{2-} \rightarrow SrSO_4(s)$$

The strontium nitrate solution contains:

$(70.0\text{ mL})(0.20\text{ mol }Sr(NO_3)_2/1{,}000\text{ mL}) \times (1\text{ mol }Sr^{2+}/1\text{ mol }Sr(NO_3)_2) = 0.014\text{ mol }Sr^{2+}$

The sodium sulfate solution contains:

$(30.0\text{ mL})(0.10\text{ mol }Na_2SO_4/1{,}000\text{ mL})(1\text{ mol }SO_4^{2-}/1\text{ mol }Na_2SO_4) = 0.0030\text{ mol }SO_4^{2-}$

The strontium and sulfate ions react in a 1:1 ratio, so 0.0030 mol of sulfate ion will combine with 0.0030 mol of strontium ion, leaving 0.011 mol of strontium in a total volume of 100.0 mL. The final strontium ion concentration is:

$$\left(\frac{0.011\text{ mol }Sr^{2+}}{100.0\text{ mL}}\right)\left(\frac{1\text{ mL}}{0.001\text{ L}}\right)$$

4. B—A (nitrous acid) and D (acetic acid) are weak acids. Weak acids and bases are weak electrolytes. C (ethanol) is a nonelectrolyte. Potassium nitrate (B) is a water-soluble ionic compound, which is normally a strong electrolyte.

5. B—The number of moles of chloride ion needed is:

$(400\text{ mL})(1.0\text{ mol }Cl^-/1{,}000\text{ mL}) = 0.40\text{ mol }Cl^-$

The initial number of moles of chloride ion in the solution is:

$(400\text{ mL})(0.30\text{ mol }MgCl_2/1{,}000\text{ mL})(2\text{ mol }Cl^-/\text{mol }MgCl_2) = 0.24\text{ mol }Cl^-$

The number of moles needed = $[(0.40 - 0.24)\text{ mol }Cl^-](1\text{ mol }CaCl_2/2\text{ mol }Cl^-) = 0.080\text{ mol}$

6. D—Both of the acids are strong acids and yield 1 mol of H^+ each. Calculate the number of moles of H^+ produced by each of the acids. Divide the total number of moles by the final volume. $(30.0\text{ mL})(0.50\text{ mol }H^+/1{,}000\text{ mL}) + (70.0\text{ mL}) \times (1.00\text{ mol }H^+/1{,}000\text{ mL}) = 0.085\text{ mol }H^+$

$$\left(\frac{0.085\text{ mol }H^+}{100.0\text{ mL}}\right)\left(\frac{1\text{ mL}}{0.001\text{ L}}\right)$$

7. C—To produce a molar solution of any type, the final volume must be the desired volume. This eliminates answer D. B involves mass of water instead of volume. A calculation of the required mass will allow a decision between A and C.

$$(3.0\ \text{L})(0.20\ \text{mol}\ K_3PO_4/\text{L})(212\ \text{g}\ K_3PO_4/1\ \text{mol}\ K_3PO_4) = 130\ \text{g}\ K_3PO_4$$

8. D—$(5.0\ \text{mol}\ MgSO_4/1{,}000\ \text{mL})(100.0\ \text{mL}) \times (120.4\ \text{g}\ MgSO_4/\text{mol}\ MgSO_4) = 60.0\ \text{g}\ MgSO_4$

9. D—This is a dilution problem. $V_{before} = (M_{after})(V_{after})/(M_{before})$

$$(6.0\ \text{M}\ HNO_3)(0.500\ \text{L})(1{,}000\ \text{mL}/1\ \text{L})/(16.0\ \text{M}\ HNO_3) = 190\ \text{mL}$$

10. A—To calculate the molarity, the moles of urea and the volume of the solution are necessary. The density of the solution and the mass of the solution give the volume of the solution (it may be necessary to convert to liters). The mass of urea and the molecular weight of urea give the moles of urea.

11. C—The strong electrolyte with the greatest concentration of ions is the best conductor. D is a weak electrolyte, not a strong electrolyte. The number of ions for the strong electrolytes may be found by simply counting the ions: A – 2, B – 2, C – 4. The best conductor has the greatest value when the molarity is multiplied by the number of ions.

12. D—This is a dilution problem. $V_{after} = (M_{before}\ V_{before})/(M_{after})$

$$(10.0\ \text{M}\ HNO_3 \times 50.0\ \text{mL})/(4.0\ \text{M}\ HNO_3) = 125\ \text{mL}$$

The final volume is 125 mL. Since the original volume was 50.0 mL, an additional 75.0 mL must be added.

13. A—Solutions cannot be separated by titrations or filtering. Electrolysis of the solution would produce hydrogen and oxygen gas. Chromatography might achieve a minimal separation.

14. C—The solubility of a gas is increased by increasing the partial pressure of the gas and by lowering the temperature.

Free-Response Questions

Question 1

You have 5 minutes to answer the following two-part question. You may use a calculator and the tables in the back of the book.

Five beakers each containing 100.0 mL of an aqueous solution are on a lab bench. The solutions are all at 25°C. Solution 1 contains 0.20 M KNO_3. Solution 2 contains 0.10 M $BaCl_2$. Solution 3 contains 0.15 M $C_2H_4(OH)_2$. Solution 4 contains 0.20 M $(NH_4)_2SO_4$. Solution 5 contains 0.25 M $KMnO_4$.

(a) Which solution has the lowest pH? Explain.

(b) Which solution would be the poorest conductor of electricity? Explain.

Question 2

You have 15 minutes to answer the following four-part question. You may use a calculator and the tables in the back of the book.

Five beakers are placed in a row on a countertop. Each beaker is half filled with a 0.20 M aqueous solution. The solutes, in order, are: (1) potassium sulfate, K_2SO_4, (2) methyl alcohol, CH_3OH, (3) sodium carbonate, Na_2CO_3, (4) ammonium chromate, $(NH_4)_2CrO_4$, and (5) barium chloride, $BaCl_2$. The solutions are all at 25°C. Answer the following questions with respect to the five solutions listed above.

(a) Which solution will form a precipitate when ammonium chromate is added to it?

(b) Which solution is the most basic? Explain.

(c) Which solution would be the poorest conductor of electricity? Explain.

(d) Which solution is colored?

Answers and Explanations for the Free-Response Questions

Question 1

(a) Solution 4, because the ammonium ion is a weak acid

You get 1 point for picking solution 4 and 1 point for saying the ammonium ion (NH_4^+) is a weak acid, or that it undergoes hydrolysis.

(b) Solution 3, because the solute is a nonelectrolyte

Give yourself 1 point for picking solution 3 and 1 point for saying it is a nonelectrolyte or that it does not ionize.

Question 2

(a) The ammonium ion, from the ammonium chromate, will not form a precipitate since most ammonium compounds are water soluble. Therefore, the precipitate must contain the chromate ion combined with a cation from one of the solutions. Solution (2) is a nonelectrolyte; therefore, there are no cations present to combine with the chromate ion. The potassium and sodium ions, from solutions (1) and (3), give soluble salts like the ammonium ion. This only leaves solution (5) barium chloride that will give a precipitate. The formula of the precipitate is $BaCrO_4$.

You get 1 point for picking the correct solution.

(b) Solution (3) sodium carbonate is the most basic. Since the carbonate ion is the conjugate base of a weak acid, it will undergo significant hydrolysis to produce a basic solution.

You get 1 point for picking the correct solution and 1 point for the correct explanation.

(c) Methyl alcohol is a nonelectrolyte, so its solutions do not conduct electricity. The remaining solutions contain ionic salts, which in general are electrolytes in solution.

You get 1 point for picking the correct solution and 1 point for the correct formula for the explanation.

(d) Solution (4) ammonium chromate is yellow. Most solutions containing a transition metal ion are colored.

You get 1 point for picking the correct solution.

› Rapid Review

- A solution is a homogeneous mixture composed of a solvent and one or more solutes. A solute is a substance that dissolves in the solvent and is normally present in smaller amount.
- The general rule of solubility is "like dissolves like." This means that polar solvents dissolve polar solutes and nonpolar solvents dissolve nonpolar solutes. Remember, however, that simply quoting this rule will not be sufficient as an explanation in the free-response section.
- A saturated solution is one in which the maximum amount of solute is dissolved for a given amount of solvent at a given temperature. Any solution with less than the maximum solute is called unsaturated. A solution with greater than maximum solute is supersaturated (an unstable state).

- For the chemist the most useful unit of concentration is molarity (M), which is the moles of solute per liter of solution. Know how to work molarity problems. Be careful not to confuse molarity, M or [ ], with moles, *n* or mol.
- Electrolytes conduct an electrical current when melted or dissolved in a solvent, whereas nonelectrolytes do not.
- A colloid is a mixture in which the solute particle size is intermediate between a true solution and a suspension. If a light is shone through a colloid, the light beam is visible. This is the Tyndall effect.

CHAPTER 14

Kinetics

IN THIS CHAPTER

Summary: Thermodynamics often can be used to predict whether a reaction will occur spontaneously, but it gives very little information about the speed at which a reaction occurs. **Kinetics** is the study of the speed of reactions and is largely an experimental science. Some general qualitative ideas about reaction speed may be developed, but accurate quantitative relationships require that experimental data to be collected.

For a chemical reaction to occur, there must be a collision between the reactant particles. That collision is necessary to transfer kinetic energy, to break reactant chemical bonds and reform product ones. If the collision doesn't transfer enough energy, no reaction will occur. And the collision must take place with the proper orientation at the correct place on the molecule, the reactive site.

Five factors affect the rates of chemical reaction:

1. **Nature of the reactants**—Large, complex molecules tend to react more slowly than smaller ones because statistically there is a greater chance of collisions occurring somewhere else on the molecule, rather than at the reactive site.
2. **The temperature**—Temperature is a measure of the average kinetic energy of the molecules. The higher the temperature, the higher the kinetic energy and the greater the chance that enough energy will be transferred to cause the reaction. Also, the higher the temperature, the greater the number of collisions and the greater the chance of a collision at the reactive site.
3. **The concentration of reactants**—The higher the concentration of reactants, the greater the chance of collision and (normally) the greater the reaction rate. For gaseous reactants, the pressure is directly related to the concentration; the greater the pressure, the greater the reaction rate.

4. **Physical state of reactants**—When reactants are mixed in the same physical state, the reaction rates should be higher than if they are in different states, because there is a greater chance of collision. Also, gases and liquids tend to react faster than solids because of the increase in surface area. The more chance for collision, the faster the reaction rate.
5. **Catalysts**—A **catalyst** is a substance that speeds up the reaction rate and is (at least theoretically) recoverable at the end of the reaction in an unchanged form. Catalysts accomplish this by reducing the activation energy of the reaction. **Activation energy** is that minimum amount of energy that must be supplied to the reactants in order to initiate or start the reaction. Many times the activation energy is supplied by the kinetic energy of the reactants.

Keywords and Equations

$\ln[A]_t - \ln[A]_0 = -kt$ (first order)

$$\frac{1}{[A]_t} - \frac{1}{[A]_0} = kt \text{ (second order)}$$

$$t_{1/2} = \frac{\ln 2}{k} = \frac{0.693}{k}$$

$$\ln k = \frac{-E_a}{R}\left(\frac{1}{T}\right) + \ln A$$

t = time (seconds)
E_a = activation energy
k = rate constant
A = frequency factor
Gas constant, $R = 8.314/\text{J mol}^{-1}\ \text{K}^{-1}$

Rates of Reaction

The rate (or speed) of reaction is related to the change in concentration of either a reactant or product with time. Consider the general reaction: $2A + B \rightarrow C + 3D$. As the reaction proceeds, the concentrations of reactants A and B will decrease and the concentrations of products C and D will increase. Thus, the rate can be expressed in the following ways:

$$\text{Rate} = -\frac{1}{2}\frac{\Delta[A]}{\Delta t} = -\frac{\Delta[B]}{\Delta t} = \frac{\Delta[C]}{\Delta t} = \frac{1}{3}\frac{\Delta[D]}{\Delta t}$$

The first two expressions involving the reactants are negative, because their concentrations will decrease with time. The square brackets represent moles per liter concentration (molarity).

The rate of reaction decreases during the course of the reaction. The rate that is calculated above can be expressed as the average rate of reaction over a given time frame or, more commonly, as the initial reaction rate—the rate of reaction at the instant the reactants are mixed.

The Rate Equation

The rate of reaction may depend upon reactant concentration, product concentration, and temperature. Cases in which the product concentration affects the rate of reaction are rare and are not covered on the AP exam. Therefore, we will not address those reactions. We will discuss temperature effects on the reaction later in this chapter. For the time being, let's just consider those cases in which the reactant concentration may affect the speed of reaction. For the general reaction: a A + b B + . . . → c C + d D + . . . where the lower-case letters are the coefficients in the balanced chemical equation; the upper-case letters stand for the reactant; and product chemical species and initial rates are used, the rate equation (rate law) is written:

$$\text{Rate} = k[A]^m[B]^n \ldots$$

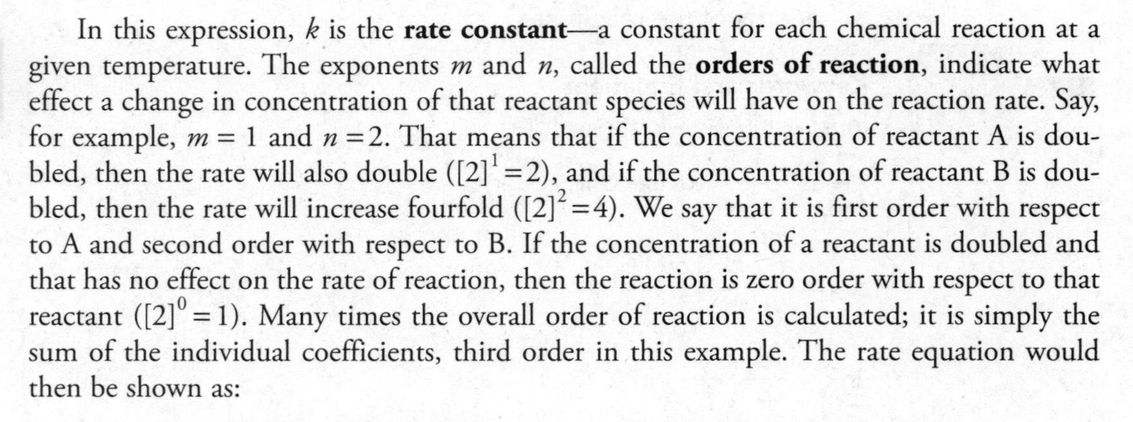

In this expression, k is the **rate constant**—a constant for each chemical reaction at a given temperature. The exponents m and n, called the **orders of reaction**, indicate what effect a change in concentration of that reactant species will have on the reaction rate. Say, for example, $m = 1$ and $n = 2$. That means that if the concentration of reactant A is doubled, then the rate will also double ($[2]^1 = 2$), and if the concentration of reactant B is doubled, then the rate will increase fourfold ($[2]^2 = 4$). We say that it is first order with respect to A and second order with respect to B. If the concentration of a reactant is doubled and that has no effect on the rate of reaction, then the reaction is zero order with respect to that reactant ($[2]^0 = 1$). Many times the overall order of reaction is calculated; it is simply the sum of the individual coefficients, third order in this example. The rate equation would then be shown as:

$$\text{Rate} = k[A][B]^2 \text{ (If the exponent is 1, it is generally not shown.)}$$

It is important to realize that the rate law (the rate, the rate constant, and the orders of reaction) is determined experimentally. Do not use the balanced chemical equation to determine the rate law.

The rate of reaction may be measured in a variety of ways, including taking the slope of the concentration versus time plot for the reaction. Once the rate has been determined, the orders of reaction can be determined by conducting a series of reactions in which the reactant species concentrations are changed one at a time, and mathematically determining the effect on the reaction rate. Once the orders of reaction have been determined, it is easy to calculate the rate constant.

For example, consider the reaction:

$$2\ NO(g) + O_2(g) \rightarrow 2\ NO_2(g)$$

The following kinetics data were collected:

Experiment	Initial [NO]	Initial [O_2]	Rate of NO_2 formation (M/s)
1	0.01	0.01	0.05
2	0.02	0.01	0.20
3	0.01	0.02	0.10

There are a couple of ways to interpret the data to generate the rate equation. If the numbers involved are simple (as above and on most tests, including the AP exam), you can reason out the orders of reaction. You can see that in going from experiment 1 to experiment 2, the [NO] was doubled, ([O_2] held constant), and the rate increased fourfold. This means that the reaction is second order with respect to NO. Comparing experiments 1 and 3, you see that the [O_2] was doubled, ([NO] was held constant), and the rate doubled.

Therefore, the reaction is first order with respect to O_2 and the rate equation can be written as:

$$\text{Rate} = k[\text{NO}]^2[\text{O}_2]$$

The rate constant can be determined by substituting the values of the concentrations of NO and O_2 from any of the experiments into the rate equation above and solving for k. Using experiment 1:

$$0.05\ \text{M/s} = k(0.01\ \text{M})^2(0.01\ \text{M})$$

$$k = (0.05\ \text{M/s})(0.01\ \text{M})^2(0.01\ \text{M})$$

$$k = 5 \times 10^4/\text{M}^2\text{s}$$

Sometimes, because of the numbers' complexity, you must set up the equations mathematically. The ratio of the rate expressions of two experiments will be used in determining the reaction orders. The equations will be chosen so that the concentration of only one reactant has changed while the others remain constant. In the example above, the ratio of experiments 1 and 2 will be used to determine the effect of a change of the concentration of NO on the rate, and then experiments 1 and 3 will be used to determine the effect of O_2. Experiments 2 and 3 cannot be used, because both chemical species have changed concentration.

Remember: In choosing experiments to compare, choose two in which the concentration of only one reactant has changed while the others have remained constant.

Comparing experiments 1 and 2:

$$\frac{0.05\ \text{M/s} = k[0.01]^m[0.01]^n}{0.20\ \text{M/s} = [0.02]^m[0.01]^n}$$

Canceling the rate constants and the $[0.01]^n$ and simplifying:

$$\frac{1}{4} = \left(\frac{1}{2}\right)^m$$

$$m = 2 \text{ (use logarithms to solve for } m)$$

Comparing experiments 1 and 3:

$$\frac{0.05\ \text{M/s} = k[0.01]^m[0.01]^n}{0.10\ \text{M/s} = [0.01]^m[0.02]^n}$$

Canceling the rate constants and the $[0.01]^n$ and simplifying:

$$\frac{1}{2} = \left(\frac{1}{2}\right)^n$$

$$n = 1$$

Writing the rate equation:

$$\text{Rate} = k[\text{NO}]^2[\text{O}_2]$$

Again, the rate constant k could be determined by choosing any of the three experiments, substituting the concentrations, rate, and orders into the rate expression, and then solving for k.

Integrated Rate Laws

Thus far, only cases in which instantaneous data are used in the rate expression have been shown. These expressions allow us to answer questions concerning the speed of the reaction at a particular moment, but not questions about how long it might take to use up a certain reactant, etc. If changes in the concentration of reactants or products over time are taken into account, as in the **integrated rate laws**, these questions can be answered. Consider the following reaction:

$$A \rightarrow B$$

Assuming that this reaction is first order, then the rate of reaction can be expressed as the change in concentration of reactant A with time:

$$\text{Rate} = -\frac{\Delta[A]}{\Delta t}$$

and also as the rate law:

$$\text{Rate} = k[A]$$

Setting these terms equal to each other gives:

$$-\frac{\Delta[A]}{\Delta t} = k[A]$$

and integrating over time gives:

$$\ln[A]_t - \ln[A]_0 = -kt$$

where ln is the natural logarithm, $[A]_0$ is the concentration of reactant A at time = 0, and $[A]_t$ is the concentration of reactant A at some time t.

If the reaction is second order in A, then the following equation can be derived using the same procedure:

$$\frac{1}{[A]_t} - \frac{1}{[A]_0} = kt$$

Consider the following problem: Hydrogen iodide, HI, decomposes through a second-order process to the elements. The rate constant is 2.40×10^{-21}/M s at 25°C. How long will it take for the concentration of HI to drop from 0.200 M to 0.190 M at 25°C?

Answer:

1.10×10^{20} s. In this problem, $k = 2.40 \times 10^{-21}$/M s, $[A]_0 = 0.200$ M, and $[A]_1 = 0.190$ M. You can simply insert the values and solve for t, or you first can rearrange the equation to give $t = [1/[A]_t - 1/[A]_0]/k$. You will get the same answer in either case. If you get a negative answer, you interchanged $[A]_t$ and $[A]_0$. A common mistake is to use the first-order equation instead of the second-order equation. The problem will always give you the information needed to determine whether the first-order or second-order equation is required.

The order of reaction can be determined graphically through the use of the integrated rate law. If a plot of the ln[A] versus time yields a straight line, then the reaction is first order with respect to reactant A. If a plot of $\frac{1}{[A]}$ versus time yields a straight line, then the reaction is second order with respect to reactant A.

The reaction **half-life,** $t_{1/2}$, is the amount of time that it takes for a reactant concentration to decrease to one-half its initial concentration. For a first-order reaction, the half-life is a constant, independent of reactant concentration, and can be shown to have the following mathematical relationship:

$$t_{1/2} = \frac{\ln 2}{k} = \frac{0.693}{k}$$

For second-order reactions, the half-life does depend on the reactant concentration and can be calculated using the following formula:

$$t_{1/2} = \frac{1}{k[A]_0}$$

This means that as a second-order reaction proceeds, the half-life increases.

Radioactive decay is a first-order process, and the half-lives of the radioisotopes are well documented (see the chapter on Nuclear Chemistry for a discussion of half-lives with respect to nuclear reactions).

Consider the following problem: The rate constant for the radioactive decay of thorium-232 is 5.0×10^{-11}/year. Determine the half-life of thorium-232.

Answer: 1.4×10^{10} yr.

This is a radioactive decay process. Radioactive decay follows first-order kinetics. The solution to the problem simply requires the substitution of the k-value into the appropriate equation:

$$t_{1/2} = 0.693/k = 0.693/5.0 \times 10^{-11}\ \text{yr}^{-1} = 1.386 \times 10^{10}\text{yr}$$

which rounds (correct significant figures) to the answer reported.

Consider another case: Hydrogen iodide, HI, decomposes through a second-order process to the elements. The rate constant is 2.40×10^{-21}/M s at 25°C. What is the half-life for this decomposition for a 0.200 M of HI at 25°C?

Answer: 2.08×10^{21} s.

The problem specifies that this is a second-order process. Thus, you must simply enter the appropriate values into the second-order half-life equation:

$$t_{1/2} = 1/k[A]_0 = 1/(2.40 \times 10^{-21}/\text{M s})(0.200\ \text{M}) = 2.08333 \times 10^{21}\ \text{seconds}$$

which rounds to the answer reported.

If you are unsure about your work in either of these problems, just follow your units. You are asked for time, so your answer must have time units only and no other units.

Activation Energy

A change in the temperature at which a reaction is taking place affects the rate constant k. As the temperature increases, the value of the rate constant increases and the reaction is faster. The Swedish scientist Arrhenius derived a relationship in 1889 that related the rate constant and temperature. The Arrhenius equation has the form: $k = Ae^{-Ea/RT}$ where k is the rate constant, A is a term called the frequency factor that accounts for molecular orientation, e is the natural logarithm base, R is the universal gas constant 8.314 J mol K^{-1}, T is the Kelvin temperature, and E_a is the **activation energy**, the minimum amount of energy that is needed to initiate or start a chemical reaction.

The Arrhenius equation is most commonly used to calculate the activation energy of a reaction. One way this can be done is to plot the ln k versus $1/T$. This gives a straight line whose slope is $-E_a/R$. Knowing the value of R allows the calculation of the value of E_a.

Normally, high activation energies are associated with slow reactions. Anything that can be done to lower the activation energy of a reaction will tend to speed up the reaction.

Reaction Mechanisms

In the introduction to this chapter we discussed how chemical reactions occurred. Recall that before a reaction can occur there must be a collision between one reactant with the proper orientation at the reactive site of another reactant that transfers enough energy to provide the activation energy. However, many reactions do not take place in quite this simple a way. Many reactions proceed from reactants to products through a sequence of reactions. This sequence of reactions is called the **reaction mechanism**. For example, consider the reaction

$$A + 2B \rightarrow E + F$$

Most likely, E and F are not formed from the simple collision of an A and two B molecules. This reaction might follow this reaction sequence:

$$A + B \rightarrow C$$

$$C + B \rightarrow D$$

$$D \rightarrow E + F$$

If you add together the three equations above, you will get the overall equation $A + 2B \rightarrow E + F$. C and D are called **reaction intermediates**, chemical species that are produced and consumed during the reaction, but that do not appear in the overall reaction.

Each individual reaction in the mechanism is called an **elementary step** or **elementary reaction**. Each reaction step has its own rate of reaction. One of the reaction steps is slower than the rest and is the **rate-determining step**. The rate-determining step limits how fast the overall reaction can occur. Therefore, the rate law of the rate-determining step is the rate law of the overall reaction.

The rate equation for an elementary step can be determined from the reaction stoichiometry, unlike the overall reaction. The reactant coefficients in the elementary step become the reaction orders in the rate equation for that elementary step.

Many times a study of the kinetics of a reaction gives clues to the reaction mechanism. For example, consider the following reaction:

$$NO_2(g) + CO(g) \rightarrow NO(g) + CO_2(g)$$

It has been determined experimentally that the rate law for this reaction is: Rate = $k[NO_2]^2$. This rate law indicates that the reaction does not occur with a simple collision between NO_2 and CO. A simple collision of this type would have a rate law of Rate = $k[NO_2][CO]$. The following mechanism has been proposed for this reaction:

$$NO_2(g) + NO_2(g) \rightarrow NO_3(g) + NO(g)$$

$$NO_3(g) + CO(g) \rightarrow NO_2(g) + CO_2(g)$$

Notice that if you add these two steps together, you get the overall reaction. The first step has been shown to be the slow step in the mechanism, the rate-determining step. If we write the rate law for this elementary step, it is: Rate = $k[NO_2]^2$, which is identical to the experimentally determined rate law for the overall reaction.

Also note that both of the steps in the mechanism are **bimolecular reactions**, reactions that involve the collision of two chemical species. In **unimolecular reactions** a single chemical species decomposes or rearranges. Both bimolecular and unimolecular reactions are common, but the collision of three or more chemical species is quite rare. Therefore, in developing or assessing a mechanism, it is best to consider only unimolecular or bimolecular elementary steps.

Catalysts

A **catalyst** is a substance that speeds up the rate of reaction without being consumed in the reaction. A catalyst may take part in the reaction and even be changed during the reaction, but at the end of the reaction it is at least theoretically recoverable in its original form. It will not produce more of the product, but it allows the reaction to proceed more quickly. In equilibrium reactions (see the chapter on Equilibrium), the catalyst speeds up both the forward and reverse reactions. Catalysts speed up the rates of reaction by providing a different mechanism that has a lower activation energy. The higher the activation energy of a reaction, the slower the reaction will proceed. Catalysts provide an alternate pathway that has a lower activation energy and thus will be faster. In general, there are two distinct types of catalyst.

Homogeneous Catalysts

Homogeneous catalysts are catalysts that are in the same phase or state of matter as the reactants. They provide an alternate reaction pathway (mechanism) with a lower activation energy.

The decomposition of hydrogen peroxide is a slow, one-step reaction, especially if the solution is kept cool and in a dark bottle:

$$2\ H_2O_2 \rightarrow 2\ H_2O + O_2$$

However, if ferric ion is added, the reaction speeds up tremendously. The proposed reaction sequence for this new reaction is:

$$2\ Fe^{3+} + H_2O_2 \rightarrow 2\ Fe^{2+} + O_2 + 2\ H^+$$

$$2\ Fe^{2+} + H_2O_2 + 2\ H^+ \rightarrow 2\ Fe^{3+} + 2\ H_2O$$

Notice that in the reaction the catalyst, Fe^{3+}, was reduced to the ferrous ion, Fe^{2+}, in the first step of the mechanism, but in the second step it was oxidized back to the ferric ion. Overall, the catalyst remained unchanged. Notice also that although the catalyzed reaction is a two-step reaction, it is significantly faster than the original uncatalyzed one-step reaction.

Heterogeneous Catalysts

A **heterogeneous catalyst** is in a different phase or state of matter from the reactants. Most commonly, the catalyst is a solid and the reactants are liquids or gases. These catalysts lower the activation energy for the reaction by providing a surface for the reaction, and also by providing a better orientation of one reactant so its reactive site is more easily hit by the other reactant. Many times these heterogeneous catalysts are finely divided metals. The Haber process, by which nitrogen and hydrogen gases are converted into ammonia, depends upon an iron catalyst, while the hydrogenation of vegetable oil to margarine uses a nickel catalyst.

Experiments

Unlike other experiments, a means of measuring time is essential to all kinetics experiments. This may be done with a clock or a timer. The initial concentration of each reactant must be determined. Often this is done through a simple dilution of a stock solution.

The experimenter must then determine the concentration of one or more substances later, or record some measurable change in the solution. Unless there will be an attempt to measure the activation energy, the temperature should be kept constant. A thermometer is needed to confirm this.

"Clock" experiments are common kinetics experiments. They do not require a separate experiment to determine the concentration of a substance in the reaction mixture. In clock experiments, after a certain amount of time, the solution suddenly changes color. This occurs when one of the reactants has disappeared, and another reaction involving a color change can begin.

In other kinetics experiments, the volume or pressure of a gaseous product is monitored. Again, it is not necessary to analyze the reaction mixture. Color changes in a solution may be monitored with a spectrophotometer. Finally, as a last resort, a sample of the reaction mixture may be removed at intervals and analyzed.

The initial measurement and one or more later measurements are required. (Remember, you measure times; you calculate changes in time [Δt]). Glassware, for mixing and diluting solutions, and a thermometer are the equipment needed for a clock experiment. Other kinetics experiments will use additional equipment to measure volume, temperature, etc. Do not forget: In all cases you measure a property, then calculate a change. You never measure a change.

Common Mistakes to Avoid

1. When working mathematical problems, be sure your units cancel to give you the desired unit in your answer.
2. Be sure to round your answer off to the correct number of significant figures.
3. In working rate law problems, be sure to use molarity for your concentration unit.
4. In writing integrated rate laws, be sure to include the negative sign with the change in *reactant* concentration, since it will be decreasing with time.
5. Remember that the rate law for an overall reaction must be derived from experimental data.
6. In mathematically determining the rate law, be sure to set up the ratio of two experiments such that the concentration of only one reactant has changed.
7. Remember that in most of these calculations the base e logarithm (ln) is used and not the base 10 logarithm (log).

› Review Questions

Use these questions to review the content of this chapter and practice for the AP Chemistry exam. First are 12 multiple-choice questions similar to what you will encounter in Section I of the AP Chemistry exam. Following those is a multipart free-response question like the ones in Section II of the exam. To make these questions an even more authentic practice for the actual exam, time yourself following the instructions provided.

Multiple-Choice Questions

Answer the following questions in 20 minutes. You may not use a calculator. You may use the periodic table and the equation sheet at the back of this book.

1. A reaction follows the rate law: Rate = $k[A]^2$. Which of the following plots will give a straight line?

(A) 1/[A] versus 1/time
(B) $[A]^2$ versus time
(C) 1/[A] versus time
(D) ln[A] versus time

2. For the following reaction: $NO_2(g) + CO(g) \rightarrow NO(g) + CO_2(g)$, the rate law is: Rate = $k[NO_2]^2$. If a small amount of gaseous carbon monoxide (CO) is added to a reaction mixture that was 0.10 molar in NO_2 and 0.20 molar in CO, which of the following statements is true?

(A) Both k and the reaction rate remain the same.
(B) Both k and the reaction rate increase.
(C) Both k and the reaction rate decrease.
(D) Only k increases, the reaction rate will remain the same.

3. The specific rate constant, k, for radioactive Beryllium-11 is 0.049 s^{-1}. What mass of a 0.500 mg sample of beryllium-11 remains after 28 seconds?

(A) 0.250 mg
(B) 0.125 mg
(C) 0.0625 mg
(D) 0.375 mg

4. The slow rate of a particular chemical reaction might be attributed to which of the following?

(A) a low activation energy
(B) a high activation energy
(C) the presence of a catalyst
(D) the temperature is high

5. The steps below represent a proposed mechanism for the catalyzed oxidation of CO by O_3.

Step 1: $NO_2(g) + CO(g) \rightarrow NO(g) + CO_2(g)$

Step 2: $NO(g) + O_3(g) \rightarrow NO_2(g) + O_2(g)$

What are the overall products of the catalyzed reaction?

(A) CO_2 and O_2
(B) NO and CO_2
(C) NO_2 and O_2
(D) NO and O_2

6. The decomposition of ammonia to the elements is a first-order reaction with a half-life of 200 s at a certain temperature. How long will it take the partial pressure of ammonia to decrease from 0.100 atm to 0.00625 atm?

(A) 200 s
(B) 400 s
(C) 800 s
(D) 1,000 s

7. The energy difference between the reactants and the transition state is:

(A) the free energy
(B) the heat of reaction
(C) the activation energy
(D) the kinetic energy

8. The purpose of striking a match against the side of a box to light the match is:

(A) to supply the free energy for the reaction
(B) to supply the activation energy for the reaction
(C) to supply the heat of reaction
(D) to supply the kinetic energy for the reaction

9. The following table gives the initial concentrations and rate for three experiments.

EXPERIMENT	INITIAL [CO] (mol L^{-1})	INITIAL [Cl_2] (mol L^{-1})	INITIAL RATE OF FORMATION OF $COCl_2$ (mol L^{-1} min^{-1})
1	0.200	0.100	3.9×10^{-25}
2	0.100	0.200	3.9×10^{-25}
3	0.200	0.200	7.8×10^{-25}

The reaction is $CO(g) + Cl_2(g) \rightarrow COCl_2(g)$. What is the rate law for this reaction?

(A) Rate = $k[CO]$
(B) Rate = $k[CO]^2[Cl_2]$
(C) Rate = $k[CO][Cl_2]$
(D) Rate = $k[CO][Cl_2]^2$

10. The reaction $(CH_3)_3CBr(aq) + H_2O(l) \rightarrow (CH_3)_3COH(aq) + HBr(aq)$ follows the rate law: Rate = $k[(CH_3)_3CBr]$. What will be the effect of decreasing the concentration of $(CH_3)_3CBr$?

(A) The rate of the reaction will increase.
(B) More HBr will form.
(C) The rate of the reaction will decrease.
(D) The reaction will shift to the left.

11. When the concentration of $H^+(aq)$ is doubled for the reaction $H_2O_2(aq) + 2\ Fe^{2+}(aq) + 2\ H^+(aq) \rightarrow 2\ Fe^{3+}(aq) + 2\ H_2O(g)$, there is no change in the reaction rate. This indicates:

(A) the H^+ is a spectator ion.
(B) the rate-determining step does not involve H^+.
(C) the reaction mechanism does not involve H^+.
(D) the H^+ is a catalyst.

12. The following mechanism has been proposed for the reaction of $CHCl_3$ with Cl_2.

Step 1: $Cl_2(g) \rightarrow 2\ Cl(g)$ fast

Step 2: $Cl(g) + CHCl_3(g) \rightarrow CCl_3(g) + HCl(g)$ slow

Step 3: $CCl_3(g) + Cl(g) \rightarrow CCl_4(g)$ fast

Which of the following rate laws is consistent with this mechanism?

(A) Rate = $k[Cl_2]$
(B) Rate = $k[CHCl_3][Cl_2]$
(C) Rate = $k[CHCl_3]$
(D) Rate = $k[CHCl_3][Cl_2]^{1/2}$

Answers and Explanations for the Multiple-Choice Questions

1. C—The "2" exponent means this is a second-order rate law. Second-order rate laws give a straight-line plot for $1/[A]$ versus t.

2. A—The value of k remains the same unless the temperature is changed or a catalyst is added. Only materials that appear in the rate law, in this case NO_2, will affect the rate. Adding NO_2 would increase the rate, and removing NO_2 would decrease the rate. CO has no effect on the rate.

3. B—The half-life is $0.693/k = 0.693/0.049\ s^{-1} = 14$ s. The time given, 28 s, represents two half-lives. The first half-life uses one-half of the beryllium, and the second half-life uses one-half of the remaining material, so only one-fourth of the original material remains.

4. B—Slow reactions have high activation energies. High activation energies are often attributed to strong bonds within the reactant molecules. All the other choices give faster rates.

5. A—Add the two equations together:

$$NO_2(g) + CO(g) + NO(g) + O_3(g) \rightarrow NO(g) + CO_2(g) + NO_2(g) + O_2(g)$$

Then cancel identical species that appear on opposite sides:

$$CO(g) + O_3(g) \rightarrow CO_2(g) + O_2(g)$$

6. C—The value will be decreased by one-half for each half-life. Using the following table:

Half-lives	Remaining
0	0.100
1	0.0500
2	0.0250
3	0.0125
4	0.00625

Four half-lives = 4(200 s) = 800 s

7. C—This is the definition of the activation energy.

8. B—The friction supplies the energy needed to start the reaction. The energy needed to start the reaction is the activation energy.

9. C—Beginning with the generic rate law: Rate = $k[CO]^m[Cl_2]^n$, it is necessary to determine the values of m and n (the orders). Comparing Experiments 2 and 3, the rate doubles when the concentration of CO is doubled. This direct change means the reaction is first order with respect to CO. Comparing Experiments 1 and 3, the rate doubles when the concentration of Cl_2 is doubled. Again, this direct change means the reaction is first order. This gives: Rate = $k[CO]^1[Cl_2]^1 = k[CO][Cl_2]$.

10. C—The compound appears in the rate law, so a change in its concentration will change the rate. The reaction is first order in $(CH_3)_3CBr$, so the rate will change directly with the change in concentration of this reactant.

11. B—All substances involved, directly or indirectly, in the rate-determining step will change the rate when their concentrations are changed. The ion is required in the balanced chemical equation, so it cannot be a spectator ion, and it must appear in the mechanism. Catalysts will change the rate of a reaction. Since H^+ does not affect the rate, the reaction is zero order with respect to this ion.

12. D—The rate law depends on the slow step of the mechanism. The reactants in the slow step are Cl and $CHCl_3$ (one of each). The rate law is first order with respect to each of these. The Cl is half of the original reactant molecule Cl_2. This replaces the [Cl] in the rate law with $[Cl_2]^{1/2}$. Do not make the mistake of using the overall reaction to predict the rate law.

Free-Response Question

You have 15 minutes to answer the following multipart question. You may use a calculator and the tables in the back of the book.

Question 1

$$2\ ClO_2(aq) + 2\ OH^-(aq) \rightarrow ClO_3^-(aq) + ClO_2^-(aq) + H_2O(l)$$

A series of experiments were conducted to study the above reaction. The initial concentrations and rates are in the following table.

EXPERIMENT	INITIAL CONCENTRATIONS (mol/L)		INITIAL RATE OF FORMATION OF ClO_3^- (mol/L min)
	$[OH^-]$	$[ClO_2]$	
1	0.030	0.020	0.166
2	0.060	0.020	0.331
3	0.030	0.040	0.661

(a) i. Determine the order of the reaction with respect to each reactant. Make sure you explain your reasoning.

ii. Give the rate law for the reaction.

(b) Determine the value of the rate constant, making sure the units are included.
(c) Calculate the initial rate of disappearance of ClO_2 in experiment 1.
(d) The following is the proposed mechanism for this reaction.

Step 1: $ClO_2 + ClO_2 \rightarrow Cl_2O_4$

Step 2: $Cl_2O_4 + OH^- \rightarrow ClO_3^- + HClO_2$

Step 3: $HClO_2 + OH^- \rightarrow ClO_2^- + H_2O$

Which step is the rate-determining step? Show that this mechanism is consistent with both the rate law for the reaction and with the overall stoichiometry.

Answer and Explanation for the Free-Response Question

(a) i. This part of the problem begins with a generic rate equation: Rate = $k[ClO_2]^m[OH]^n$. The values of the exponents, the orders, must be determined. It does not matter which exponent is done first. If you want to begin with ClO_2, you must pick two experiments from the table where its concentration changes but the OH^- concentration does not change. These are experiments 1 and 3. Experiment 3 has twice the concentration of ClO_2 as experiment 1. This doubling of the ClO_2 concentration has quadrupled the rate. The relationship between the concentration (× 2) and the rate (× 4 = × 2^2) indicates that the order for ClO_2 is 2 (= m). Using experiments 1 and 2 (only the OH^- concentration changes), we see that doubling the concentration simply doubles the rate. Thus, the order for OH^- is 1 (= n).
Give yourself 1 point for each order you got correct.

ii. Inserting the orders into the generic rate law gives: Rate = $k[ClO_2]^2 [OH^-]^1$, which is usually simplified to: Rate = $k[ClO_2]^2[OH^-]$.
Give yourself 1 point if you got this equation correct.

(b) Any one of the three experiments may be used to calculate the rate constant. If the problem asked for an average rate constant, you would need to calculate a value for each of the experiments and then average the values.

The rate law should be rearranged to k = Rate/$[ClO_2]^2[OH^-]$. Then the appropriate values are entered into the equation. Using experiment 1 as an example:

$$k = (0.166 \text{ mol/L min})/[(0.020 \text{ M})^2(0.030 \text{ M})]$$

$$= 1.3833 \times 10^4 \text{ M/M}^3 \text{ min} = 1.4 \times 10^4/\text{M}^2 \text{ min}$$

The answer could also be reported as 1.4×10^4 L^2/mol^2 min. You should not forget that M = mol/L.

Give yourself 1 point for the correct numerical value. Give yourself 1 point for the correct units. If you had the wrong rate law in part **a. ii**, and use it correctly in part **b**, you will still get the points.

(c) The coefficients from the equation say that for every mole of ClO_3^- that forms, 2 mol of ClO_2 reacted. Thus, the rate of ClO_2 is twice the rate of ClO_3^-. Do not forget that since ClO_3^- is forming, it has a positive rate, and since ClO_2 is reacting, it has a negative rate. Therefore:

Rearranging and inserting the rate from experiment 1 gives: $\Delta[ClO]/\Delta t$ = $-2(0.166$ mol/L min)] $=-8.332$ mol/L min

Give yourself 2 points if you got the entire answer correct. You get only 1 point if the sign or units are missing.

(d) The rate-determining step must match the rate law. One approach is to determine the rate law for each step in the mechanism. This gives:

Step 1: Rate $= k[ClO_2]^2$
Step 2: Rate $= k[Cl_2O_4][OH^-] = k[Cl_2O]^2[OH^-]$
Step 3: Rate $= k[HClO_2][OH^-] = k[ClO_2][OH^-]^2$

For steps 2 and 3, it is necessary to replace the intermediates with reactants. Step 2 gives a rate law matching the one derived in part **a**.

Give yourself 1 point if you picked step 2, or if you picked a step with a rate law that matches a wrong answer for part **a**. Give yourself 1 more point if you explained the substitution of reactants for intermediates.

To see if the stoichiometry is correct, simply add the three steps together and cancel the intermediates (materials that appear on both sides of the reaction arrow).

Step 1: $ClO_2 + ClO_2 \rightarrow Cl_2O_4$
Step 2: $Cl_2O_4 + OH^- \rightarrow ClO_3^- + HClO_2$
Step 3: $HClO_2 + OH^- \rightarrow ClO_2^- + H_2O$
Total: $2\ ClO_2 + Cl_2O_4 + 2\ OH^- + HClO_2 \rightarrow$
$Cl_2O_4 + ClO_3^- + HClO_2 + ClO_2^- + H_2O$

After removing the intermediates (Cl_2O_4 and $HClO_2$):

$$2\ ClO_2 + 2\ OH^- \rightarrow ClO_3^- + ClO_2^- + H_2O$$

As this matches the original reaction equation, the mechanism fulfills the overall stoichiometry requirement.

Give yourself 1 point for summing the equations and proving the overall equation is consistent.

The total is 10 points for this question. Subtract 1 point if any answer has an incorrect number of significant figures.

› Rapid Review

- Kinetics is a study of the speed of a chemical reaction.
- The five factors that can affect the rates of chemical reaction are the nature of the reactants, the temperature, the concentration of the reactants, the physical state of the reactants, and the presence of a catalyst.
- The rate equation relates the speed of reaction to the concentration of reactants and has the form: Rate $= k[A]^m[B]^n$ . . . where k is the rate constant and m and n are the orders of reaction with respect to that specific reactant.
- The rate law must be determined from experimental data. Review how to determine the rate law from kinetics data.
- When mathematically comparing two experiments in the determination of the rate equation, be sure to choose two in which all reactant concentrations except one remain constant.
- Rate laws can be written in the integrated form.

- If a reaction is first order, it has the rate law of Rate = $k[A]$; $\ln [A]_t - \ln [A]_0 = -kt$; a plot of ln[A] versus time gives a straight line.
- If a reaction is second order, it has the form of Rate = $k[A]^2$; $\frac{1}{[A]_t} - \frac{1}{[A]_0} = kt$ (integrated rate law); a plot of $\frac{1}{[A]}$ versus time gives a straight line.
- The reaction half-life is the amount of time that it takes the reactant concentration to decrease to one-half its initial concentration.
- The half-life can be related to concentration and time by these two equations (first and second order, respectively): $t_{1/2} = \frac{\ln 2}{k} = \frac{0.693}{k}$ and $t_{1/2} = 1/k[A]_0$ to apply these equations.
- The activation energy is the minimum amount of energy needed to initiate or start a chemical reaction.
- Many reactions proceed from reactants to products by a series of steps called elementary steps. All these steps together describe the reaction mechanism, the pathway by which the reaction occurs.
- The slowest step in a reaction mechanism is the rate-determining step. It determines the rate law.
- A catalyst is a substance that speeds up a reaction without being consumed in the reaction.
- A homogeneous catalyst is in the same phase as the reactants, whereas a heterogeneous catalyst is in a different phase from the reactants.

CHAPTER 15

Equilibrium

IN THIS CHAPTER

Summary: We've been discussing chemical reactions for several chapters. In the Kinetics chapter you saw how chemical reactions take place and some of the factors that affect the reactions' speed. In this chapter we will discuss another aspect of chemical reactions: equilibrium.

A few chemical reactions proceed to completion, using up one or more of the reactants and then stopping. However, most reactions behave in a different way. Consider the general reaction:

$$aA + bB \rightarrow cC + dD$$

Reactants A and B are forming C and D. Then C and D start to react to form A and B:

$$cC + dD \rightarrow aA + bB$$

These two reactions proceed until the two rates of reaction become equal. That is, the speed of production of C and D in the first reaction is equal to the speed of production of A and B in the second reaction. Since these two reactions are occurring simultaneously in the same container, the amounts of A, B, C, and D become constant. A **chemical equilibrium** has been reached, in which two exactly opposite reactions are occurring at the same place, at the same time, and with the same rates of reaction. When a system reaches the equilibrium state, the reactions do not stop. A and B are still reacting to form C and D; C and D are still reacting to form A and B. But because the reactions proceed at the same rate, the amounts of each chemical species are constant. This state is sometimes called a **dynamic** equilibrium state to emphasize the fact that the reactions are still occurring—it is a dynamic, not a static state. An equilibrium state is indicated by a double arrow instead of a single arrow. For the reaction above it would be shown as:

$$aA + bB \rightleftharpoons cC + dD$$

It is important to remember that at equilibrium the concentrations of the chemical species are constant, not necessarily equal. There may be a lot of C and D and a little A and B, or vice versa. The concentrations are constant, unchanging, but not necessarily equal.

At any point during the preceding reaction, a relationship may be defined called the **reaction quotient,** Q. It has the following form:

$$Q = \frac{[C]^c[D]^d}{[A]^a[B]^b}$$

The reaction quotient is a fraction. In the numerator is the product of the chemical species on the right-hand side of the equilibrium arrow, each raised to the power of that species' coefficient in the balanced chemical equation. It is called the Q_c in this case, because molar concentrations are being used. If this was a gas-phase reaction, gas pressures could be used and it would become a Q_p.

Remember: products over reactants.

Keywords and Equations

Q = reaction quotient

$$Q = \frac{[C]^c[D]^d}{[A]^a[B]^b}, \text{ where } aA + bB \rightarrow cC + dD$$

Equilibrium Constants:

K = equilibrium constant

K_a (weak acid) K_b (weak base) K_w (water) K_p (gas pressure)

K_c (molar concentrations)

$$K_a = \frac{[H^+][A^-]}{[HA]} \qquad K_b = \frac{[OH^-][HB^+]}{[B]}$$

$K_w = [OH^-][H^+] = 1.0 \times 10^{-14} = K_a \times K_b$ at 25°C

$pH = -\log [H^+]$, $pOH = -\log [OH^-]$

$14 = pH + pOH$

$$pH = pK_a + \log\frac{[A^-]}{[HA]}$$

$$pOH = pK_b + \log\frac{[HB^+]}{[B]}$$

$pK_a = -\log K_a$, $pK_b = -\log K_b$

$K_p = K_c(RT)^{\Delta n}$, where Δn = moles product gas − moles reactant gas

Gas constant, $R = 0.0821$ L atm mol^{-1} K^{-1}

Equilibrium Expressions

The reactant quotient can be written at any point during the reaction, but the most useful point is when the reaction has reached equilibrium. At equilibrium, the reaction quotient becomes the **equilibrium constant,** K_c (or K_p if gas pressures are being used). Usually this equilibrium constant is expressed simply as a number without units, since it is a ratio of concentrations or pressures. In addition, the concentrations of solids or pure liquids (not in solution) that appear in the equilibrium expression are assumed to be 1, since their concentrations do not change.

Consider the Haber process for the production of ammonia:

$$N_2(g) + 3H_2(g) \rightleftharpoons 2NH_3(g)$$

The equilibrium constant expression would be written as:

$$K_c = \frac{[NH_3]^2}{[N_2][H_2]^3}$$

If the partial pressures of the gases were used, then K_p would be written in the following form:

$$K_p = \frac{P_{NH_3}{}^2}{P_{N_2} \times P_{H_2}{}^3}$$

There is a relationship between K_c and K_p: $K_p = K_c(RT)^{\Delta n}$, where R is the ideal gas constant (0.0821 L atm/mol K) and Δn is the change in the number of moles of gas in the reaction.

Remember: Be sure that your value of R is consistent with the units chosen for the partial pressures of the gases.

For the following equilibrium $K_p = 1.90$: $C(s) = CO_2(g) \rightleftharpoons 2\ CO(g)$. Calculate K_c for this equilibrium at 25°C.

$$C(s) + CO_2(g) \rightleftharpoons 2CO(g) \quad K_p = 1.90$$

$$K_p = K_c(RT)^{\Delta n}$$

$$1.90 = K_c \frac{[(0.0826\ \text{L atm})(298\text{K})]^{(2-1)}}{[(\text{mol K})]}$$

$$K_c = 0.0777$$

The numerical value of the equilibrium constant can give an indication of the extent of the reaction after equilibrium has been reached. If the value of K_c is large, that means the numerator is much larger than the denominator and the reaction has produced a relatively large amount of products (reaction lies far to the right). If K_c is small, then the numerator is much smaller than the denominator and not much product has been formed (reaction lies far to the left).

Le Châtelier's Principle

At a given temperature, a reaction will reach equilibrium with the production of a certain amount of product. If the equilibrium constant is small, that means that not much product will be formed. But is there anything that can be done to produce more? Yes, there is—through the application of **Le Châtelier's principle**. Le Châtelier, a French scientist, discovered that if a chemical system at equilibrium is stressed (disturbed) it will reestablish equilibrium by shifting the reactions involved. This means that the amounts of the reactants and products will change, but the final ratio will remain the same. The equilibrium may be stressed in a number of ways: changes in concentration, pressure, and temperature. Many times the use of a catalyst is mentioned. However, a catalyst will have no effect on the equilibrium amounts, because it affects both the forward and reverse reactions equally. It will, however, cause the reaction to reach equilibrium faster.

Changes in Concentration

If the equilibrium system is stressed by a change in concentration of one of the reactants or products, the equilibrium will react to remove that stress. If the concentration of a chemical species is decreased, the equilibrium will shift to produce more of it. In doing so, the concentration of chemical species on the other side of the reaction arrows will be decreased. If the concentration of a chemical species is increased, the equilibrium will shift to consume it, increasing the concentration of chemical species on the other side of the reaction arrows.

For example, again consider the Haber process:

$$N_2(g) + 3H_2(g) \rightleftharpoons 2NH_3(g)$$

If one increases the concentration of hydrogen gas, then the equilibrium shifts to the right to consume some of the added hydrogen. In doing so, the concentration of ammonia (NH_3) will increase and the concentration of nitrogen gas will decrease. On the other hand, if the concentration of nitrogen gas was decreased, the equilibrium would shift to the left to form more, the concentration of ammonia would decrease, and the concentration of hydrogen would increase.

Again, remember that the concentrations may change, but the value of K_c or K_p would remain the same.

Changes in Pressure

Changes in pressure are significant only if gases are involved. The pressure may be changed by changing the volume of the container or by changing the concentration of a gaseous species (although this is really a change in concentration and can be treated as a concentration effect, as above). If the container becomes smaller, the pressure increases because there is an increased number of collisions on the inside walls of the container. This stresses the equilibrium system, and it will shift to reduce the pressure. This can be accomplished by shifting the equilibrium toward the side of the equation that has the lesser number of moles of gas. If the container size is increased, the pressure decreases and the equilibrium will shift to the side containing more moles of gas to increase the pressure. If the number of moles of gas is the same on both sides, changing the pressure will not affect the equilibrium.

Once again, consider the Haber reaction:

$$N_2(g) + 3H_2(g) \rightleftharpoons 2NH_3(g)$$

Note that there are 4 mol of gas (1 of nitrogen and 3 of hydrogen) on the left side and 2 mol on the right. If the container is made smaller, the pressure will increase and the equilibrium will shift to the right because 4 mol would be converted to 2 mol. The concentrations of nitrogen and hydrogen gases would decrease, and the concentration of ammonia would increase.

Remember: Pressure effects are only important for gases.

Changes in Temperature

Changing the temperature changes the value of the equilibrium constant. It also changes the amount of heat in the system and can be treated as a concentration effect. To treat it this way, one must know which reaction, forward or reverse, is exothermic (releasing heat).

One last time, let's consider the Haber reaction:

$$N_2(g) + H_2(g) \rightleftharpoons 2NH_3(g)$$

The formation of ammonia is exothermic (liberating heat), so the reaction could be written as:

$$N_2(g) + 3H_2(g) \rightleftharpoons 2NH_3(g) + \text{heat}$$

If the temperature of the reaction mixture were increased, the amount of heat would be increased and the equilibrium would shift to the left to consume the added heat. In doing so, the concentration of nitrogen and hydrogen gases would increase and the concentration of ammonia gas would decrease. If you were in the business of selling ammonia, you would probably want to operate at a reduced temperature, in order to shift the reaction to the right.

Consider the following equilibrium (endothermic as written), and predict what changes, if any, would occur if the following stresses were applied after equilibrium was established.

$$CaCO_3(s) \rightleftharpoons CaO(s) + CO_2(g)$$

a. add CO_2
b. remove CO_2
c. add CaO
d. increase T
e. decrease V
f. add a catalyst

Answers:

a. Left—the equilibrium shifts to remove some of the excess CO_2.
b. Right—the equilibrium shifts to replace some of the CO_2.
c. No change—solids do not shift equilibria unless they are totally removed.
d. Right—endothermic reactions shift to the right when heated.
e. Left—a decrease in volume, or an increase in pressure, will shift the equilibrium toward the side with less gas.
f. No change—catalysts do not affect the position of an equilibrium.

Acid–Base Equilibrium

In the Reactions and Periodicity chapter we introduced the concept of acids and bases. Recall that acids are proton (H^+) donors and bases are proton acceptors. Also recall that acids and bases may be strong or weak. **Strong acids** completely dissociate in water; **weak acids** only partially dissociate. For example, consider two acids HCl (strong) and CH_3COOH (weak). If each is added to water to form aqueous solutions the following reactions take place:

$$HCl(aq) + H_2O(l) \rightarrow H_3O^+(aq) + Cl^-(aq)$$
$$CH_3COOH(aq) + H_2O(l) \rightleftharpoons H_3O^+(aq) + CH_3COO^-(aq)$$

The first reaction essentially goes to completion—there is no HCl left in solution. The second reaction is an equilibrium reaction—there are appreciable amounts of both reactants and products left in solution.

There are generally only two strong bases to consider: the hydroxide and the oxide ion (OH^- and O^{2-}, respectively). All other common bases are weak. **Weak bases**, like weak acids, also establish an equilibrium system, as in aqueous solutions of ammonia:

$$NH_3(aq) + H_2O(l) \rightleftharpoons OH^-(aq) + NH_4^+(aq)$$

In the Brønsted–Lowry acid–base theory, there is competition for an H^+. Consider the acid–base reaction between acetic acid, a weak acid, and ammonia, a weak base:

$$CH_3COOH(aq) + NH_3(aq) \rightleftharpoons CH_3COO^-(aq) + NH_4^+(aq)$$

Acetic acid donates a proton to ammonia in the forward (left-to-right) reaction of the equilibrium to form the acetate and ammonium ions. But in the reverse (right-to-left) reaction, the ammonium ion donates a proton to the acetate ion to form ammonia and acetic acid. The ammonium ion is acting as an acid, and the acetate ion as a base. Under the Brønsted–Lowry system, acetic acid (CH_3COOH) and the acetate ion (CH_3COO^-) are called a conjugate acid–base pair. **Conjugate acid–base pairs** differ by only a single H^+. Ammonia (NH_3) and the ammonium ion (NH_4^+) are also a conjugate acid–base pair. In this reaction there is a competition for the H^+ between acetic acid and the ammonium ion. To predict on which side the equilibrium will lie, this general rule applies: *The equilibrium will favor the side in which the weaker acid and base are present.* Figure 15.1 shows the relative strengths of the conjugate acid–base pairs.

In Figure 15.1 you can see that acetic acid is a stronger acid than the ammonium ion and ammonia is a stronger base than the acetate ion. Therefore, the equilibrium will lie to the right.

The reasoning above allows us to find good qualitative answers, but in order to be able to do quantitative problems (how much is present, etc.), the extent of the dissociation of the weak acids and bases must be known. That is where a modification of the equilibrium constant is useful.

K_a—the Acid Dissociation Constant

Strong acids completely dissociate (ionize) in water. Weak acids partially dissociate and establish an equilibrium system. But as shown in Figure 15.1 there is a large range of weak acids based upon their ability to donate protons. Consider the general weak acid, HA, and its reaction when placed in water:

$$HA(aq) + H_2O(l) \rightleftharpoons H_3O^+(aq) + A^-(aq)$$

An equilibrium constant expression can be written for this system:

$$K_c = \frac{[H_3O^+][A^-]}{[HA]}$$

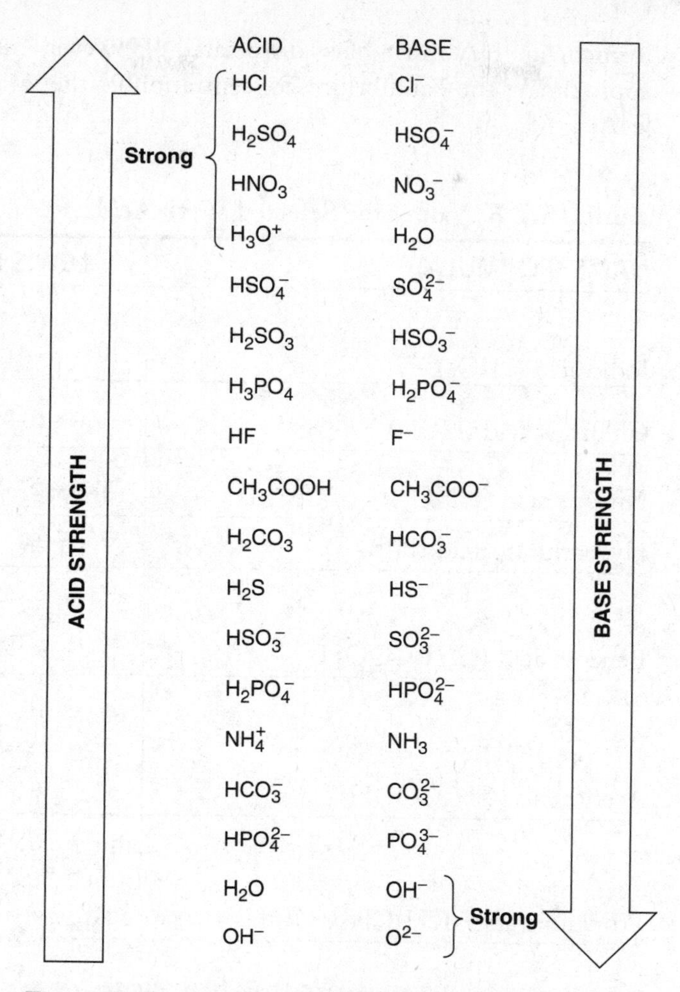

Figure 15.1 Conjugate acid–base pair strengths.

The $[H_2O]$ is assumed to be a constant and is incorporated into the K_a value. It is not shown in the equilibrium constant expression.

Since this is the equilibrium constant associated with a weak acid dissociation, this particular K_c is most commonly called the **acid dissociation constant**, K_a. The K_a expression is then:

$$K_a = \frac{[H_3O^+][A^-]}{[HA]}$$

Many times the weak acid dissociation reaction will be shown in a shortened notation, omitting the water:

$$HA(aq) \rightleftharpoons H^+(aq) + A^-(aq) \qquad \text{with } K_a = \frac{[H^+][A^-]}{[HA]}$$

The greater the amount of dissociation is, the larger the value of K_a. Table 15.1, on the next page, shows the K_a values of some common weak acids.

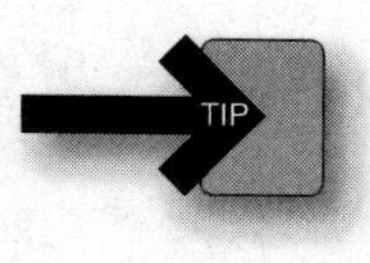

Here are a couple of tips: For every H^+ formed, an A^- is formed, so the numerator of the K_a expression can be expressed as $[H^+]^2$ (or $[A^-]^2$, although it is rarely done this way). Also, the [HA] is the equilibrium molar concentration of the undissociated weak acid, not its initial concentration. The exact expression would then be $[HA] = M_{initial} - [H^+]$, where $M_{initial}$ is the initial concentration of the weak acid. This is true because for every H^+ that is formed, an HA must have dissociated. However, many times if K_a is small, you can approximate the equilibrium concentration of the weak acid by its initial concentration, $[HA] = M_{initial}$.

Table 15.1 K_a Values for Selected Weak Acids

NAME (FORMULA)	LEWIS STRUCTURE	K_a
Iodic acid (HIO_3)	H—Ö—Ï=Ö, with =:O: on I	1.6×10^{-1}
Chlorous acid ($HClO_2$)	H—Ö—Cl=Ö	1.12×10^{-2}
Nitrous acid (HNO_2)	H—Ö—N=Ö	7.1×10^{-4}
Hydrofluoric acid (HF)	H—F:	6.8×10^{-4}
Benzoic acid (C_6H_5COOH)	C₆H₅—C(=:O:)—Ö—H	6.3×10^{-5}
Acetic acid (CH_3COOH)	H—C(H)(H)—C(=:O:)—Ö—H	1.8×10^{-5}
Propanoic acid (CH_3CH_2COOH)	H—C(H)(H)—C(H)(H)—C(=:O:)—Ö—H	1.3×10^{-5}
Hypochlorous acid (HClO)	H—Ö—Cl:	2.9×10^{-8}
Hypobromous acid (HBrO)	H—Ö—Br:	2.3×10^{-9}
Phenol (C_6H_5OH)	C₆H₅—Ö—H	1.0×10^{-10}
Hypoiodous acid (HIO)	H—Ö—I:	2.3×10^{-11}

If the initial molarity and K_a of the weak acid are known, the $[H^+]$ (or $[A^-]$) can be calculated easily. And if the initial molarity and $[H^+]$ are known, K_a can be calculated.

For example, calculate the $[H^+]$ of a 0.300 M acetic acid solution.

$$K_a = 1.8 \times 10^{-5}$$

$$HC_2H_3O_2(aq) \rightleftharpoons H^+(aq) + C_2H_3O_2^-(aq)$$

$$0.300 - x \qquad x \qquad x$$

$$K_a = \frac{[H^+][C_2H_3O_2^-]}{[HC_2H_3O_2]} = 1.8 \times 10^{-5}$$

$$= \frac{(x)(x)}{0.300 - x} = 1.8 \times 10^{-5}$$

$$x = [H^+] = 2.3 \times 10^{-3} \text{ M}$$

For **polyprotic acids**, acids that can donate more than one proton, the K_a for the first dissociation is much larger than the K_a for the second dissociation. If there is a third K_a, it is much smaller still. For most practical purposes you can simply use the first K_a.

K_w—the Water Dissociation Constant

Before examining the equilibrium behavior of aqueous solutions of weak bases, let's look at the behavior of water itself. In the initial discussion of acid–base equilibrium above, we showed water acting both as an acid (proton donor when put with a base) and a base (proton acceptor when put with an acid). Water is **amphoteric**, it will act as either an acid or a base, depending on whether the other species is a base or acid. But in pure water the same amphoteric nature is noted. In pure water a very small amount of proton transfer is taking place:

$$H_2O(l) + H_2O(l) \rightleftharpoons H_3O^+(aq) + OH^-(aq)$$

This is commonly written as:

$$H_2O(l) \rightleftharpoons H^+(aq) + OH^-(aq)$$

There is an equilibrium constant, called the **water dissociation constant, K_w**, which has the form:

$$K_w = [H^+][OH^-] = 1.0 \times 10^{-14} \text{ at } 25°C$$

Again, the concentration of water is a constant and is incorporated into K_w.

The numerical value of K_w of 1.0×10^{-14} is true for the product of the $[H^+]$ and $[OH^-]$ in pure water and for aqueous solutions of acids and bases.

In the discussion of weak acids, we indicated that the $[H^+] = [A^-]$. However, there are two sources of H^+ in the system: the weak acid and water. The amount of H^+ that is due to the water dissociation is very small and can be easily ignored.

pH

Because the concentration of the hydronium ion, H_3O^+, can vary tremendously in solutions of acids and bases, a scale to easily represent the acidity of a solution was developed. It is called the pH scale and is related to the $[H_3O^+]$:

$$pH = -\log [H_3O^+] \text{ or } -\log [H^+] \text{ using the shorthand notation}$$

Remember that in pure water $K_w = [H_3O^+][OH^-] = 1.0 \times 10^{-14}$. Since both the hydronium ion and hydroxide ions are formed in equal amounts, the K_w expression can be expressed as:

$$[H_3O^+]^2 = 1.0 \times 10^{-14}$$

Solving for $[H_3O^+]$ gives us $[H_3O^+] = 1.0 \times 10^{-7}$. If you then calculate the pH of pure water:

$$pH = -\log[H_3O^+] = -\log\ [1.0 \times 10^{-7}] = -(-7.00) = 7.00$$

The pH of pure water is 7.00. On the pH scale this is called **neutral**. A solution whose $[H_3O^+]$ is greater than in pure water will have a pH less than 7.00 and is called **acidic**. A solution whose $[H_3O^+]$ is less than in pure water will have a pH greater than 7.00 and is called **basic**. Figure 15.2, on the next page, shows the pH scale and the pH values of some common substances.

The pOH of a solution can also be calculated. It is defined as $pOH = -\log[OH^-]$. The pH and the pOH are related:

$$pH + pOH = pK_w = 14.00 \text{ at } 25°C$$

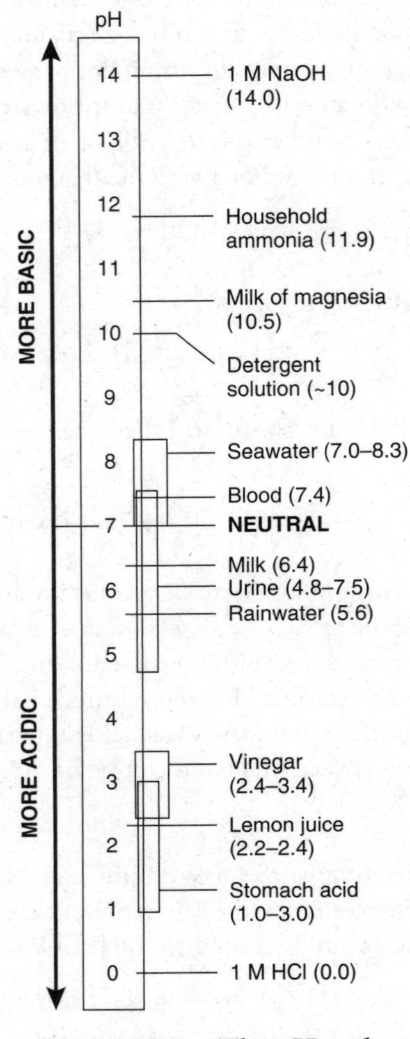

Figure 15.2 The pH scale.

In any of the problems above in which $[H^+]$ or $[OH^-]$ was calculated, you can now calculate the pH or pOH of the solution.

You can estimate the pH of a solution by looking at its $[H^+]$. For example, if a solution has an $[H^+] = 1 \times 10^{-5}$, its pH would be 5. This value was determined from the value of the exponent in the $[H^+]$.

K_b—the Base Dissociation Constant

Weak bases (B), when placed into water, also establish an equilibrium system much like weak acids:

$$3(aq) + H_2O(l) \rightleftharpoons HB^+(aq) + OH^-(aq)$$

The equilibrium constant expression is called the weak **base dissociation constant, K_b**, and has the form:

$$K_b = \frac{[HB^+][OH^-]}{[HB]}$$

The same reasoning that was used in dealing with weak acids is also true here: $[HB^+] = [OH^-]$; $[HB] \approx M_{initially}$; the numerator can be represented as $[OH^-]^2$; and knowing the initial molarity and K_b of the weak base, the $[OH^-]$ can easily be calculated. And if the initial molarity and $[OH^-]$ are known, K_b can be calculated.

For example, a 0.500 M solution of ammonia has a pH of 11.48. What is the K_b of ammonia?

$$pH = 11.48$$

$$[H^+] = 10^{-11.48}$$

$$[H^+] = 3.3 \times 10^{-12}\ M$$

$$K_w = [H^+][OH^-] = 1.0 \times 10^{-14}$$

$$[OH^-] = 3.0 \times 10^{-3}\ M$$

$$NH_3 + H_2O \rightleftharpoons NH_4^+ + OH^-$$

$$0.500 - x \qquad x \qquad x$$

$$K_b = \frac{[NH_4^+][OH^-]}{[NH_3]}$$

$$[OH^-] = [NH_4^+] = 3.0 \times 10^{-3}\ M$$

$$[NH_3] = 0.500 - 3.0 \times 10^{-3} = 0.497\ M$$

$$K_b = \frac{(3.0 \times 10^{-3})^2}{(0.497)} = 1.8 \times 10^{-5}$$

The K_a and K_b of conjugate acid–base pairs are related through the K_w expression:

$$K_a \times K_b = K_w$$

This equation shows an inverse relationship between K_a and K_b for any conjugate acid–base pair.

This relationship may be used in problems such as: Determine the pH of a solution made by adding 0.400 mol of strontium acetate to sufficient water to produce 2.000 L of solution.

Solution:

The initial molarity is 0.400 mol/2.000 L = 0.200 M.

When a salt is added to water dissolution will occur:

$$Sr(C_2H_3O_2)_2 \rightarrow Sr^{2+}(aq) + 2C_2H_3O_2^-(aq)$$

The resultant solution, since strontium acetate is soluble, has 0.200 M Sr^{2+} and 0.400 M $C_2H_3O_2^-$.

Ions such as Sr^{2+}, which come from strong acids or strong bases, may be ignored in this type of problem. Ions such as $C_2H_3O_2^-$, from weak acids or bases, will undergo hydrolysis. The acetate ion is the conjugate BASE of acetic acid ($K_a = 1.8 \times 10^{-5}$). Since acetate is not a strong base this will be a K_b problem, and OH^- will be produced. The equilibrium is:

$$\begin{array}{lllll} C_2H_3O_2^- & + H_2O \rightleftharpoons & OH^- & + & HC_2H_3O_2 \\ 0.400 - x & & +x & & +x \end{array}$$

Determining K_b from K_a (using $K_w = K_aK_b = 1.0 \times 10^{-14}$) gives:

$$\frac{[x][x]}{0.400 - x} = K_b = 5.6 \times 10^{-10}$$

with $x = 1.5 \times 10^{-5} = [OH^-]$, and $pH = 9.180$

Acidic/Basic Properties of Salts

The behavior of a salt will depend upon the acid–base properties of the ions present in the salt. The ions may lead to solutions of the salt being acidic, basic, or neutral. The pH of a solution depends on hydrolysis, a generic term for a variety of reactions with water. Some ions will undergo hydrolysis and this changes the pH.

The reaction of an acid and a base will produce a salt. The salt will contain the cation from the base and the anion from the acid. In principle, the cation of the base is the conjugate acid of the base, and the anion from the acid is the conjugate base of the acid. Thus, the salt contains a conjugate acid and a conjugate base. This is always true in principle. In some cases, one or the other of these ions is not a true conjugate base or a conjugate acid. Just because the ion is not a true conjugate acid or base does not mean that we cannot use the ion as if it were.

The conjugate base of any strong acid is so weak that it will not undergo any significant hydrolysis; the conjugate acid of any strong base is so weak that it, too, will not undergo any significant hydrolysis. Ions that do not undergo any significant hydrolysis will have no effect upon the pH of a solution and will leave the solution neutral. The presence of the following conjugate bases Cl^-, Br^-, I^-, NO_3^-, ClO_3^-, and ClO_4^- will leave the solution neutral. The cations from the strong bases, Li^+, Na^+, K^+, Rb^+, Cs^+ Ca^{2+}, Sr^{2+}, and Ba^{2+}, while not true conjugate acids, will also leave the solution neutral. Salts containing a combination of only these cations and anions are neutral.

The conjugate base from any weak acid is a strong base and will undergo hydrolysis in aqueous solution to produce a basic solution. If the conjugate base (anion) of a weak acid is in a salt with the conjugate of a strong base (cation), the solution will be basic, because only the anion will undergo any significant hydrolysis. Salts of this type are basic salts. All salts containing the cation of a strong base and the anion of a weak acid are basic salts.

The conjugate acid of a weak base is a strong acid, and it will undergo hydrolysis in an aqueous solution to make the solution acidic. If the conjugate acid (cation) of a weak base is in a salt with the conjugate base of a strong acid (anion), the solution will be acidic, because only the cation will undergo any significant hydrolysis. Salts of this type are acidic salts. All salts containing the cation of a weak base and the anion of a strong acid are acidic salts.

There is a fourth category, consisting of salts that contain the cation of a weak base with the anion of a weak acid. Prediction of the acid–base character of these salts is less obvious, because both ions undergo hydrolysis. The two equilibria not only alter the pH of the solution, but also interfere with each other. Predictions require a comparison of the K values for the two ions. The larger K value predominates. If the larger value is K_a, the solution is acidic. If the larger value is K_b, the solution is basic. In the rare case where the two values are equal, the solution would be neutral.

The following table summarizes this information:

CATION FROM	ANION FROM	SOLUTION
Strong Base	Strong Acid	Neutral
Strong Base	Weak Acid	Basic
Weak Base	Strong Acid	Acidic
Weak Base	Weak Acid	Must be determined by comparing K values

For example, suppose you are asked to determine if a solution of sodium carbonate, Na_2CO_3, is acidic, basic, or neutral. Sodium carbonate is the salt of a strong base (NaOH) and a weak acid (HCO_3^-). Salts of strong bases and weak acids are basic salts. As a basic salt, we know the final answer must be basic (pH above 7).

Buffers

Buffers are solutions that resist a change in pH when an acid or base is added to them. The most common type of buffer is a mixture of a weak acid and its conjugate base. The weak acid will neutralize any base added, and the weak base of the buffer will neutralize any acid added to the solution. The hydronium ion concentration of a buffer can be calculated using an equation derived from the K_a expression:

$$[H_3O^+] = K_a \times \frac{[HA]}{[A^-]}$$

Taking the negative log of both sides yields the **Henderson–Hasselbalch equation**, which can be used to calculate the pH of a buffer:

$$pH = pK_a + \log\frac{[A^-]}{[HA]}$$

The weak base K_b expression can also be used giving:

$$[OH^-] = K_b \times \frac{[B]}{[HB^+]} \quad \text{and} \quad pOH = pK_b + \log\frac{[HB^+]}{[B]}$$

These equations allow us to calculate the pH or pOH of the buffer solution knowing K of the weak acid or base and the concentrations of the conjugate weak acid and its conjugate base. Also, if the desired pH is known, along with K, the ratio of base to acid can be calculated. The more concentrated these species are, the more acid or base can be neutralized and the less the change in buffer pH. This is a measure of the **buffer capacity**, the ability to resist a change in pH.

Let's calculate the pH of a buffer. What is the pH of a solution containing 2.00 mol of ammonia and 3.00 mol of ammonium chloride in a volume of 1.00 L?

$$K_b = 1.81 \times 10^{-5}$$

$$NH_3 + H_2O \rightleftharpoons NH_4^+ + OH^-$$

There are two ways to solve this problem.

$$K_b = \frac{[NH_4^+][OH^-]}{[NH_3]} = \frac{(3.00+x)(x)}{(2.00-x)} = 1.81\times10^{-5}$$

Assume x is small:

$$1.81\times10^{-5} = \frac{3.00x}{2.00}$$
$$x = 1.21\times10^{-5}$$
$$pOH = 4.918$$
$$pH = 14.000 - 4.918 = 9.082$$

Alternate solution:

$$pOH = -\log 1.81\times10^{-5} + \log\frac{[NH_4^+]}{[NH_3]}$$
$$= 4.742 + \log\frac{3.00}{2.00}$$
$$= 4.918 \quad pH = 9.082$$

Titration Equilibria

An acid–base **titration** is a laboratory procedure commonly used to determine the concentration of an unknown solution. A base solution of known concentration is added to an acid solution of unknown concentration (or vice versa) until an acid–base **indicator** visually signals that the **endpoint** of the titration has been reached. The **equivalence point** is the point at which a stoichiometric amount of the base has been added to the acid. Both chemists and chemistry students hope that the equivalence point and the endpoint are close together.

If the acid being titrated is a weak acid, then there are equilibria which will be established and accounted for in the calculations. Typically, a plot of pH of the weak acid solution being titrated versus the volume of the strong base added (the **titrant**) starts at a low pH and gradually rises until close to the equivalence point, where the curve rises dramatically. After the equivalence point region, the curve returns to a gradual increase. This is shown in Figure 15.3.

In many cases, one may know the initial concentration of the weak acid, but may be interested in the pH changes during the titration. To study the changes one can divide the titration curve into four distinctive areas in which the pH is calculated.

1. Calculating the initial pH of the weak acid solution is accomplished by treating it as a simple weak acid solution of known concentration and K_a.
2. As base is added, a mixture of weak acid and conjugate base is formed. This is a buffer solution and can be treated as one in the calculations. Determine the moles of acid consumed from the moles of titrant added—that will be the moles of conjugate base formed. Then calculate the molar concentration of weak acid and conjugate base, taking into consideration the volume of titrant added. Finally, apply your buffer equations.
3. At the equivalence point, all the weak acid has been converted to its conjugate base. The conjugate base will react with water, so treat it as a weak base solution and calculate the $[OH^-]$ using K_b. Finally, calculate the pH of the solution.
4. After the equivalence point, you have primarily the excess strong base that will determine the pH.

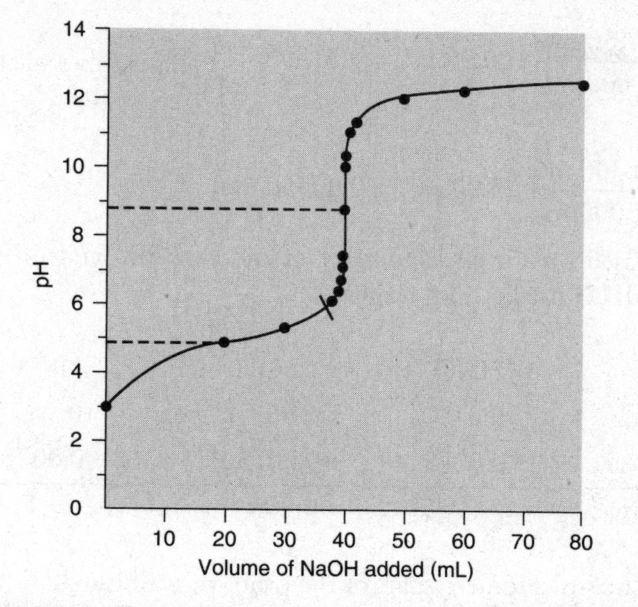

Figure 15.3 The titration of a weak acid with a strong base.

Let's consider a typical titration problem. A 100.0 mL sample of 0.150 M nitrous acid (pK_a = 3.35) was titrated with 0.300 M sodium hydroxide. Determine the pH of the solution after the following quantities of base have been added to the acid solution:

a. 0.00 mL
b. 25.00 mL
c. 49.50 mL
d. 50.00 mL
e. 55.00 mL
f. 75.00 mL

a. 0.00 mL. Since no base has been added, only HNO_2 is present. HNO_2 is a weak acid, so this can only be a K_a problem.

$$\begin{array}{lll} HNO_2 \rightleftharpoons & H^+(aq) + & NO_2^- \\ 0.150 - x & x & x \end{array}$$

$$K_a = 10^{-3.35} = 4.5 \times 10^{-4} = \frac{(x)(x)}{0.150 - x}$$

Quadratic needed: $x^2 + 4.47 \times 10^{-5}x - 6.70 \times 10^{-5} = 0$

(extra sig. figs.)

$$x = [H^+] = 8.0 \times 10^{-3}M \quad pH = 2.10$$

b. 25.00 mL. Since both an acid and a base are present (and they are not conjugates), this must be a stoichiometry problem. Stoichiometry requires a balanced chemical equation and moles.

$$HNO_2 + NaOH \rightarrow Na^+ + NO_2^- + H_2O$$

$Na^+ + NO_2^-$ could be written as $NaNO_2$, but the separated ions are more useful.

Acid: $\frac{0.150 \text{ mol}}{1{,}000 \text{ mL}} 100.00 \text{ mL} = 0.0150 \text{ mol}$ (This number will be used in all remaining steps.)

Base: $\frac{0.300 \text{ mol}}{1{,}000 \text{ mL}} 25.00 \text{ mL} = 0.00750 \text{ mol}$

Based on the stoichiometry of the problem, and on the moles of acid and base, NaOH is the limiting reagent.

	HNO_2 +	NaOH →	Na^+ +	NO_2^- + H_2O
init.	0.0150	0.00750 mol	0	0
react.	−0.0148	−0.00750	+0.00750	+0.00750
final	0.00750	0.000	—	0.00750

The stoichiometry part of the problem is finished.

The solution is no longer HNO_2 and NaOH, but HNO_2 and NO_2^- (a conjugate acid–base pair).

Since a CA/CB pair is present, this is now a buffer problem, and the Henderson–Hasselbalch equation may be used.

$$pH + pK_a + \log (CB/CA) = 3.35 - \log (0.00750/0.00750) = 3.35$$

Note the simplification in the CB/CA concentrations. Both moles are divided by exactly the same volume (since they are in the same solution), so the identical volumes cancel.

$$\frac{\left[\frac{0.00750 \text{ mol base}}{0.12500 \text{ L solution}}\right]}{\left[\frac{0.00750 \text{ mol acid}}{0.12500 \text{ L solution}}\right]}$$

c. 49.50 mL. Since both an acid and a base are present (and they are not conjugates), this must be a stoichiometry problem again. Stoichiometry requires a balanced chemical equation and moles.

$$HNO_2 + NaOH \rightarrow Na^+ + NO_2^- + H_2O$$

Base: $\frac{0.300 \text{ mol}}{1000 \text{ mL}} 49.50 \text{ mL} = 0.0148 \text{ mole}$

Based on the stoichiometry of the problem, and on the moles of acid and base, NaOH is the limiting reagent.

	HNO_2 +	NaOH →	Na^+ +	NO_2^- + H_2O
init.	0.0150	0.0148 mol	0	0
react.	−0.0148	−0.0148	+0.0148	+0.0148
final	0.0002	0.000	—	0.0148

The stoichiometry part of the problem is finished.

The solution is no longer HNO_2 and NaOH, but HNO_2 and NO_2^- (a conjugate acid–base pair).

Since a CA/CB pair is present, this is now a buffer problem, and the Henderson–Hasselbalch equation may be used.

$$pH = pK_a + \log(CB/CA) = 3.35 + \log(0.0148/0.0002) = 5.2$$

d. 50.00 mL. Since both an acid and a base are present (and they are not conjugates), this must be a stoichiometry problem. Stoichiometry requires a balanced chemical equation and moles.

$$HNO_2 + NaOH \rightarrow Na^+ + NO_2^- + H_2O$$

$$\text{Base: } \frac{0.300\ \text{mol}}{1{,}000\ \text{mL}} 50.00\ \text{mL} = 0.0150\ \text{mol}$$

Based on the stoichiometry of the problem, and on the moles of acid and base, both are limiting reagents.

	$HNO_2 +$	$NaOH \rightarrow$	$Na^+ +$	$NO_2^- + H_2O$
init.	0.0150	0.0150 mol	0	0
react.	−0.0150	−0.0150	+0.0150	+0.0150
final	0.0000	0.000	—	0.0150

$$[NO_2] = 0.0150\ \text{mol} / 0.150\text{L} = 0.100\text{M}$$

The stoichiometry part of the problem is finished.

The solution is no longer HNO_2 and NaOH, but an NO_2^- solution (a conjugate base of a weak acid).

Since the CB of a weak acid is present, this is a K_b problem.

$$pK_b = 14.000 - pK_a = 14.000 - 3.35 = 10.65$$

$$NO_2^- + H_2O \rightleftharpoons OH^- + HNO_2$$
$$0.100 - x \qquad x \qquad x$$

$$K_b = 10^{-10.65} = 2.24 \times 10^{-11} = \frac{(x)(x)}{0.100 - x} \text{ neglect } x$$

$$x = [OH^-] = 1.50 \times 10^{-6}\text{M} \quad pOH = 5.82$$

$$pH = 14.00 - pOH^- = 14.00 - 5.82 = 8.18$$

e. 55.00 mL. Since both an acid and a base are present (and they are not conjugates), this must be a stoichiometry problem. Stoichiometry requires a balanced chemical equation and moles.

$$HNO_2 + NaOH \rightarrow Na^+ + NO_2^- + H_2O$$

$$\text{Base: } \frac{0.300\ \text{mol}}{1{,}000\ \text{mL}} 55.00\ \text{mL} = 0.0165\ \text{mol}$$

Based on the stoichiometry of the problem, and on the moles of acid and base, the acid is now the limiting reagent.

	$HNO_2 +$	$NaOH \rightarrow$	$Na^+ +$	$NO_2^- + H_2O$
init.	0.0150	0.0165 mol	0	0
react.	−0.0150	−0.0150	+0.0150	+0.0150
final	0.0000	0.000	—	0.0150

The strong base will control the pH.

$$[OH^-] = 0.0015 \text{ mol}/0.155 \text{ L} = 9.7 \times 10^{-3} \text{ M}$$

The stoichiometry part of the problem is finished.
Since this is now a solution of a strong base, it is now a simple pOH/pH problem.

$$pOH = -\log 9.7 \times 10^{-3} = 2.01$$

$$pH = 14.00 - pOH = 14.00 - 2.01 = 11.99$$

f. 75.00 mL. Since both an acid and a base are present (and they are not conjugates), this must be a stoichiometry problem. Stoichiometry requires a balanced chemical equation and moles.

$$HNO_2 + NaOH \rightarrow Na^+ + NO_2^- + H_2O$$

$$\text{Base: } \frac{0.300 \text{ mol}}{1{,}000 \text{ mL}} 75.00 \text{ ML} = 0.0225 \text{ mol}$$

Based on the stoichiometry of the problem, and on the moles of acid and base, the acid is now the limiting reagent.

	$HNO_2 +$	$NaOH \rightarrow$	$Na^+ +$	$NO_2^- + H_2O$
init.	0.0150	0.0225 mol	0	0
react.	−0.0150	−0.0150	+0.0150	+0.0150
final	0.0000	0.0075	—	0.0150

The strong base will control the pH.

$$[OH^-] = 0.0075 \text{ mol}/0.175 \text{ L} = 4.3 \times 10^{-2} \text{ M}$$

The stoichiometry part of the problem is finished.
Since this is now a solution of a strong base, it is now a simple pOH/pH problem.

$$pOH = -\log 4.3 \times 10^{-2} = 1.37$$

$$pH = 14.00 - pOH = 14.00 - 1.37 = 12.63$$

Solubility Equilibria

Many salts are soluble in water, but some are only slightly soluble. These salts, when placed in water, quickly reach their solubility limit and the ions establish an equilibrium system with the undissolved solid. For example, $PbSO_4$, when dissolved in water, establishes the following equilibrium:

$$PbSO_4(s) \rightleftharpoons Pb^{2+}(aq) + SO_4^{2-}(aq)$$

The equilibrium constant expression for systems of slightly soluble salts is called the **solubility product constant, K_{sp}**. It is the product of the ionic concentrations, each one raised to the power of the coefficient in the balanced chemical equation. It contains no denominator since the concentration of a solid is, by convention, 1 and does not appear in the equilibrium constant expressions. (Some textbooks will say that the concentrations of solids, liquids, and solvents are included in the equilibrium constant.) The K_{sp} expression for the $PbSO_4$ system would be:

$$K_{sp} = [Pb^{2+}][SO_4^{2-}]$$

For this particular salt the numerical value of K_{sp} is 1.6×10^{-8} at 25°C. Note that the Pb^{2+} and SO_4^{2-} ions are formed in equal amounts, so the right-hand side of the equation could be represented as $[x]^2$. If the numerical value of the solubility product constant is known, then the concentration of the ions can be determined. And if one of the ion concentrations can be determined, then K_{sp} can be calculated.

For example, the K_{sp} of magnesium fluoride in water is 8×10^{-8}. How many grams of magnesium fluoride will dissolve in 0.250 L of water?

$$MgF_2(s) \rightleftharpoons Mg^{2+} + 2F^-$$

$$K_{sp} = [Mg^{2+}][F^-]^2 = 8 \times 10^{-8}$$

$$= (x)(2x)^2 = 4x^3 = 8 \times 10^{-8}$$

$$x = 3 \times 10^{-3} = [Mg^{2+}]$$

$$\frac{(3 \times 10^{-3}\,\text{mol Mg}^{2+})}{(\text{L})}(0.250\,\text{L})\frac{(1\,\text{mol MgF}_2)(62.3\text{g MgF}_2)}{(1\,\text{mol Mg}^{2+})(1\,\text{mol MgF}_2)}$$

$$= 0.05\text{ g}$$

If a slightly soluble salt solution is at equilibrium and a solution containing one of the ions involved in the equilibrium is added, the solubility of the slightly soluble salt is decreased. For example, let's again consider the $PbSO_4$ equilibrium:

$$PbSO_4(s) \rightleftharpoons Pb^{2+}(aq) + SO_4^{2-}(aq)$$

Suppose a solution of Na_2SO_4 is added to this equilibrium system. The additional sulfate ion will disrupt the equilibrium, by Le Châtelier's principle, and shift it to the left, decreasing the solubility. The same would be true if you tried to dissolve $PbSO_4$ in a solution of Na_2SO_4 instead of pure water—the solubility would be lower. This application of Le Châtelier's principle to equilibrium systems of slightly soluble salts is called the **common-ion effect**. Calculations like the ones above involving finding concentrations and K_{sp} can still be done, but the concentration of the additional common ion will have to be inserted into the solubility product constant expression. Sometimes, if K_{sp} is very small and the common ion concentration is large, the concentration of the common ion can simply be approximated by the concentration of the ion added.

For example, calculate the silver ion concentration in each of the following solutions:

a. $Ag_2CrO_4(s)$ + water
b. $Ag_2CrO_4(s)$ + 1.00 MNa_2CrO_4

$K_{sp} = 1.9 \times 10^{-12}$

a. $Ag_2CrO_4(s) \rightleftharpoons 2Ag^+(aq) + CrO_4^{2-}(aq)$

$$ 2x \qquad\qquad x$$

$K_{sp} = (2x)^2(x) = 1.9 \times 10^{-12} = 4x^3$

$x = 7.8 \times 10^{-5}$

$[Ag^+] = 2x = 1.6 \times 10^{-4}$ M

b. 1.00 M $Na_2CrO_4 \rightarrow$ 1.00 M CrO_4^{2-} (common ion)

$$Ag_2CrO_4(s) \rightleftharpoons 2Ag^+(aq) + CrO_4^{2-}(aq)$$
$$2x \qquad 1.00 + x$$

$$K_{sp} = (2x)^2(1.00 + x) = 1.9 \times 10^{-12} = 4x^2 \text{ (neglect } x\text{)}$$
$$x = 6.9 \times 10^{-6} \text{ M}$$
$$[Ag^+] = 2x = 1.4 \times 10^{-6} \text{ M}$$

Knowing the value of the solubility product constant can also allow us to predict whether a precipitate will form if two solutions, each containing an ion component of a slightly soluble salt, are mixed. The **ion-product**, sometimes represented as Q (same form as the solubility product constant), is calculated taking into consideration the mixing of the volumes of the two solutions, and this ion-product is compared to K_{sp}. If it is greater than K_{sp}, precipitation will occur until the ion concentrations have been reduced to the solubility level.

If 10.0 mL of a 0.100 M $BaCl_2$ solution is added to 40.0 mL of a 0.0250 M Na_2SO_4 solution, will $BaSO_4$ precipitate? K_{sp} for $BaSO_4 = 1.1 \times 10^{-10}$

To answer this question, the concentrations of the barium ion and the sulfate ion *before* precipitation must be used. These may be determined simply from $M_{dil} = M_{con} V_{con}/V_{dil}$.

$$\text{For } Ba^{2+}\text{: } M_{dil} = (0.100 \text{ M})(10.0 \text{ mL})/(10.0 + 40.0 \text{ mL}) = 0.0200 \text{ M}$$

$$\text{For } SO_4^{2-}\text{: } M_{dil} = (0.0250 \text{ M})(40.0 \text{ mL})/(50.0 \text{ mL}) = 0.0200 \text{ M}$$

Entering these values into the following relation produces:

$$Q = [Ba^{2+}][SO_4^{2-}] = (0.0200)(0.0200) = 0.000400$$

Since Q is greater than K_{sp}, precipitation will occur.

Other Equilibria

Other types of equilibria can be treated in much the same way as the ones discussed above. For example, there is an equilibrium constant associated with the formation of complex ions. This equilibrium constant is called the **formation constant, K_f**. $Zn(H_2O)_4^{2+}$ reacts with ammonia to form the $Zn(NH_3)_4^{2+}$ complex ion according to the following equation:

$$Zn(H_2O)_4^{2+}(aq) + 4NH_3(aq) \rightleftharpoons Zn(NH_3)_4^{2+} + 4H_2O(l)$$

The K_f of $Zn(NH_3)_4^{2+}(aq)$ is 7.8×10^8, indicating that the equilibrium lies to the right.

Experiments

Equilibrium experiments such as 10, 11, and 19 in Chapter 19 (Experimental Investigations), directly or indirectly involve filling a table like the following:

Reactants and Products
Initial amount
Change
Equilibrium amount

The initial amounts—concentrations or pressures—are normally zero for the products, and a measured or calculated value for the reactants. Once equilibrium has been established, the amount of at least one of the substances is determined. Based on the change in this one substance and the stoichiometry, the amounts of the other materials may be calculated.

Measurements may include the pressure, the mass (to be converted to moles), the volume (to be used in calculations), and the pH (to be converted into either the hydrogen ion or hydroxide ion concentration). Some experiments measure the color intensity (with a spectrophotometer), which may be converted to a concentration.

Do not make the mistake of "measuring" a change. Changes are never measured; they are always calculated.

Common Mistakes to Avoid

1. Be sure to check the units and significant figures of your final answer.
2. When writing equilibrium constant expressions, use products over reactants. Each concentration is raised to the power of the coefficient in the balanced chemical equation.
3. In converting from K_c to K_p be sure to use the ideal gas constant, R, whose units are consistent with the units of the partial pressures of the gases.
4. Remember, in working Le Châtelier problems, pressure effects are important only for gases that are involved in the equilibrium.
5. Be sure, when working weak-base problems, to use K_b and not K_a.
6. In titration problems, make sure you compensate for dilution when mixing two solutions together.
7. A K_a expression must have $[H^+]$ in the numerator, and a K_b expression must have $[OH^-]$ in the numerator.

❯ Review Questions

Use these questions to review the content of this chapter and practice for the AP Chemistry exam. First are 20 multiple-choice questions similar to what you will encounter in Section I of the AP Chemistry exam. Following those is a multipart free-response question like the ones in Section II of the exam. To make these questions an even more authentic practice for the actual exam, time yourself following the instructions provided.

Multiple-Choice Questions

Answer the following questions in 25 minutes. You may not use a calculator. You may use the periodic table and the equation sheet at the back of this book.

1. A 0.1 molar solution of acetic acid (CH_3COOH) has a pH of about:

(A) 1
(B) 3
(C) 7
(D) 10

2.

Acid	K_a, acid dissociation constant
H_3PO_4	7.2×10^{-3}
$H_2PO_4^-$	6.3×10^{-8}
HPO_4^{2-}	4.2×10^{-13}

Using the given information, choose the best answer for preparing a pH = 8 buffer.

(A) $K_2HPO_4 + KH_2PO_4$
(B) H_3PO_4
(C) $K_2HPO_4 + K_3PO_4$
(D) K_3PO_4

Use the following information for questions 3–4.

Ionization Constants:

HCOOH	$K_a = 1.8 \times 10^{-4}$
CH_3NH_2	$K_b = 4.4 \times 10^{-4}$
H_3PO_2	$K_{a1} = 3 \times 10^{-2}$
	$K_{a2} = 1.7 \times 10^{-7}$

3. What is a solution with an initial KCOOH concentration of 1 M, and an initial K_2HPO_2 concentration of 1 M?

(A) a solution with a pH > 7, which is a buffer
(B) a solution with a pH < 7, which is not a buffer
(C) a solution with a pH < 7, which is a buffer
(D) a solution with a pH > 7, which is not a buffer

4. What is a solution with an initial H_3PO_2 concentration of 1 M and an initial KH_2PO_2 concentration of 1 M?

(A) a solution with a pH > 7, which is a buffer
(B) a solution with a pH < 7, which is not a buffer
(C) a solution with a pH < 7, which is a buffer
(D) a solution with a pH > 7, which is not a buffer

5. A solution of a weak base is titrated with a solution of a standard strong acid. The progress of the titration is followed with a pH meter. Which of the following observations would occur?

(A) The pH of the solution gradually decreases throughout the experiment.
(B) Initially the pH of the solution drops slowly, and then it drops much more rapidly.
(C) At the equivalence point, the pH is 7.
(D) After the equivalence point, the pH becomes constant because this is the buffer region.

6. What is the ionization constant, K_a, for a weak monoprotic acid if a 0.30 molar solution has a pH of 4.0?

(A) 3.3×10^{-8}
(B) 4.7×10^{-2}
(C) 1.7×10^{-6}
(D) 3.0×10^{-4}

7. Phenol, C_6H_5OH, has $K_a = 1.0 \times 10^{-10}$. What is the pH of a 0.010 M solution of phenol?

(A) between 3 and 7
(B) 10
(C) 2
(D) between 7 and 10

8. You are given equimolar solutions of each of the following. Which has the lowest pH?

(A) NH_4Cl
(B) NaCl
(C) K_3PO_4
(D) Na_2CO_3

9. When sodium nitrite dissolves in water:

(A) The solution is acidic because of hydrolysis of the sodium ion.
(B) The solution is basic because of hydrolysis of the NO_2^- ion.
(C) The solution is basic because of hydrolysis of the sodium ion.
(D) The solution is acidic because of hydrolysis of the NO_2^- ion.

10. Which of the following solutions has a pH nearest 7?

(A) 1 M $H_2C_2O_4$ (oxalic acid) and 1 M KHC_2O_4 (potassium hydrogen oxalate)
(B) 1 M KNO_3 (potassium nitrate) and 1 M HNO_3 (nitric acid)
(C) 1 M NH_3 (ammonia) and 1 M NH_4NO_3 (ammonium nitrate)
(D) 1 M CH_3NH_2 (methylamine) and 1 M $HC_2H_3O_2$ (acetic acid)

11. Determine the $OH^-(aq)$ concentration in 1.0 M aniline ($C_6H_5NH_2$) solution. (The K_b for aniline is 4.0×10^{-10}.)

(A) 2.0×10^{-5} M
(B) 4.0×10^{-10} M
(C) 3.0×10^{-6} M
(D) 5.0×10^{-7} M

12. A student wishes to reduce the zinc ion concentration in a saturated zinc iodate solution to 1×10^{-6} M. How many moles of solid KIO_3 must be added to 1.00 L of solution? [K_{sp} $Zn(IO_3)_2 = 4 \times 10^{-6}$ at 25°C]

(A) 1 mol
(B) 0.5 mol
(C) 2 mol
(D) 4 mol

13. At constant temperature, a change in volume will NOT affect the moles of substances present in which of the following?

(A) $H_2(g) + I_2(g) \rightleftarrows 2\ HI(g)$
(B) $CO(g) + Cl_2(g) \rightleftarrows COCl_2(g)$
(C) $PCl_5(g) \rightleftarrows PCl_3(g) + Cl_2(g)$
(D) $N_2(g) + 3\ H_2(g) \rightleftarrows 2\ NH_3(g)$

14. The equilibrium constant for the hydrolysis of $C_2O_4^{2-}$ is best represented by which of the following?

(A) $K = [OH^-][C_2O_4^{2-}]/[HC_2O_4^-]$
(B) $K = [H_3O^+][C_2O_4^{2-}]/[HC_2O_4^-]$
(C) $K = [HC_2O_4^-][OH^-]/[C_2O_4^{2-}]$
(D) $K = [C_2O_4^{2-}]/[HC_2O_4^-][OH^-]$

15. $ZnS(s) + 2H^+(aq) \rightleftarrows Zn^{2+}(aq) + H_2S(aq)$

What is the equilibrium constant for the above reaction? The successive acid dissociation constants for H_2S are 9.5×10^{-8} (K_{a1}) and 1×10^{-19} (K_{a2}). The K_{sp}, the solubility product constant, for ZnS equals 1.6×10^{-24}.

(A) $1.6 \times 10^{-24}/9.5 \times 10^{-8}$
(B) $1 \times 10^{-79}/1.6 \times 10^{-24}$
(C) $1.6 \times 10^{-24}/9.5 \times 10^{-27}$
(D) $9.5 \times 10^{-8}/1.6 \times 10^{-24}$

16. $C(s) + H_2O(g) \rightleftarrows CO(g) + H_2O(g)$ endothermic

An equilibrium mixture of the reactants is placed in a sealed container at 150°C. The amount of the products may be increased by which of the following changes?

(A) increasing the volume of the container
(B) raising the temperature of the container and increasing the volume of the container
(C) raising the temperature of the container
(D) increasing the volume of the container and adding 1 mol of C(s) to the container

17. $CH_4(g) + CO_2(g) \rightleftarrows 2\ CO(g) + 2\ H_2(g)$

A 1.00 L flask is filled with 0.30 mol of CH_4 and 0.40 mol of CO_2, and allowed to come to equilibrium. At equilibrium, there are 0.20 mol of CO in the flask. What is the value of K_c, the equilibrium constant, for the reaction?

(A) 1.2
(B) 0.027
(C) 0.30
(D) 0.060

18. $NO_2(g) \rightleftarrows 2\ NO(g) + O_2(g)$

The above materials were sealed in a flask and allowed to come to equilibrium at a certain temperature. A small quantity of $O_2(g)$ was added to the flask, and the mixture was allowed to return to equilibrium at the same temperature. Which of the following has increased over its original equilibrium value?

(A) the quantity of $NO_2(g)$ present
(B) the quantity of NO(g) present
(C) the equilibrium constant, *K*, increases
(D) the rate of the reaction

19. $2CH_4(g) + O_2(g) \rightleftarrows 2CO(g) + 4H_2(g) \quad \Delta H < 0$

In order to increase the value of the equilibrium constant, *K*, which of the following changes must be made to the above equilibrium?

(A) increase the temperature
(B) increase the volume
(C) decrease the temperature
(D) add CO(g)

20. The addition of nitric acid increases the solubility of which of the following compounds?

(A) KCl(s)
(B) $Pb(CN)_2(s)$
(C) $Cu(NO_3)_2(s)$
(D) $NH_4NO_3(s)$

Answers and Explanations for the Multiple-Choice Questions

1. B—An acid, any acid, will give a pH below 7; thus, answers C and D are eliminated. A 0.1 molar solution of a strong acid would have a pH of 1. Acetic acid is not a strong acid, and this eliminates answer A.

2. A—The K nearest 10^{-8} will give a pH near 8. The answer must involve the $H_2PO_4^-$ ion.

3. D—The two substances are not a conjugate acid–base pair, so this is not a buffer. Both compounds are salts of a strong base and a weak acid; such salts are basic (pH > 7).

4. C—The two substances constitute a conjugate acid–base pair, so this is a buffer. The pH should be near $-\log K_{a1}$. This is about 2 (acid).

5. B—Any time an acid is added, the pH will drop. The reaction of the weak base with the acid produces the conjugate acid of the weak base. The combination of the weak base and its conjugate is a buffer, so the pH will not change very much until all the base is consumed. After all the base has reacted, the pH will drop much more rapidly. The equivalence point of a weak base–strong acid titration is always below 7 (only strong base–strong acid titrations will give a pH of 7 at the equivalence point). The value of pOH is equal to pK_b halfway to the equivalence point.

6. A—If pH = 4.0, then $[H^+] = 1 \times 10^{-4} = [A^-]$, and $[HA] = 0.30 - 1 \times 10^{-4}$. The generic K_a is $[H^+][A^-]/[HA]$, and when the values are entered into this equation: $(1 \times 10^{-4})2/0.30 = 3.3 \times 10^{-8}$. Since you can estimate the answer, no actual calculations are necessary.

7. A—This is an acid-dissociation constant, thus the solution must be acidic (pH < 7). The pH of a 0.010 M strong acid would be 2.0. This is not a strong acid, so the pH must be above 2.

8. A—A is the salt of a strong acid and a weak base; it is acidic. B is a salt of a strong acid and a strong base; they are neutral. C and D are salts of a weak acid and a strong base; they are basic. The lowest pH would be the acidic choice.

9. B—Sodium nitrite is a salt of a weak acid and a strong base. Ions from strong bases, Na^+ in this case, do not undergo hydrolysis, and do not affect the pH. Ions from weak acids, NO_2^- in this case, undergo hydrolysis to produce basic solutions.

10. D—The weak acid and the weak base partially cancel each other to give a nearly neutral solution.

11. A—The equilibrium constant expression is: $K_b = 4.0 \times 10^{-10} = [OH^-][C_6H_5NH_3^+]/[C_6H_5NH_2]$. This expression becomes: $(x)(x)/(1.0 - x) = 4.0 \times 10^{-10}$, which simplifies to: $x^2/1.0 = 4.0 \times 10^{-10}$. Taking the square root of each side gives: $x = 2.0 \times 10^{-5} = [OH^-]$. Since you can estimate the answer, no actual calculations are necessary.

12. C—The solubility-product constant expression is: $K_{sp} = [Zn^{2+}][IO_3^-]^2 = 4 \times 10^{-6}$. This may be rearranged to: $[IO_3^-]^2 = 4 \times 10^{-6}/[Zn^{2+}]$. Inserting the desired zinc ion concentration gives: $[IO_3^-]^2 = 4 \times 10^{-6}/(1 \times 10^{-6}) = 4$. Taking the square root of each side leaves a desired IO_3^- concentration of 2 M. Two moles of KIO_3 must be added to 1.00 L of solution to produce this concentration. Since you can estimate the answer, no actual calculations are necessary.

13. A—When dealing with gaseous equilibriums, volume changes are important when there is a difference in the total number of moles of gas on opposite sides of the equilibrium arrow. All the answers, except A, have differing numbers of moles of gas on opposite sides of the equilibrium arrow.

14. C—Hydrolysis of any ion begins with the interaction of that ion with water. Thus, both the ion and water must be on the left side of the equilibrium arrow, and hence in the denominator of the equilibrium-constant expression (water, as all solvents, will be left out of the expression). The oxalate ion is the conjugate base of a weak acid. As a base it will produce OH^- in solution along with the conjugate acid ($HC_2O_4^-$) of the base. The equilibrium reaction is: $C_2O_4^{2-}(aq) + H_2O(l) \rightleftarrows OH^-(aq) + HC_2O_4^-(aq)$.

15. C—The equilibrium given is actually the sum of the following three equilibriums:

$ZnS(s) \rightleftarrows Zn^{2+}(aq) + S^{2-}(aq)$

$$K_{sp} = 1.6 \times 10^{-24}$$

$S^{2-}(aq) + H^{+}(aq) \rightleftarrows HS^{-}(aq)$

$$K = 1/K_{a2} = 1/1 \times 10^{-19}$$

$HS^{-}(aq) + H^{+}(aq) \rightleftarrows H_2S(aq)$

$$K' = 1/K_{a1} = 1/9.5 \times 10^{-8}$$

Summing these equations means you need to multiply the equilibrium constants:

$$K_{sum} = K_{sp}KK' = K_{sp}/K_{a2}K_{a1}$$
$$= 1.6 \times 10^{-24}/[(1 \times 10^{-19})\ (9.5 \times 10^{-8})]$$

16. B—The addition or removal of some solid, as long as some remains present, will not change the equilibrium. An increase in volume will cause the equilibrium to shift toward the side with more moles of gas (right). Raising the temperature of an endothermic process will shift the equilibrium to the right. Any shift to the right will increase the amounts of the products.

17. B—Using the following table:

	$[CH_4]$	$[CO_2]$	$[CO]$	$[H_2]$
Initial	0.30	0.40	0	0
Change	$-x$	$-x$	$+2x$	$+2x$
Equilibrium	$0.30 - x$	$0.40 - x$	$2x$	$2x$

The presence of 0.20 mol of CO (0.20 M) at equilibrium means that $2x = 0.20$ and that $x = 0.10$. Using this value for x, the bottom line of the table becomes:

	$[CH_4]$	$[CO_2]$	$[CO]$	$[H_2]$
Equilibrium	0.20	0.30	0.20	0.20

The equilibrium expression is: $K = [CO]^2[H_2]^2/[CH_4][CO_2]$. Entering the equilibrium values into the equilibrium expression gives: $K = (0.20)^2(0.20)^2/(0.20)(0.30)$

18. A—The addition of a product will cause the equilibrium to shift to the left. The amounts of all the reactants will increase, and the amounts of all the products will decrease (the O_2 will not go below its earlier equilibrium value since excess was added). The value of K is constant, unless the temperature is changed. The rates of the forward and reverse reactions are equal at equilibrium.

19. C—The only way to change the value of K is to change the temperature. For an exothermic process ($\Delta H < 0$), K is increased by a decrease in temperature.

20. B—Nitric acid, being an acid, will react with a base. In addition to obvious bases containing OH^-, the salts of weak acids are also bases. All of the anions, except CN^-, are from strong acids.

Free-Response Question

You have 15 minutes to answer the following multipart question. You may use a calculator and the tables in the back of the book.

Question 1

An aqueous solution is prepared that is initially 0.100 M in CdI_4^{2-}. After equilibrium is established, the solution is found to be 0.013 M in Cd^{2+}. The products of the equilibrium are $Cd^{2+}(aq)$ and $I^-(aq)$.

(a) Derive the expression for the dissociation equilibrium constant, K_d, for the equilibrium and determine the value of the constant.

(b) What will be the cadmium ion concentration arising when 0.400 mol of KI is added to 1.00 L of the solution in part **a**?

(c) A solution is prepared by mixing 0.500 L of the solution from part **b** and 0.500 L of 2.0×10^{-5} M NaOH. Will cadmium hydroxide, $Cd(OH)_2$, precipitate? The K_{sp} for cadmium hydroxide is 2.2×10^{-14}.

(d) When the initial solution is heated, the cadmium ion concentration increases. Is the equilibrium an exothermic or an endothermic process? Explain how you arrived at your conclusion.

Answer and Explanation for the Free-Response Question

(a) $K_d = [Cd^{2+}][I^-]^4/[CdI_4^{2-}]$

Give yourself 1 point for this expression.

Using the following table:

	$CdI_4^{2-}(aq)$	$Cd^{2+}(aq)$	$I^-(aq)$
Initial	0.100 M	0	0
Change	$-x$	$+x$	$+4x$
Equilibrium	$0.100 - x$	x	$4x$

The value of $[Cd^{2+}]$ is given (= 0.013), and this is x. This changes the last line of the table to

	$CdI_4^{2-}(aq)$	$Cd^{2+}(aq)$	$I^-(aq)$
Equilibrium	$0.100 - x = 0.087$	$x = 0.013$	$4x = 0.052$

Entering these values into the K_d expression gives: $(0.013)(0.052)^4/(0.087) = 1.1 \times 10^{-6}$.

Give yourself 1 point for this answer. You can also get 1 point if you correctly put your values into the wrong equation.

(b) The table in part **a** changes to the following:

	$CdI_4^{2-}(aq)$	$Cd^{2+}(aq)$	$I^-(aq)$
Initial	0.100 M	0	0.400
Change	$-x$	$+x$	$+4x$
Equilibrium	$0.100 - x$	x	$0.400 + 4x$

$$K_d = [Cd^{2+}][I]^4/[CdI_4^{2-}] = 1.1 \times 10^{-6} = (x)(0.400 + 4x)^4/(0.100 - x)$$

$$= (x)(0.400)^4/(0.100)$$

$$x = 4.3 \times 10^{-6}\ M = [Cd^{2+}]$$

Give yourself 1 point for the correct setup and 1 point for the correct answer. If you got the wrong value for K in part **a**, you can still get one or both points for using it correctly.

(c) The equilibrium is $Cd(OH)_2 \rightleftarrows Cd^{2+}(aq) + 2\ OH^-(aq)$

The dilution reduces both the Cd^{2+} and OH^- concentration by a factor of 2. This gives:

$$[Cd^{2+}] = 4.3 \times 10^{-6}/2 = 2.2 \times 10^{-6} \text{ and } [OH^-] = 2.0 \times 10^{-5}/2 = 1.0 \times 10^{-5}$$

The reaction quotient is $Q = [Cd^{2+}][OH^{-2}] = (2.2 \times 10^{-6})(1.0 \times 10^{-5})^2 = 2.2 \times 10^{-16}$. This value is less than the K_{sp}, so no precipitate will form.

Give yourself 1 point for a correct calculation, and another point for the correct conclusion. If you correctly use a wrong $[Cd^{2+}]$ you calculated in part **b**, you can still get both points.

(d) Since the cadmium ion concentration increases, the equilibrium must shift to the right. Endothermic processes shift to the right when they are heated. This is in accordance with Le Châtelier's principle.

Give yourself 1 point for endothermic. Give yourself 1 point for mentioning Le Châtelier's principle.

Your score is based on a total of 8 points. Subtract 1 point if any answer has an incorrect number of significant figures.

› Rapid Review

- A chemical equilibrium is established when two exactly opposite reactions occur in the same container at the same time and with the same rates of reaction.
- At equilibrium the concentrations of the chemical species become constant, but not necessarily equal.
- For the reaction aA + bB $\rightleftharpoons$ cC + dD, the equilibrium constant expression would be:

  $K_c = \frac{[C]^c[D]^d}{[A]^a[B]^b}$. Know how to apply this equation.
- Le Châtelier's principle says that if an equilibrium system is stressed, it will reestablish equilibrium by shifting the reactions involved. A change in concentration of a species will cause the equilibrium to shift to reverse that change. A change in pressure or temperature will cause the equilibrium to shift to reverse that change.
- Strong acids completely dissociate in water, whereas weak acids only partially dissociate.
- Weak acids and bases establish an equilibrium system.
- Under the Brønsted–Lowry acid–base theory, acids are proton (H^+) donors and bases are proton acceptors.
- Conjugate acid–base pairs differ only in a single H^+; the one that has the extra H^+ is the acid.
- The equilibrium for a weak acid is described by K_a, the acid dissociation constant.

  It has the form: $K_a = \frac{[H^+][A^-]}{[HA]}$. Know how to apply this equation.
- Most times the equilibrium concentration of the weak acid, [HA], can be approximated by the initial molarity of the weak acid.

- Knowing K_a and the initial concentration of the weak acid allows the calculation of the $[H^+]$.
- Water is an amphoteric substance, acting either as an acid or a base.
- The product of the $[H^+]$ and $[OH^-]$ in a solution or in pure water is a constant, K_w, called the water dissociation constant, 1.0×10^{-14}. $K_w = [H^+][OH^-] = 1.0 \times 10^{-14}$ at 25°C. Know how to apply this equation.
- The pH is a measure of the acidity of a solution. $pH = -\log[H^+]$. Know how to apply this equation and estimate the pH from the $[H^+]$.
- On the pH scale 7 is neutral; pH > 7 is basic; and pH < 7 is acidic.
- $pH + pOH = pK_w = 14.00$. Know how to apply this equation.
- K_b is the ionization constant for a weak base. $K_b = \frac{[HB^+][OH^-]}{[HB]}$. Know how to apply this equation.
- $K_a \times K_b = K_w$ for conjugate acid–base pairs. Know how to apply this equation.
- Buffers are solutions that resist a change in pH by neutralizing either an added acid or an added base.
- The Henderson–Hasselbalch equation allows the calculation of the pH of a buffer solution: $pH = pK_a + \log\frac{[A^-]}{[HA]}$. Know how to apply this equation.
- The buffer capacity is a quantitative measure of the ability of a buffer to resist a change in pH. The more concentrated the acid–base components of the buffer, the higher its buffer capacity.
- A titration is a laboratory technique to determine the concentration of an acid or base solution.
- An acid–base indicator is used in a titration and changes color in the presence of an acid or base.
- The equivalence point or endpoint of a titration is the point at which an equivalent amount of acid or base has been added to the base or acid being neutralized.
- Know how to determine the pH at any point of an acid–base titration.
- The solubility product constant, K_{sp}, is the equilibrium constant expression for sparingly soluble salts. It is the product of the ionic concentration of the ions, each raised to the power of the coefficient of the balanced chemical equation.
- Know how to apply ion-products and K_{sp} values to predict precipitation.
- Formation constants describe complex ion equilibria.

CHAPTER 16

Electrochemistry

IN THIS CHAPTER

Summary: Electrochemistry is the study of chemical reactions that produce electricity, and chemical reactions that take place because electricity is supplied. Electrochemical reactions may be of many types. Electroplating is an electrochemical process. So are the electrolysis of water, the production of aluminum metal, and the production and storage of electricity in batteries. All these processes involve the transfer of electrons and redox reactions.

Keywords and Equations

A table of half-reactions is given in the exam booklet and in the back of this book.

I = current (amperes)
q = charge (coulombs)
$E°$ = standard reduction potential
K = equilibrium constant
$G°$ = standard free energy
Faraday's constant, F = 96,500 coulombs per mole of electrons
Gas constant, R = 8.31 volt coulomb mol^{-1} K^{-1}

$$I = \frac{q}{t} \qquad \log K = \frac{nE°}{0.0592} \qquad \Delta G° = -nFE°$$

$$E_{cell} = E°_{cell} - \left(\frac{RT}{nF}\right)\ln Q = E°_{cell} - \left(\frac{0.0592}{n}\right)\log Q \text{ at } 25°C$$

$$Q = \frac{[C]^c[D]^d}{[A]^a[B]^b} \quad \text{where } aA + bB \rightarrow cC + dD$$

Redox Reactions

Electrochemical reactions involve redox reactions. In the chapter on Reactions and Periodicity we discussed redox reactions, but here is a brief review: *Redox* is a term that stands for reduction and oxidation. Reduction is the gain of electrons, and oxidation is the loss of electrons. For example, suppose a piece of zinc metal is placed in a solution containing Cu^{2+}. Very quickly, a reddish solid forms on the surface of the zinc metal. That substance is copper metal. At the molecular level the zinc metal is losing electrons to form Zn^{2+} and Cu^{2+} is gaining electrons to form copper metal. These two processes can be shown as:

$$Zn(s) \rightarrow Zn^{2+}(aq) + 2e^{-} \qquad \text{(oxidation)}$$

$$Cu^{2+}(aq) + 2e^{-} \rightarrow Cu(s) \qquad \text{(reduction)}$$

The electrons that are being lost by the zinc metal are the same electrons that are being gained by the cupric ion. The zinc metal is being oxidized, and the cupric cation is being reduced.

Something must cause the oxidation (taking of the electrons), and that substance is called the oxidizing agent (the reactant being reduced). In the example above, the oxidizing agent is Cu^{2+}. The reactant undergoing oxidation is called the reducing agent, because it is furnishing the electrons used in the reduction half-reaction. Zinc metal is the reducing agent above. The two half-reactions, oxidation and reduction, can be added together to give you the overall redox reaction. The electrons must cancel—that is, there must be the same number of electrons lost as electrons gained:

$$Zn(s) + Cu^{2+}(aq) + 2e^{-} \rightarrow Zn^{2+}(aq) + 2e^{-} + Cu(s) \qquad \text{or}$$

$$Zn(s) + Cu^{2+}(aq) \rightarrow Zn^{2+}(aq) + Cu(s)$$

In these redox reactions, like the electrochemical reactions we will show you, there is a simultaneous loss and gain of electrons. In the oxidation reaction (commonly called a half-reaction) electrons are being lost, but in the reduction half-reaction those very same electrons are being gained. So, in redox reactions electrons are being exchanged as reactants are being converted into products. This electron exchange may be direct, as when copper metal plates out on a piece of zinc, or it may be indirect, as in an electrochemical cell (battery). In this chapter, we will show you both processes and the calculations associated with each.

The balancing of redox reactions is beginning to appear on the AP exam, so we have included the half-reaction method of balancing redox reactions in the Appendixes, just in case you are having trouble with the technique in your chemistry class.

The definitions for oxidation and reduction given above are the most common and the most useful ones. A couple of others might also be useful: Oxidation is the gain of oxygen or loss of hydrogen and involves an increase in oxidation number. Reduction is the gain of hydrogen or loss of oxygen and involves a decrease in oxidation number.

Electrochemical Cells

In the example above, the electron transfer was direct, that is, the electrons were exchanged directly from the zinc metal to the cupric ions. But such a direct electron transfer doesn't allow for any useful work to be done by the electrons. Therefore, in order to use these electrons, indirect electron transfer must be done. The two half-reactions are physically separated and connected by a wire. The electrons that are lost in the oxidation half-reaction are allowed to flow through the wire to get to the reduction half-reaction. While those electrons

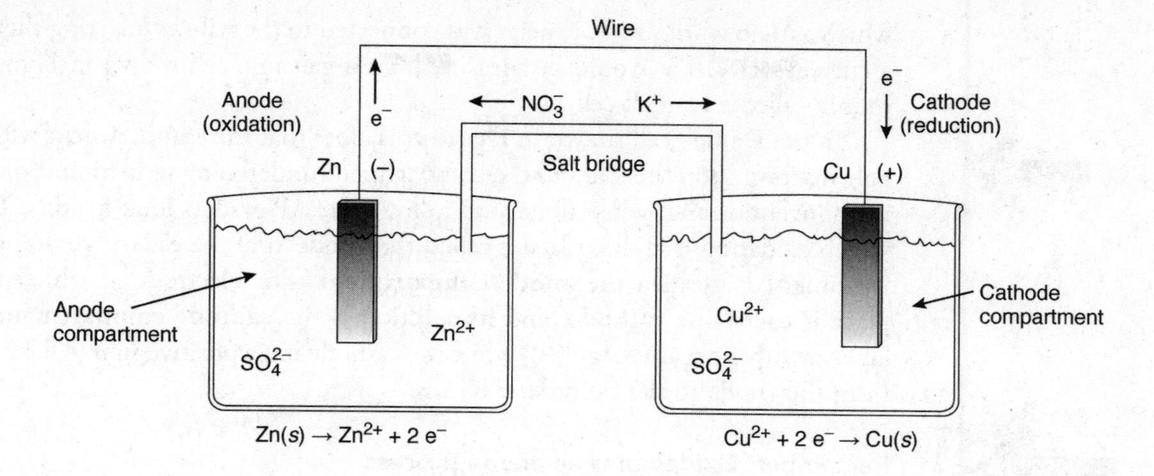

Figure 16.1 A galvanic cell.

are flowing through the wire they can do useful work, like powering a calculator or a pacemaker. **Electrochemical cells** use indirect electron transfer to produce electricity by a redox reaction, or they use electricity to produce a desired redox reaction.

Galvanic (Voltaic) Cells

Galvanic (voltaic) cells produce electricity by using a redox reaction. Let's take that zinc/copper redox reaction that we studied before (the direct electron transfer reaction) and make it a galvanic cell by separating the oxidation and reduction half-reactions. (See Figure 16.1.)

Instead of one container, as before, two will be used. A piece of zinc metal will be placed in one, a piece of copper metal in another. A solution of aqueous zinc sulfate will be added to the beaker containing the zinc electrode and an aqueous solution of copper(II) sulfate will be added to the beaker containing the copper metal. The zinc and copper metals will form the **electrodes** of the cell, the solid portion of the cell that conducts the electrons involved in the redox reaction. The solutions in which the electrodes are immersed are called the **electrode compartments.** The electrodes are connected by a wire and . . . nothing happens. If the redox reactions were to proceed, the beaker containing the zinc metal would build up a positive charge due to the zinc cations being produced in the oxidation half-reaction. The beaker containing the copper would build up a negative charge due to the loss of the copper(II) ions. The solutions (compartments) must maintain electrical neutrality. To accomplish this, a salt bridge will be used. A **salt bridge** is often an inverted U-tube that holds a gel containing a concentrated electrolyte solution, such as KNO_3 in this example. Any electrolyte could be used as long as it does not interfere with the redox reaction. The anions in the salt bridge will migrate through the gel into the beaker containing the zinc metal, and the salt-bridge cations will migrate in the opposite direction. In this way electrical neutrality is maintained. In electrical terms, the circuit has been completed and the redox reaction can occur. The zinc electrode is being oxidized in one beaker, and the copper(II) ions in the other beaker are being reduced to copper metal. The same redox reaction is happening in this indirect electron transfer as happened in the direct one:

$$Zn(s) + Cu^{2+}(aq) \rightarrow Zn^{2+}(aq) + Cu(s)$$

The difference is that the electrons are now flowing through a wire from the oxidation half-reaction to the reduction half-reaction. And electrons flowing through a wire is electricity, which can do work. If a voltmeter was connected to the wire connecting the two electrodes, a current of 1.10 V would be measured. This galvanic cell shown in Figure 16.1 is commonly called a Daniell cell.

In the Daniell cell shown in Figure 16.1 note that the compartment with the oxidation half-reaction is on the left and the compartment undergoing reduction is on the right. This is a convention that you will have to follow. The AP graders look for this. The electrode at which oxidation is taking place is called the **anode**, and the electrolyte solution in which it is immersed is called the **anode compartment**. The electrode at which reduction takes place is called the **cathode**, and its solution is the **cathode compartment**. The anode is labeled with a negative sign (−), while the cathode has a positive sign (+). The electrons flow from the anode to the cathode.

Remember: Oxidation is an anode process.

Sometimes the half-reaction(s) involved in the cell lack a solid conductive part to act as the electrode, so an **inert (inactive) electrode**, a solid conducting electrode that does not take part in the redox reaction, is used. Graphite and platinum are commonly used as inert electrodes.

Note: The electrode must be a conductor on to which a wire may be attached. It can never be an ion in solution.

Cell Notation

Cell notation is a shorthand notation representing a galvanic cell. To write the cell notation in Figure 16.1:

1. Write the chemical formula of the anode: $\mathbf{Zn(s)}$
2. Draw a single vertical line to represent the phase boundary between the solid anode and the solution: $\mathbf{Zn(s)|}$
3. Write the reactive part of the anode compartment with its initial concentration (if known) in parentheses (assume 1M in this case): $\mathbf{Zn(s)|Zn^{2+}(1M)}$
4. Draw a double vertical line to represent the salt bridge connecting the two electrode compartments: $\mathbf{Zn(s)|Zn^{2+}(1M)||}$
5. Write the reactive part of the cathode compartment with its initial concentration (if known) shown in parentheses: $\mathbf{Zn(s)|Zn^{2+}(1M)||Cu^{2+}(1M)}$
6. Draw a single vertical line representing the phase boundary between the solution and the solid cathode: $\mathbf{Zn(s)|Zn^{2+}(1M)||Cu^{2+}(1M)|}$
7. Finally, write the chemical formula of the cathode: $\mathbf{Zn(s)|Zn^{2+}(1M)||Cu^{2+}(1M)|Cu(s)}$

If an inert electrode is used because one or both redox half-reactions do not have a suitable conducting electrode material associated with the reaction, the inert electrode is shown with its phase boundary. If the electrode components are in the same phase, they are separated by commas; if not, a vertical phase boundary line. For example, consider the following redox reaction:

$$Ag^{+}(aq) + Fe^{2+}(aq) \rightarrow Fe^{3+}(aq) + Ag(s)$$

The oxidation of the ferrous ion to ferric doesn't involve a suitable electrode material, so an inert electrode, such as platinum, would be used. The cell notation would then be:

$$Pt(s)|Fe^{2+}(aq), Fe^{3+}(aq)||Ag^{+}(aq)|Ag(s)$$

Cell Potential

In the discussion of the Daniell cell we indicated that this cell produces 1.10 volts. This voltage is really the difference in potential between the two half-cells. There are half-cell potentials associated with all half-cells A list of all possible combinations of half-cells would be tremendously long. Therefore, a way of combining desired half-cells has been developed. The cell potential (really the half-cell potentials) depends on concentration and temperature, but initially we'll simply look at the half-cell potentials at the standard temperature of 298 K (25°C) and all components in their standard states (1 M concentration of all solutions, 1 atmosphere pressure for any gases, and pure solid electrodes). All the half-cell potentials are tabulated as the reduction potentials, that is, the potentials associated with the reduction reaction. The hydrogen half-reaction has been defined as the standard and has been given a value of exactly 0.00 V. All the other half-reactions have been measured relative to it, some positive and some negative. The table of **standard reduction potentials** provided on the AP exam is shown in Table 16.1, on the next page, and in the back of this book.

Here are some things to be aware of in looking at this table:

- All reactions are shown in terms of the reduction reaction relative to the standard hydrogen electrode.
- The more positive the value of the voltage associated with the half-reaction ($E°$), the more readily the reaction occurs.
- The strength of the oxidizing agent increases as the value becomes more positive, and the strength of the reducing agent increases as the value becomes more negative.

This table of standard reduction potentials can be used to write the overall cell reaction and to calculate the **standard cell potential ($E°$)**, the potential (voltage) associated with the cell at standard conditions. There are a few things to remember when using these standard reduction potentials to generate the cell reaction and cell potential:

1. The standard cell potential for a galvanic cell is a positive value, $E° > 0$.
2. Because one half-reaction must involve oxidation, one of the half-reactions shown in the table of reduction potentials must be reversed to indicate the oxidation. If the half-reaction is reversed, the sign of the standard reduction potential must be reversed. However, this is not necessary to calculate the standard cell potential.
3. Because oxidation occurs at the anode and reduction at the cathode, the standard cell potential can be calculated from the standard reduction potentials of the two half-reactions involved in the overall reaction by using the equation:

$$E^\circ_{cell} = E^\circ_{cathode} - E^\circ_{anode} > 0$$

But remember, both $E^\circ_{cathode}$ and E°_{anode} are shown as reduction potentials, used directly from the table without reversing.

Once the standard cell potential has been calculated, the reaction can be written by reversing the half-reaction associated with the anode and adding the half-reactions together, using appropriate multipliers if needed to ensure that the numbers of electrons lost and gained are equal.

Suppose a galvanic cell was to be constructed utilizing the following two half-reactions taken from Table 16.1:

$$Ni^{2+}(aq) \rightarrow Ni(s) \qquad E° = -0.25V$$
$$Ag^{+}(aq) \rightarrow Ag(s) \qquad E° = 0.80V$$

Table 16.1 Standard Reduction Potentials in Aqueous Solution at 25°C

HALF-REACTION			E°(V)
$Li^+ + e^-$	→	Li(s)	−3.05
$Cs^+ + e^-$	→	Cs(s)	−2.92
$K^+ + e^-$	→	K(s)	−2.92
$Rb^+ + e^-$	→	Rb(s)	−2.92
$Ba^{2+} + 2\ e^-$	→	Ba(s)	−2.90
$Sr^{2+} + 2\ e^-$	→	Sr(s)	−2.89
$Ca^{2+} + 2\ e^-$	→	Ca(s)	−2.87
$Na^+ + e^-$	→	Na(s)	−2.71
$Mg^{2+} + 2\ e^-$	→	Mg(s)	−2.37
$Be^{2+} + 2\ e^-$	→	Be(s)	−1.70
$Al^{3+} + 3\ e^-$	→	Al(s)	−1.66
$Mn^{2+} + 2\ e^-$	→	Mn(s)	−1.18
$Zn^{2+} + 2\ e^-$	→	Zn(s)	−0.76
$Cr^{3+} + 3\ e^-$	→	Cr(s)	−0.74
$Fe^{2+} + 2\ e^-$	→	Fe(s)	−0.44
$Cr^{3+} + e^-$	→	Cr^{2+}	−0.41
$Cd^{2+} + 2\ e^-$	→	Cd(s)	−0.40
$Tl^+ + e^-$	→	Tl(s)	−0.34
$Co^{2+} + 2\ e^-$	→	Co(s)	−0.28
$Ni^{2+} + 2\ e^-$	→	Ni(s)	−0.25
$Sn^{2+} + 2\ e^-$	→	Sn(s)	−0.14
$Pb^{2+} + 2\ e^-$	→	Pb(s)	−0.13
$2\ H^+ + 2\ e^-$	→	$H_2(g)$	0.00
$S(s) + 2\ H^+ + 2\ e^-$	→	$H_2S(g)$	0.14
$Sn^{4+} + 2\ e^-$	→	Sn^{2+}	0.15
$Cu^{2+} + e^-$	→	Cu^+	0.15
$Cu^{2+} + 2\ e^-$	→	Cu(s)	0.34
$Cu^+ + e^-$	→	Cu(s)	0.52
$I_2(s) + 2\ e^-$	→	$2I^-$	0.53
$Fe^{3+} + e^-$	→	Fe^{2+}	0.77
$Hg_2^{2+} + 2\ e^-$	→	2 Hg(l)	0.79
$Ag^+ + e^-$	→	Ag(s)	0.80
$Hg^{2+} + 2\ e^-$	→	Hg(l)	0.85
$2\ Hg^{2+} + 2\ e^-$	→	Hg_2^{2+}	0.92
$Br_2(l) + 2\ e^-$	→	$2Br^-$	1.07
$O_2(g) + 4\ H^+ + 4\ e^-$	→	$2\ H_2O(l)$	1.23
$Cl_2(g) + 2\ e^-$	→	$2Cl^-$	1.36
$Au^{3+} + 3\ e^-$	→	Au(s)	1.50
$Co^{3+} + e^-$	→	Co^{2+}	1.82
$F_2(g) + 2\ e^-$	→	$2F^-$	2.87

First, the cell voltage can be calculated using:

$$E^\circ_{cell} = E^\circ_{cathode} - E^\circ_{anode} > 0$$

Since the cell potential must be positive (a galvanic cell), there is only one arrangement of −0.25 and 0.80 volts than can result in a positive value:

$$E^\circ_{cell} = 0.80\text{ V} - (-0.25\text{ V}) = 1.05\text{ V}$$

TIP

This means that the Ni electrode is the anode and must be involved in oxidation, so its reduction half-reaction must be reversed, changing the sign of the standard half-cell potential, and added to the silver half-reaction. **Note that the silver half-reaction must be multiplied by two to equalize electron loss and gain, but the half-cell potential remains the same:**

$$\begin{array}{ll} Ni(s) \rightarrow Ni^{2+}(aq) + 2e^- & E^\circ = 0.25\text{ V} \\ 2 \times (Ag^+(aq) + e^- \rightarrow Ag(s)) & E^\circ = 0.80\text{ V} \\ \hline Ni(s) + 2Ag^+(aq) \rightarrow Ni^{2+}(aq) + Ag(s) & E^\circ_{cell} = 1.05\text{ V} \end{array}$$

Note that the same cell potential is obtained as using: $E^\circ_{cell} = E^\circ_{cathode} - E^\circ_{anode} > 0$.

If, for example, you are given the cell notation, you could use this method to determine the cell potential. In this case, the cell notation would be: $Ni \mid Ni^{2+} \mid\mid Ag^+ \mid Ag \mid$.

Electrolytic Cells

Electrolytic cells use electricity from an external source to produce a desired redox reaction. Electroplating and the recharging of an automobile battery are examples of electrolytic cells.

Figure 16.2, on the next page, shows a comparison of a galvanic cell and an electrolytic cell for the Sn/Cu system. On the left-hand side of Figure 16.2, the galvanic cell is shown for this system. Note that this reaction produces 0.48 V. But what if we wanted the reverse reaction to occur, the nonspontaneous reaction? This can be accomplished by applying a voltage in excess of 0.48 V from an external electrical source. This is shown on the right-hand side of Figure 16.2. In this electrolytic cell, electricity is being used to produce the nonspontaneous redox reaction.

Quantitative Aspects of Electrochemistry

One of the most widely used applications of electrolytic cells is in **electrolysis**, the decomposition of a compound. Water may be decomposed into hydrogen and oxygen. Aluminum oxide may be electrolyzed to produce aluminum metal. In these situations, several questions may be asked: *How long* will it take; *how much* can be produced; *what current* must be used? Given any two of these quantities, the third may be calculated. To answer these questions, the balanced half-reaction must be known. Then the following relationships can be applied:

1 Faraday = 96,500 coulombs per mole of electrons

(F = 96,500 C/mol e^- or 96,500 J/V)

1 ampere = 1 coulomb/second (A = C/s)

Knowing the amperage and how long it is being applied (seconds), the coulombs can be calculated. Then the coulombs can be converted into moles of electrons, and the moles of electrons can be related to the moles (and then grams) of material being electrolyzed through the balanced half-reaction.

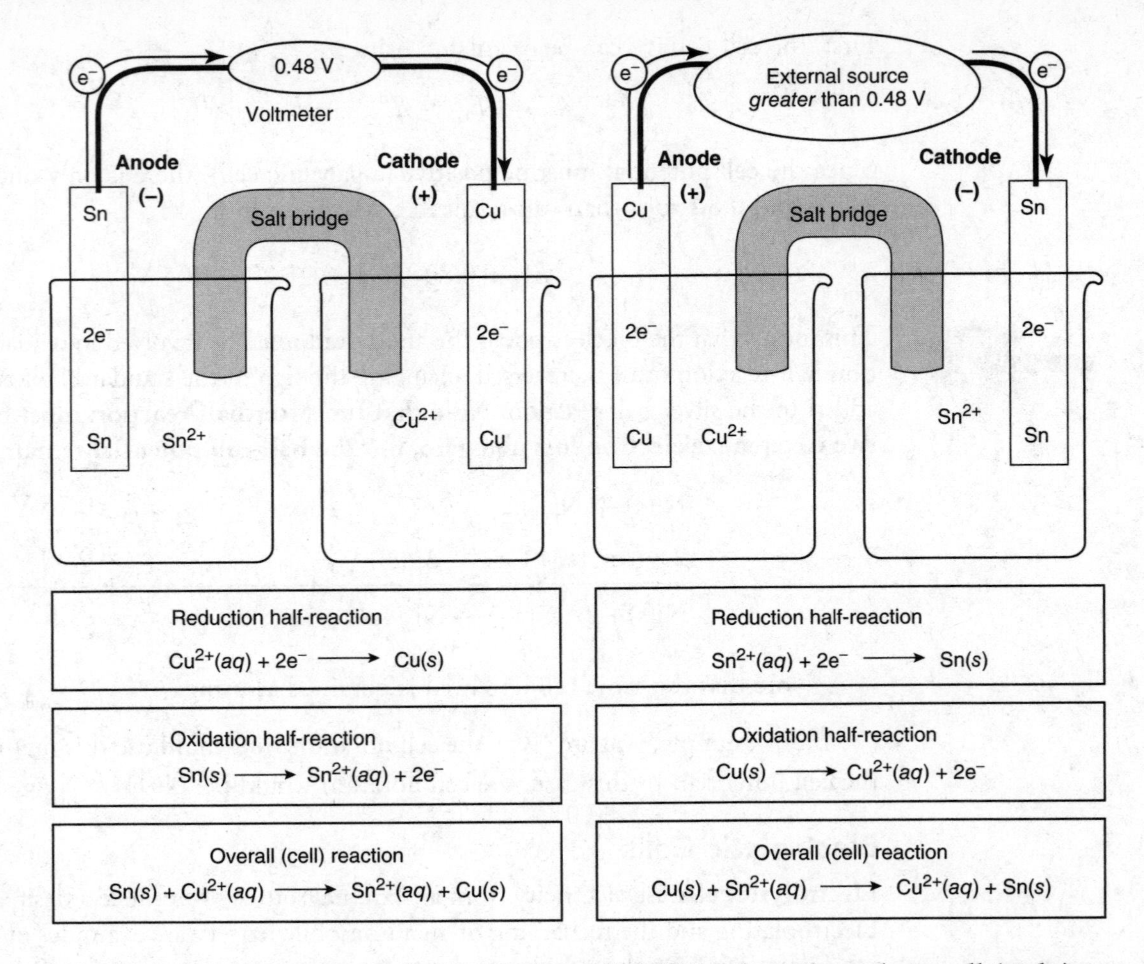

Figure 16.2 Comparison of a galvanic cell (left) and an electrolytic cell (right).

For example, if liquid titanium(IV) chloride (acidified with HCl) is electrolyzed by a current of 1.000 amp for 2.000 h, how many grams of titanium will be produced?

Answer:

$$TiCl_4(l) \rightarrow Ti(s) + 2Cl_2(g) \text{ (not necessary)}$$

$$Ti^{4+} + 4e^- \rightarrow Ti \text{ (necessary)}$$

$$(2.000\ \text{h})\left(\frac{3{,}600\ \text{s}}{\text{h}}\right)\left(\frac{1.000\ \text{C}}{\text{s}}\right)\left(\frac{1\ \text{mole e}^-}{96{,}485\ \text{C}}\right)\left(\frac{1\ \text{mole Ti}}{4\ \text{mole e}^-}\right)\left(\frac{47.90\ \text{g Ti}}{\text{mol Ti}}\right)$$

$$= 0.8936\ \text{g Ti}$$

Calculation of E°_{cell} also allows for the calculation of two other useful quantities—the Gibbs free energy (ΔG°) and the equilibrium constant (K).

The Gibbs free energy is the best single thermodynamic indicator of whether a reaction will be spontaneous (review the Thermodynamics chapter). The Gibbs free energy for a reaction can be calculated from the E° of the reaction using the following equation:

$$\Delta G^\circ = -nFE^\circ_{cell}$$

where F is Faraday's constant of 96,500 C/mol e^- = 96,500 J/V.

If the redox reaction is at equilibrium, $E° = 0$, the equilibrium constant may be calculated by:

$$E^\circ_{\text{cell}} = \frac{0.0592\,\text{V}}{n}\log K \quad \text{or} \quad \log K = \frac{nE^\circ_{\text{cell}}}{0.0592\,\text{V}}$$

Let's apply these relationships. Determine $\Delta G°$ and K for the following reaction:

$$\text{Ni(s)} + 2\text{Ag}^+\text{(aq)} \rightarrow \text{Ni}^{2+}\text{(aq)} + \text{Ag(s)} \ \ E^\circ_{\text{cell}} = 1.05\,\text{V}$$

Answer:

For this reaction, two electrons are transferred from the Ni to the Ag. Thus, n is 2 for this reaction. The value of F (96,500 J/V) is given on the exam, so you will not need to memorize it.

The first answer is:

$$\Delta G° = -nFE^\circ_{\text{cell}} = -2(96{,}500\ \text{J/V})(1.05\,\text{V}) = -2.03 \times 10^5\ \text{J}$$

The second answer is:

$$\log K = \frac{nE^\circ_{\text{cell}}}{0.0592\ \text{V}} = \frac{2(1.05\ \text{V})}{0.0592\ \text{V}} = 35.5$$

This gives a K of about 10^{35} (actually $K = 3 \times 10^{35}$). In many cases, the approximate value will be all you need for the AP exam.

Nernst Equation

Thus far, all of our calculations have been based on the standard cell potential or standard half-cell potentials—that is, the standard state conditions that were defined previously. However, many times the cell is not at standard conditions—commonly the concentrations are not 1 M. The actual cell potential, E_{cell}, can be calculated by the use of the **Nernst equation**:

$$E_{\text{cell}} = E^\circ_{\text{cell}} - \left(\frac{RT}{nF}\right)\ln Q = E^\circ_{\text{cell}} - \left(\frac{0.0592}{n}\right)\log Q \text{ at } 25°\text{C}$$

where R is the ideal gas constant, T is the Kelvin temperature, n is the number of electrons transferred, F is Faraday's constant, and Q is the reaction quotient discussed in the Equilibrium chapter. The second form, involving log Q, is the more useful form. If one knows the cell reaction, the concentrations of ions, and E°_{cell}, then the actual cell potential can be calculated. Another useful application of the Nernst equation is in calculating the concentration of one of the reactants from cell-potential measurements. Knowing the actual cell potential and E°_{cell} allows calculation of Q, the reaction quotient. Knowing Q and all but one of the concentrations allows the calculation of the unknown concentration. Another application of the Nernst equation is in concentration cells. A **concentration cell** is an electrochemical cell in which the same chemical species is used in both cell compartments, but differing in concentration. Because the half-reactions are the same, $E^\circ_{\text{cell}} = 0.00$ V. Simply substituting the appropriate concentrations into the reaction quotient allows calculation of the actual cell potential.

When using the Nernst equation on a cell reaction in which the overall reaction is not supplied, only the half-reactions and concentrations, there are two equivalent methods to work the problem. The first way is to write the overall redox reaction based upon $E°$ values, and then apply the Nernst equation. If E_{cell} turns out to be negative, it indicates that the reaction is not a spontaneous one (an electrolytic cell), or that the reaction is written backwards if it supposed to be a galvanic cell. If it is supposed to be a galvanic cell, all you need to do is reverse the overall reaction and change the sign on E_{cell} to positive. The other method involves using the Nernst equation with the individual half-reactions, then combining them depending on whether or not it is a galvanic cell. The only disadvantage to the second method is that you must use the Nernst equation twice. Either method should lead you to the correct answer.

Let's practice. Calculate the potential of a half-cell containing 0.10 M $K_2Cr_2O_7(aq)$, 0.20 M $Cr^{3+}(aq)$, and 1.0×10^{-4} M $H^+(aq)$.

Answer:

The following half-reaction is given on the AP exam:

$$Cr_2O_7^{2-}(aq) + 14H^+(aq) + 6e^- \rightarrow 2Cr^{3+}(aq) + 7H_2O(l) \quad E° = 1.33\,V$$

$$E = E° - \frac{0.0592}{n} \log \frac{[Cr^{3+}]^2}{[Cr_2O_7^{2-}][H^+]^{14}} \quad \text{ignore } H_2O$$

$$= 1.33\text{ V} - \frac{0.0592}{6} \log \frac{[.20]^2}{[.10][1.0 \times 10^{-4}]^{14}}$$

$$= 0.78\,V$$

Experiments

Electrochemical experiments are normally concerned with standard cell voltages. Measurements of the cell potential are essential and require a voltmeter (potentiometer). These measurements may be taken from different combinations of half-cells, or from measurements before and after changes of some aspect of the cell were made.

Using measurements of different half-cell combinations, a set of "standard" reduction potentials may be constructed. This set will be similar to a table of standard reduction potentials. The solutions used in the half-cells must be of known concentration. These solutions are produced by weighing reagents and diluting to volume. The measurements will require a balance and a volumetric flask. It is also possible to produce known concentrations by diluting solutions. This method requires a pipette and a volumetric flask. Review Chapter 13 on Solutions and Colligative Properties for solution techniques.

Common Mistakes to Avoid

1. Be sure your units cancel to give the unit wanted in your final answer.
2. Be sure to round your answer off to the correct number of significant figures.
3. Remember that oxidation is the loss of electrons and reduction the gain, and that in redox reactions the same number of electrons is lost and gained.
4. When diagramming an electrochemical cell, be sure the electrons go from anode to cathode.
5. Be sure that for a galvanic cell, the cell potential is greater than 0.
6. In cell notation, be sure to write anode, anode compartment, salt bridge, cathode compartment, cathode in this specific order.
7. When using a multiplier to equalize electron loss and gain in reduction half-cell potentials, **do not** use the multiplier on the voltage of the half-cell.

❯ Review Questions

Use these questions to review the content of this chapter and practice for the AP Chemistry exam. First are 14 multiple-choice questions similar to what you will encounter in Section I of the AP Chemistry exam. Following those is a multipart free-response question like the ones in Section II of the exam. To make these questions an even more authentic practice for the actual exam, time yourself following the instructions provided.

Multiple-Choice Questions

Answer the following questions in 20 minutes. You may not use a calculator. You may use the periodic table and the equation sheet at the back of this book.

Choose one of the following for questions 1 and 2.

(A) There is no change in the voltage.
(B) The voltage becomes zero.
(C) The voltage increases.
(D) The voltage decreases, but stays positive.

The following reaction takes place in a voltaic cell:

$$Zn(s) + Cu^{2+}(1\ M) \rightarrow Cu(s) + Zn^{2+}(1\ M)$$

The cell has a voltage that is measured and found to be +1.10 V.

1. What happens to the cell voltage when the copper electrode is made smaller?

2. What happens to the cell voltage when the salt bridge is filled with deionized water instead of 1 M KNO_3?

3. $MnO_4^-(aq) + H^+(aq) + C_2O_4^{2-}(aq) \rightarrow Mn^{2+}(aq) + H_2O(l) + CO_2(g)$

What is the coefficient of H^+ when the above reaction is balanced?

(A) 16
(B) 2
(C) 8
(D) 5

4. $S_2O_3^{2-}(aq) + OH^-(aq) \rightarrow SO_4^{2-}(aq) + H_2O(l) + e^-$

After the above half-reaction is balanced, which of the following are the respective coefficients of OH^- and SO_4^{2-} in the balanced half-reaction?

(A) 8 and 3
(B) 6 and 2
(C) 10 and 2
(D) 5 and 2

5. How many moles of Pt may be deposited on the cathode when 0.80 F of electricity is passed through a 1.0 M solution of Pt^{4+}?

(A) 1.0 mol
(B) 0.60 mol
(C) 0.20 mol
(D) 0.80 mol

6. $Cr_2O_7^{2-}(aq) + 14\ H^+(aq) + 3\ S^{2-}(aq) \rightarrow 2\ Cr^{3+}(aq) + 3\ S(s) + 7\ H_2O(l)$

For the above reaction, pick the true statement from the following.

(A) The S^{2-} is reduced by $Cr_2O_7^{2-}$.
(B) The oxidation number of chromium changes from +7 to +3.
(C) The oxidation number of sulfur remains –2.
(D) The S^{2-} is oxidized by $Cr_2O_7^{2-}$.

7. $H^+(aq) + NO_3^-(aq) + e^- \rightarrow NO(g) + H_2O(g)$

What is the coefficient for water when the above half-reaction is balanced?

(A) 3
(B) 4
(C) 2
(D) 1

8. $Co^{2+} + 2\ e^- \rightarrow Co\ E° = -0.28\ V$

$Cd^{2+} + 2\ e^- \rightarrow Cd\ E° = -0.40\ V$

Given the above standard reduction potentials, estimate the approximate value of the equilibrium constant for the following reaction:

$$Cd + Co^{2+} \rightarrow Cd^{2+} + Co$$

(A) 10^{-4}
(B) 10^{-2}
(C) 10^{4}
(D) 10^{16}

9. A sample of silver is to be purified by electrorefining. This will separate the silver from an impurity of gold. The impure silver is made into an electrode. Which of the following is the best way to set up the electrolytic cell?

(A) an impure silver cathode and an inert anode
(B) an impure silver cathode and a pure gold anode
(C) a pure silver cathode with an impure silver anode
(D) a pure gold cathode with an impure silver anode

10. $2\ Fe^{3+} + Zn \rightarrow Zn^{2+} + 2\ Fe^{2+}$

The reaction shown above was used in an electrolytic cell. The voltage measured for the cell was not equal to the calculated $E°$ for the cell. Which of the following could cause this discrepancy?

(A) The anion in the anode compartment was chloride instead of nitrate as in the cathode compartment.
(B) One or more of the ion concentrations was not 1 M.
(C) Both of the solutions were at 25°C instead of 0°C.
(D) The solution in the salt bridge was Na_2SO_4 instead of KNO_3.

11. How many grams of mercury could be produced by electrolyzing a 1.0 M $Hg(NO_3)_2$ solution with a current of 2.00 A for 3.00 h?

(A) 22.4 g
(B) 201 g
(C) 11.2 g
(D) 44.8 g

12. An electrolysis cell was constructed with two platinum electrodes in a 1.00 M aqueous solution of KCl. An odorless gas evolves from one electrode, and a gas with a distinctive odor evolves from the other electrode. Choose the correct statement from the following list.

(A) The gas with the distinctive odor was evolved at the anode.
(B) The odorless gas was oxygen.
(C) The gas with the distinctive odor was evolved at the cathode.
(D) The odorless gas was evolved at the anode.

13. $2\ BrO_3^-(aq) + 12\ H^+(aq) + 10\ e^- \rightarrow Br_2(aq) + 6\ H_2O(l)$

Which of the following statements is correct for the above reaction?

(A) The BrO_3^- undergoes oxidation at the anode.
(B) Br goes from a –1 oxidation to a 0 oxidation state.
(C) Br_2 is oxidized at the anode.
(D) The BrO_3^- undergoes reduction at the cathode.

14. $2\ M(s) + 3\ Zn^{2+}(aq) \rightarrow 2\ M^{3+}(aq) + 3\ Zn^{2+}(aq)$ $\quad E° = 0.90\ V$

$Zn^{2+}(aq) + 2e^- \rightarrow Zn(s)$ $\quad E° = -0.76\ V$

Using the above information, determine the standard reduction potential for the following reaction:

$$M^{3+}(aq) + 3\ e^- \rightarrow M(s)$$

(A) 0.90 V
(B) –1.66 V
(C) 0.00 V
(D) –0.62 V

Answers and Explanations for the Multiple-Choice Questions

1. A—The size of the electrode is not important.

2. B—The salt bridge serves as an ion source to maintain charge neutrality. Deionized water would not be an ion source, so the cell could not operate.

3. A—The balanced equation is:

$$2\ MnO_4^-(aq) + 16\ H^+(aq) + C_2O_4^{2-}(aq) \rightarrow 2\ Mn^{2+}(aq) + 8\ H_2O(l) + 10\ CO_2(g)$$

4. C—The balanced equation is:

$$S_2O_3^{2-}(aq) + 10\ OH^-(aq) \rightarrow 2\ SO_4^{2-}(aq) + 5\ H_2O(l) + 8\ e-$$

5. C—It takes 4 mol of electrons (4 F) to change the platinum ions to platinum metal. The calculation would be: (0.80 F) (1 mol Pt/4 F) = 0.20 mol Pt

6. D—The dichromate ion oxidizes the sulfide ion to elemental sulfur, as the sulfide ion reduces the dichromate ion to the chromium(III) ion. Chromium goes from +6 to +3, while sulfur goes from –2 to 0. The hydrogen remains at +1, so it is neither oxidized nor reduced.

7. C—The balanced chemical equation is:

$$4\ H^+(aq) + NO_3^-(aq) + 3\ e^- \rightarrow NO(g) + 2\ H_2O(l)$$

8. C—Using the equation:

$$\log K = \frac{nE°}{0}.0592 = \frac{2(0.12)}{0.0592} = 4.05$$

You should realize that the log $K = 4$ gives $K = 10^4$. The actual value is $K = 1.1 \times 10^4$.

9. C—The impure silver must be oxidized so it will go into solution. Oxidation occurs at the anode. Reduction is required to convert the silver ions to pure silver. Reduction occurs at the cathode. The cathode must be pure silver; otherwise, it could be contaminated with the cathode material.

10. B—If the voltage was not equal to $E°$, then the cell was not standard. Standard cells have 1 M concentrations and operate at 25°C with a partial pressure of each gas equal to 1 atm. No gases are involved in this reaction, so the cell must be operating at a different temperature or a different concentration (or both).

11. A—

$$\left(\frac{2.00\,C}{s}\right)\left(\frac{3{,}600\,s}{h}\right)(3.00\,h)\left(\frac{1\,F}{96{,}500\,C}\right)\times \left(\frac{1\,mol\,Hg}{2\,F}\right)\left(\frac{200.6\,g\,Hg}{1\,mol\,Hg}\right)$$

You can estimate the answer by replacing 96,500 with 100,000 and 200.6 with 200.

12. A—The gases produced are hydrogen (at the cathode) and chlorine (at the anode). Hydrogen is odorless, while chlorine has a distinctive odor.

13. D—The bromate ion, BrO_3^-, is gaining electrons, so it is being reduced. Reduction always occurs at the cathode.

14. B—The half-reactions giving the overall reaction must be:

$$3\ [Zn^{2+}(aq) + 2\ e^- \rightarrow Zn(s)] \qquad E° = -0.76\,V$$

$$2\ [M(s) \rightarrow M^{3+}(aq) + 3\ e^-] \qquad E° = ?$$

$$2\ M(s) + 3\ Zn^{2+}(aq) \rightarrow 2\ M^{3+}(aq) + 2\ Zn(s) \qquad E° = 0.90\,V$$

Thus, –0.76 + ? = 0.90, giving ? = 1.66 V. The half-reaction under consideration is the reverse of the one used in this combination, so the sign of the calculated voltage must be reversed. Do not make the mistake of multiplying the voltages when the half-reactions were multiplied to equalize the electrons.

Free-Response Question

You have 15 minutes to answer the following multipart question. You may use a calculator and the tables in the back of the book.

Question 1

V = voltmeter

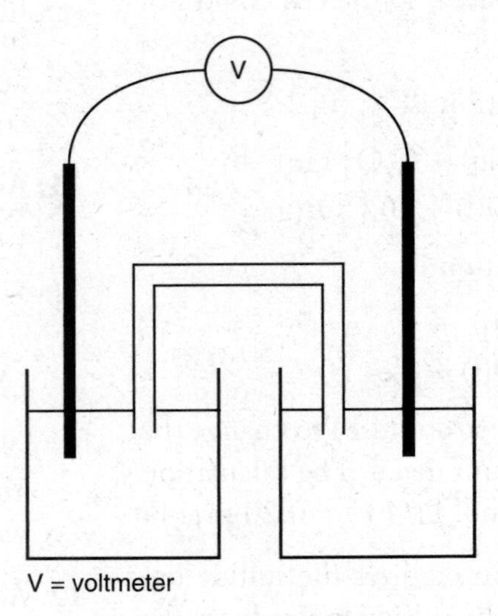

V = voltmeter

The above galvanic cell is constructed with a cobalt electrode in a 1.0 M $Co(NO_3)_2$ solution in the left compartment and a silver electrode in a 1.0 M $AgNO_3$ in the right compartment. The salt bridge contains a KNO_3 solution. The cell voltage is positive.

(a) What is the balanced net ionic equation for the reaction, and what is the cell potential?

$$Co^{2+} + 2e^- \rightarrow Co \qquad E° = -0.28\text{ V}$$

$$Ag^+ + 1e^- \rightarrow Ag \qquad E° = +0.80\text{ V}$$

(b) Which electrode is the anode? Justify your answer.

(c) If some solid $Co(NO_3)_2$ is added to the cobalt compartment, what will happen to the voltage? Justify your answer.

(d) If the cell operates until equilibrium is established, what will the potential be? Justify your answer.

Answer and Explanation for the Free-Response Question

(a) The cell reaction is:

$$Co(s) + 2\ Ag^+(aq) \rightarrow Co^{2+}(aq) + 2\ Ag(s)$$

Give yourself 1 point if you got this correct. The physical states are not necessary. The calculation of the cell potential may be done in different ways. Here is one method:

$$Co \rightarrow Co^{2+} + 2\ e^- \qquad E° = +0.28\text{ V}$$

$$2(Ag^+ + 1\ e^- \rightarrow Ag) \qquad E° = +0.80\text{ V}$$

$$Co(s) + 2\ Ag^+(aq) \rightarrow Co^{2+}(aq) + 2\ Ag(s) \qquad E° = +1.08\text{ V}$$

Give yourself 1 point for the correct answer regardless of the method used. The most common mistake is to multiply the silver voltage by 2. You do not get the point for an answer of 1 V.

(b) The cobalt is the anode.
You get 1 point for this statement.
The reason Co is the anode is because the Co is oxidized.
You get 1 point for this statement or if you state that the Co loses electrons.

(c) The voltage would decrease. The excess Co^{2+}, from the $Co(NO_3)_2$, would impede the reactions from proceeding as written.
Give yourself 1 point for saying decrease. Give yourself 1 point for the explanation.

(d) At equilibrium the cell voltage would be 0 V.
This is worth 1 point.
At equilibrium no work is done, so the potential must be zero.
Or give yourself 1 point for this answer.

There are a total of 7 points possible on this question.

› Rapid Review

- In redox reactions electrons are lost and gained. Oxidation is the loss of electrons, and reduction is the gain of electrons.
- The same number of electrons is lost and gained in redox reactions.
- Galvanic (voltaic) cells produce electricity through the use of a redox reaction.
- The anode is the electrode at which the oxidation half-reaction takes place. The anode compartment is the solution in which the anode is immersed.
- The cathode is the electrode at which reduction takes place, and the cathode compartment is the solution in which the cathode is immersed.
- A salt bridge is used in an electrochemical cell to maintain electrical neutrality in the cell compartments.
- Be able to diagram an electrochemical cell.
- The cell notation is a shorthand way of representing a cell. It has the form:

  anode|anode compartment||cathode compartment|cathode
- Standard reduction potentials are used to calculate the cell potential under standard conditions. All half-reactions are shown in the reduction form.
- For a galvanic cell $E^\circ_{cell} > 0$.
- $E^\circ_{cell} = E^\circ_{cathode} - E^0_{anode} > 0$. Know how to use this equation to calculate E°_{cell}.
- Electrolytic cells use an external source of electricity to produce a desired redox reaction.
- Review how to diagram an electrolytic cell.

- The following relationships can be used to calculate quantitative changes that occur in an electrochemical cell, especially an electrolytic one: 1 F = 96,500 C per mole of electron (F = 96,500 C/mol e^- = 96,500 J/V) and 1 amp = 1 C/s (A = C/s).
- The standard cell potetial can be used to calculate the Gibbs free energy for the reaction: $\Delta G° = -nFE°_{cell}$. Know how to use this equation.
- The standard cell potential can also be used to calculate the equilibrium constant for a reaction: $\log K = \dfrac{nE°_{cell}}{0.0592\ \text{V}}$. Know how to use this equation.

CHAPTER 17

Nuclear Chemistry

IN THIS CHAPTER

Summary: Radioactivity, the spontaneous decay of an unstable isotope to a more stable one, was first discovered by Henri Becquerel in 1896. Marie Curie and her husband expanded on his work and developed most of the concepts that are used today.

Throughout this book you have been studying traditional chemistry and chemical reactions. This has involved the transfer or sharing of electrons from the electron clouds, especially the valence electrons. Little has been said up to this point regarding the nucleus. Now we are going to shift our attention to nuclear reactions and, for the most part, ignore the electron clouds.

Keywords and Equations

No specific nuclear equations are provided, but review first-order equations in the Kinetics chapter.

Nuclear Reactions

Balancing Nuclear Reactions

Most nuclear reactions involve breaking apart the nucleus into two or more different elements or subatomic particles. If all but one of the particles is known, the unknown particle can be determined by balancing the nuclear equation. When chemical equations are balanced, coefficients are added to ensure that there are the same number of each type of atom on both sides of the reaction arrow. To balance nuclear equations, we ensure that there is the same sum of both mass numbers and atomic numbers on both the left and right of the reaction arrow. Recall that a specific isotope of an element can be represented by the following symbolization:

$${}^{A}_{Z}X$$

In this symbolization A is the mass number (sum of protons and neutrons), Z is the atomic number (number of protons), and X is the element symbol (from the periodic table). In balancing nuclear reactions, make sure the sum of all A values on the left of the arrow equals the sum of all A values to the right of the arrow. The same will be true of the sums of the Z values. Knowing that these sums must be equal allows one to predict the mass and atomic number of an unknown particle if all the others are known.

Consider the **transmutation**—creation of one element from another—of Cl-35. This isotope of chlorine is bombarded by a neutron and H-1 is created, along with an isotope of a different element. First, a partial nuclear equation is written:

$${}^{35}_{17}\mathrm{Cl} + {}^{1}_{0}\mathrm{n} \rightarrow {}^{1}_{1}\mathrm{H} + {}^{x}_{y}?$$

The sum of the mass numbers on the left of the equation is 36 = (35 + 1) and on the right is 1 $+x$. The mass number of the unknown isotope must be 35. The sum of the atomic numbers on the left is 17 = (17 + 0), and 1 + y on the right. The atomic number of the unknown must then be 16. This atomic number identifies the element as sulfur, so a complete nuclear equation can be written:

$${}^{35}_{17}\mathrm{Cl} + {}^{1}_{0}\mathrm{n} \rightarrow {}^{1}_{1}\mathrm{H} + {}^{35}_{16}\mathrm{S}$$

Sulfur-35 does not occur in nature; it is an artificially produced isotope.

Natural Radioactive Decay Modes

Three common types of radioactive decay are observed in nature, and two others are occasionally observed.

Alpha Emission

An alpha particle is a helium nucleus with two protons and two neutrons. It is represented as: ${}^{4}_{2}\mathrm{He}$ or α. As this particle is expelled from the nucleus of the radioisotope that is undergoing decay, it has no electrons and thus has a 2+ charge. However, it quickly acquires two electrons from its surroundings to form the neutral atom. Most commonly, the alpha particle is shown as the neutral particle and not the cation.

Radon-222 undergoes alpha decay according to the following equation:

$${}^{222}_{86}\mathrm{Rn} \rightarrow {}^{218}_{84}\mathrm{Po} + {}^{4}_{2}\mathrm{He}$$

Notice that in going from Rn-222 to Po-218, the atomic number has decreased by 2 and the mass number by 4.

Beta Emission

A beta particle is an electron and can be represented as either ${}^{0}_{-1}\beta$ or ${}^{0}_{-1}\mathrm{e}$. This electron comes from the nucleus, *not* the electron cloud, and results from the conversion of a neutron into a proton and an electron: ${}^{1}_{0}\mathrm{n} \rightarrow {}^{1}_{1}\mathrm{p} + {}^{0}_{-1}\mathrm{e}$.

Nickel-63 will undergo beta decay according to the following equation:

$$^{63}_{28}\text{Ni} \rightarrow ^{63}_{29}\text{Cu} + ^{0}_{-1}\text{e}.$$

Notice that the atomic number has increased by 1 in going from Ni-63 to Cu-63, but the mass number has remained unchanged.

Gamma Emission

Gamma emission is the giving off of high-energy, short-wavelength photons similar to X-rays. This radiation is commonly represented as γ. Gamma emission commonly accompanies most other types of radioactive decay, but is often not shown in the balanced nuclear equation because it has neither appreciable mass nor charge.

Alpha, beta, and gamma emissions are the most common types of natural decay mode, but positron emission and electron capture are also observed occasionally.

Positron Emission

A positron is essentially an electron that has a positive charge instead of a negative one. It is represented as $^{0}_{1}\beta$ or $^{0}_{1}\text{e}$. Positron emission results from the conversion of a proton to a neutron and a positron: $^{1}_{+1}\text{p} \rightarrow ^{1}_{0}\text{n} + ^{0}_{+1}\text{e}$. It is observed in the decay of some natural radioactive isotopes, such as K-40: $^{40}_{19}\text{K} \rightarrow ^{40}_{18}\text{Ar} + ^{0}_{+1}\text{e}$.

Electron Capture

The four decay modes described above all involve the emission or giving off of a particle; electron capture is the capturing of an electron from the energy level closest to the nucleus (1s) by a proton in the nucleus. This creates a neutron: $^{0}_{-1}\text{e} + ^{1}_{1}\text{p} \rightarrow ^{1}_{0}\text{n}$. Electron capture leaves a vacancy in the 1s energy level, and an electron from a higher energy level drops down to fill this vacancy. A cascading effect occurs as the electrons shift downward and, as they do so, energy is released. This energy falls in the X-ray part of the electromagnetic spectrum. These X-rays give scientists a clue that electron capture has taken place.

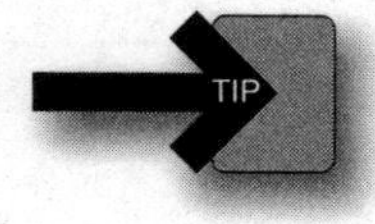

Polonium-204 undergoes electron capture: $^{204}_{84}\text{Po} + ^{0}_{-1}\text{e} \rightarrow ^{204}_{83}\text{Bi} + \text{X-rays}$. Notice that the atomic number has decreased by 1, but the mass number has remained the same. Remember that electron capture is the only decay mode that involves adding a particle to the left side of the reaction arrow.

Nuclear Stability

Predicting whether a particular isotope is stable and what type of decay mode it might undergo can be tricky. All isotopes containing 84 or more protons are unstable and will undergo nuclear decay. For these large, massive isotopes, alpha decay is observed most commonly. Alpha decay gets rid of four units of mass and two units of charge, thus helping to relieve the repulsive stress found in these nuclei. For other isotopes, with atomic numbers less than 84, stability is best predicted by the use of the neutron-to-proton (n/p) ratio.

If one plots the number of neutrons versus the number of protons for the known stable isotopes, the nuclear belt of stability is formed. At the low end of this belt of stability ($Z < 20$), the n/p ratio is 1. At the high end ($Z \approx 80$), the n/p ratio is about 1.5. One can then use the n/p ratio of the isotope under question to predict whether or not it will be stable. If it is unstable, the isotope will utilize a decay mode that will bring it back onto the belt of stability.

For example, consider Ne-18. It has 10 p and 8 n, giving an n/p ratio of 0.8. That is less than 1, so the isotope is unstable. This isotope is neutron-poor, meaning it doesn't have enough neutrons (or has too many protons) to be stable. Decay modes which increase the number of neutrons, decrease the number of protons, or both, would be favored. Both positron emission and electron capture accomplish this by converting a proton into a neutron. As a general rule, positron emission occurs with lighter isotopes and electron capture with heavier isotopes.

Isotopes that are neutron-rich, that have too many neutrons or not enough protons, lie above the belt of stability and tend to undergo beta emission because that decay mode converts a neutron into a proton.

A particular isotope may undergo a series of nuclear decays until finally a stable isotope is formed. For example, radioactive U-238 decays to stable Pb-206 in 14 steps, a majority of which are alpha emissions, as one might predict.

Nuclear Decay Calculations

A radioactive isotope may be unstable, but it is impossible to predict when a certain atom will decay. However, if a statistically large enough sample is examined, some trends become obvious. The radioactive decay follows first-order kinetics (see Chapter 14 on kinetics for a more in-depth discussion of first-order reactions and equations). If the number of radioactive atoms in a sample is monitored, it can be determined that it takes a certain amount of time for half the sample to decay; it takes the same amount of time for half the remaining sample to decay; and so on. The amount of time it takes for half the sample to decay is called the half-life of the isotope and is given the symbol $t_{1/2}$. The table below shows the percentage of radioactive isotope remaining versus half-life.

HALF-LIFE, $t_{1/2}$	PERCENT RADIOACTIVE ISOTOPE REMAINING
0	100
1	50
2	25
3	12.5
4	6.25
5	3.12
6	1.56
7	0.78
8	0.39
9	0.19
10	0.09

As a general rule, the amount of radioactivity at the end of 10 half-lives drops below the level of detection and the sample is said to be "safe."

Half-lives may be very short, 4.2×10^{-6} seconds for Po-213, or very long, 4.5×10^{9} years for U-238. The long half-lives of some waste products is a major problem with nuclear fission reactors. Remember, it takes 10 half-lives for the sample to be safe.

If only multiples of half-lives are considered, the calculations are very straightforward. For example, I-131 is used in the treatment of thyroid cancer and has a $t_{1/2}$ of 8 days. How long would it take to decay to 25% of its original amount? Looking at the chart, you see that 25% decay would occur at two half-lives or 16 days. However, since radioactive decay is not a linear process, you cannot use the chart to predict how much would still be radioactive at the end of 12 days or at some time (or amount) that is not associated with a multiple of a half-life. To solve these types of problems, one must use the mathematical relationships associated with first-order kinetics that were presented in the Kinetics chapter. In general, two equations are used:

$$(1)\ \ln [A]_t - \ln [A]_o$$

$$(2)\ t_{1/2} = \ln 2/k$$

In these equations, the ln is the natural logarithm; A_t is the amount of isotope radioactive at some time t; A_o is the amount initially radioactive; and k is the rate constant for the decay. If you know initial and final amounts and are looking for the half-life, you would use equation (1) to solve for the rate constant and then use equation (2) to solve for $t_{1/2}$.

For example: What is the half-life of a radioisotope that takes 15 min to decay to 90% of its original activity?

Using equation (1):

Using equation (1)

$$\ln 90/100 = -k(15\text{ min})$$
$$-0.1054 = -k(15\text{ min})$$
$$7.02 \times 10^{-3}\text{ min}^{-1} = k$$

Now equation (2):

$$t_{1/2} = \ln 2/7.02 \times 10^{-3}\text{ min}^{-1}$$
$$t_{1/2} = 0.693/7.02 \times 10^{-3}\text{ min}^{-1}$$
$$t_{1/2} = 98.7\text{ min}$$

If one knows the half-life and amount remaining radioactive, equation (2) can be used to calculate the rate constant k and equation (1) can then be used to solve for the time. This is the basis of C-14 dating, which is used to determine the age of objects that were once alive.

For example, suppose a wooden tool is discovered and its C-14 activity is determined to have decreased to 65% of the original. How old is the object?

The half-life of C-14 is 5,730 yr. Substituting this into equation (2):

$$5{,}730\text{ yr} = \ln 2/k$$
$$5{,}730\text{ yr} = 0.6931/k$$
$$k = 1.21 \times 10^{-4}\text{ yr}^{-1}$$

Substituting this rate constant into equation (1):

$$\ln 65/100 = -(1.21 \times 10^{-4}\text{ yr}^{-1})t$$
$$-0.4308 = -(1.21 \times 10^{-4}\text{ yr}^{-1})t$$
$$t = 3{,}600\text{ yr}$$

Mass–Energy Relationships

Whenever a nuclear decay or reaction takes place, energy is released. This energy may be in the form of heat and light, gamma radiation, or kinetic energy of the expelled particle and recoil of the remaining particle. This energy results from the conversion of a very small amount of matter into energy. (Remember that in nuclear reactions there is no conservation of matter, as in ordinary chemical reactions.) The amount of energy that is produced can be calculated by using Einstein's equation $E = mc^2$, where E is the energy produced, m is the mass converted into energy (the mass defect), and c is the speed of light. The amount of matter that is converted into energy is normally very small, but when it is multiplied by the speed of light (a very large number) squared, the amount of energy produced is very large.

For example: When 1 mol of U-238 decays to Th-234, 5×10^{-6} Kg of matter is converted to energy (the mass defect). To calculate the amount of energy released:

$$E = mc^2$$
$$E = (5 \times 10^{-6}\,\text{kg})(3.00 \times 10^{8}\,\text{m/s})^2$$
$$E = 5 \times 10^{11}\,\text{kg} \times \text{m}^2/\text{s}^2 = 5 \times 10^{11}\,\text{J}$$

If the mass is in kilograms, the answer will be in joules.

Common Mistakes to Avoid

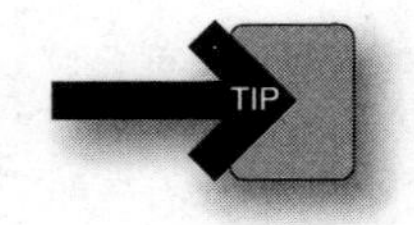

1. Make sure your answer is reasonable. Don't just write down the answer from your calculator.
2. Make sure your units cancel in your calculations, leaving the unit you want.
3. Make sure that in alpha, beta, gamma, and positron emissions the particle being emitted is on the right-hand side of the reaction arrow. In electron capture, the electron should be on the left side of the arrow.
4. In half-life problems, don't omit the minus sign. Watch your units.
5. In half-life problems, be sure to use the amount of isotope still radioactive as N_t and not the amount decayed.

› Review Questions

Use these questions to review the content of this chapter and practice for the AP Chemistry exam. Below are 6 multiple-choice questions similar to what you will encounter in Section I of the AP Chemistry exam. To make these questions an even more authentic practice for the actual exam, time yourself following the instructions provided.

Multiple-Choice Questions

Answer the following questions in 10 minutes. You may not use a calculator. You may use the periodic table and the equation sheet at the back of this book.

1. When $^{226}_{88}\text{Ra}$ decays, it emits 2 α particles, then a β particle, followed by an α particle. The resulting nucleus is:
   (A) $^{212}_{83}\text{Bi}$
   (B) $^{222}_{86}\text{Rn}$
   (C) $^{214}_{82}\text{Pb}$
   (D) $^{214}_{83}\text{Bi}$

2. The formation of $^{230}_{90}\text{Th}$ from $^{234}_{92}\text{U}$ occurs by:
   (A) electron capture
   (B) α decay
   (C) β decay
   (D) positron decay

3. Which of the following lists the types of radiation in the correct order of increasing penetrating power?
   (A) α, γ, β
   (B) β, α, γ
   (C) α, β, γ
   (D) β, γ, α

4. Which of the following statements is correct concerning β particles?

(A) They are electrons, with a mass number of zero and a charge of −1.
(B) They have a mass number of zero, a charge of −1, and are less penetrating than α particles.
(C) They are electrons with a charge of +1 and are less penetrating than α particles.
(D) They have a mass number of zero and a charge of +1.

5. An atom of $^{238}_{92}U$ undergoes radioactive decay by α emission. What is the product nuclide?

(A) $^{230}_{90}Th$
(B) $^{234}_{90}Th$
(C) $^{230}_{92}U$
(D) $^{230}_{91}Pa$

6. If 75% of a sample of pure $^{3}_{1}H$ decays in 24.6 years, what is the half-life of $^{3}_{1}H$?

(A) 24.6 yr
(B) 18.4 yr
(C) 12.3 yr
(D) 6.15 yr

Answers and Explanations for the Multiple-Choice Questions

1. D—The mass should be 226 − (4 + 4 + 0 + 4) = 214. The atomic number should be 88 − (2 + 2 − 1 + 2) = 83.

2. B—Mass difference = 234 − 230 = 4, and atomic number difference = 92 − 90 = 2. These correspond to an α particle.

3. C—Alpha particles are the least penetrating, and gamma rays are the most penetrating.

4. A—In nuclear reactions, the mass of a β particle is treated as 0, with a charge of −1. Electrons and β particles are the same thing.

5. B—Mass number = 238 − 4 = 234, and atomic number = 92 − 2 = 90.

6. C—After one half-life, 50% would remain. After another half-life this would be reduced by one-half to 25%. The total amount decayed is 75%. Thus, 24.6 years must be two half-lives of 12.3 years each.

› Rapid Review

- Know the five naturally occurring decay modes:

  1. Alpha emission, in which a helium nucleus, $^{4}_{2}He$, is emitted from the nucleus.
  2. Beta emission in which an electron, $^{0}_{-1}e$, is emitted from the nucleus. This is due to the conversion of a neutron into a proton plus the beta particle.
  3. Gamma emission, in which high-energy electromagnetic radiation is emitted from the nucleus. This commonly accompanies the other types of radioactive decay. It is due to the conversion of a small amount of matter into energy.
  4. Positron emission, in which a positron, $^{0}_{+1}e$, a particle having the same mass as an electron but a positive charge, is emitted from the nucleus. This is due to a proton converting into a neutron and the positron.
  5. Electron capture, in which an inner-shell electron is captured by a proton in the nucleus with the formation of a neutron. X-rays are emitted as the electrons cascade down to fill the vacancy in the lower energy level.

- Know that nuclear stability is best related to the neutron-to-proton ratio (n/p), which starts at about 1/1 for light isotopes and ends at about 1.5/1 for heavier isotopes with atomic numbers up to 83. All isotopes of atomic number greater than 84 are unstable and will commonly undergo alpha decay. Below atomic number 84, neutron-poor isotopes will probably undergo positron emission or electron capture, while neutron-rich isotopes will probably undergo beta emission.

- Know that the half-life, $t_{1/2}$, of a radioactive isotope is the amount of time it takes for one-half of the sample to decay. Know how to use the appropriate equations to calculate amounts of an isotope remaining at any given time, or use similar data to calculate the half-life of an isotope.

- Know how to use Einstein's equation $E = mc^2$ to calculate the amount of energy produced from a mass defect (the amount of matter that was converted into energy).

CHAPTER 18

Organic Chemistry

SPECIAL NOTE

Although organic chemistry is not specifically tested on the AP Chemistry exam, an overview of the subject will help you recognize and understand organic structures that appear in relation to other topics on the exam. In addition, this material is taught by some high school and college teachers. This is an enrichment chapter that should be considered a bonus chapter to this book.

IN THIS CHAPTER

Summary: Organic chemistry is the study of the chemistry of carbon. Almost all the compounds containing carbon are classified as organic compounds. Only a few—for example, carbonates and cyanides—are classified as inorganic. It used to be thought that all organic compounds had to be produced by living organisms, but this idea was proven wrong in 1828 when German chemist Friedrich Wöhler produced the first organic compound from inorganic starting materials. Since that time, chemists have synthesized many organic compounds found in nature and have also made many never found naturally. It is carbon's characteristic of bonding strongly to itself and to other elements in long, complex chains and rings that gives carbon the ability to form the many diverse and complex compounds needed to support life.

Keywords and Equations

No keywords or equations specific to this chapter are listed on the AP exam.

Alkanes

Alkanes are members of a family of organic compounds called hydrocarbons, compounds of carbon and hydrogen. These hydrocarbons are the simplest of organic compounds, but are extremely important to our society as fuels and raw materials for chemical industries. We heat our homes and run our automobiles through the combustion (burning) of these hydrocarbons. Paints, plastics, and pharmaceuticals are often made from hydrocarbons. **Alkanes** are hydrocarbons that contain only single covalent bonds within their molecules. They are called saturated hydrocarbons because they are bonded to the maximum number of other atoms. These alkanes may be straight-chained hydrocarbons, in which the carbons are sequentially bonded; branched hydrocarbons, in which another hydrocarbon group is bonded to the hydrocarbon "backbone"; or they may be cyclic, in which the hydrocarbon is composed entirely or partially of a ring system. The straight-chained and branched alkanes have the general formula of C_nH_{2n+2}, whereas the cyclic alkanes have the general formula of C_nH_{2n}. The *n* stands for the number of carbon atoms in the compound. The first 10 straight-chained alkanes are shown in Table 18.1.

There can be many more carbon units in a chain than are shown in Table 18.1, but these are enough to allow us to study alkane nomenclature—the naming of alkanes.

Alkane Nomenclature

The naming of alkanes is based on choosing the longest carbon chain in the structural formula, then naming the hydrocarbon branches while indicating onto which carbon that branch is attached. Here are the specific rules for naming simple alkanes:

1. Find the continuous carbon chain in the compound that contains the most carbon atoms. This will provide the base name of the alkane.
2. This base name will be modified by adding the names of the branches (substituent groups) in front of the base name. Alkane branches are named by taking the name of the alkane that contains the same number of carbon atoms, dropping the *-ane* ending and adding *-yl*. Methane becomes methyl, propane becomes propyl, etc. If there is more than one branch, list them alphabetically.

Table 18.1 The First Ten Straight-Chained Alkanes

NAME	MOLECULAR FORMULA	STRUCTURAL FORMULA
methane	CH_4	CH_4
ethane	C_2H_6	CH_3-CH_3
propane	C_3H_8	CH_3-CH_2-CH_3
butane	C_4H_{10}	CH_3-CH_2-CH_2-CH_3
pentane	C_5H_{12}	CH_3-CH_2-CH_2-CH_2-CH_3
hexane	C_6H_{14}	CH_3-CH_2-CH_2-CH_2-CH_2-CH_3
heptane	C_7H_{16}	CH_3-CH_2-CH_2-CH_2-CH_2-CH_2-CH_3
octane	C_8H_{18}	CH_3-CH_2-CH_2-CH_2-CH_2-CH_2-CH_2-CH_3
nonane	C_9H_{20}	CH_3-CH_2-CH_2-CH_2-CH_2-CH_2-CH_2-CH_2-CH_3
decane	$C_{10}H_{22}$	CH_3-CH_2-CH_2-CH_2-CH_2-CH_2-CH_2-CH_2-CH_2-CH_3

3. The position where a particular substituent is attached to the chain is indicated by a location number. These numbers are assigned by consecutively numbering the carbons of the base hydrocarbon, starting at one end of the hydrocarbon chain. Choose the end that will result in the lowest sum of location numbers for the substituent groups. Place this location number in front of the substituent name and separate it from the name by a hyphen (for example, 2-methyl).

4. Place the substituent names with their location numbers in front of the base name of the alkane in alphabetical order. If there are identical substituents (two methyl groups, for example), give the location numbers of each, separated by commas using the common Greek prefixes (di-, tri-, tetra-, etc.) to indicate the number of identical substituent groups (i.e., 2,3-dimethyl). These Greek prefixes are not considered in the alphabetical arrangement.

5. The last substituent group becomes a part of the base name as a prefix.

Studying Figures 18.1 and 18.2 may help you learn the naming of substituted alkanes.

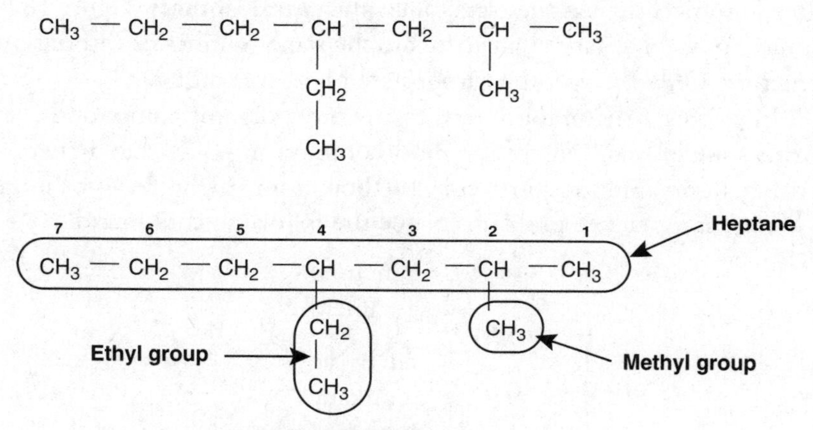

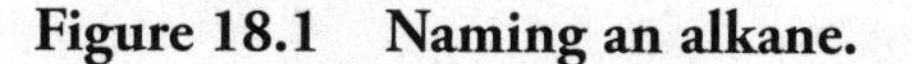

Figure 18.1 Naming an alkane.

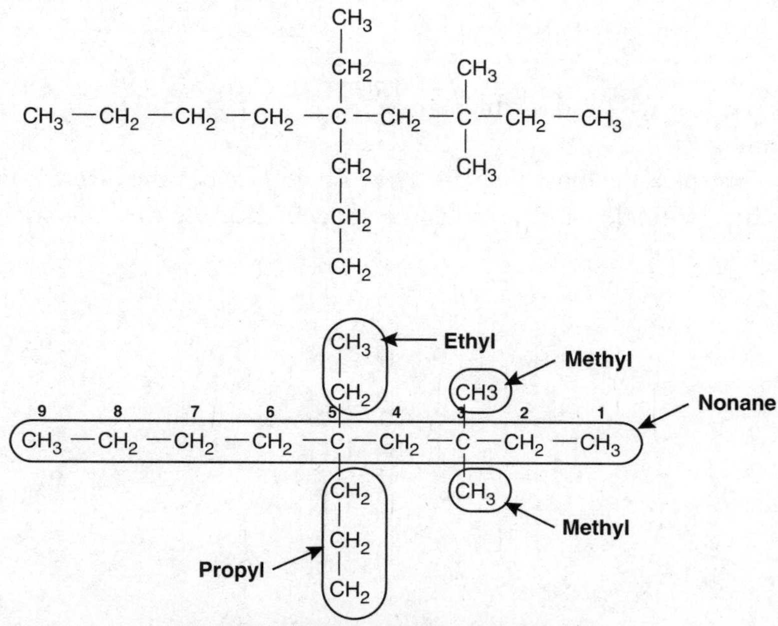

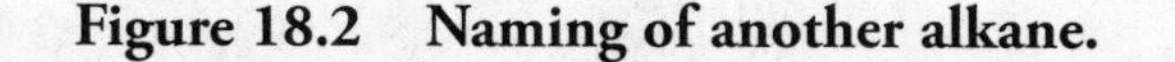

Figure 18.2 Naming of another alkane.

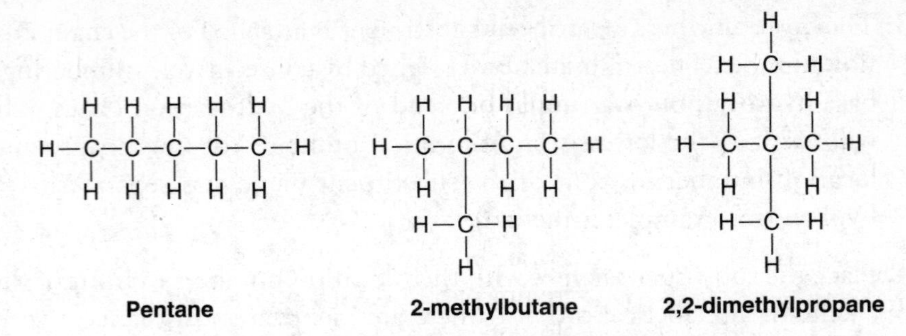

Figure 18.3 Structural isomers of C_5H_{12}.

Structural Isomerism

Compounds that have the same molecular formulas but different structural formulas are called isomers. With hydrocarbons, this applies to a different arrangement of the carbon atoms. Isomers such as these are called **structural isomers.** Figure 18.3 shows the structural isomers of C_5H_{12}. Note that there are the same number of carbons and hydrogens in each structure. Only the way the carbons are bonded is different.

In writing structural isomers, or any other organic compounds, remember that **carbon forms four bonds.** One of the most common mistakes that a chemistry student makes is writing an organic structure with a carbon atom having fewer or more than four bonds.

Here is a practice problem. Name the following compound:

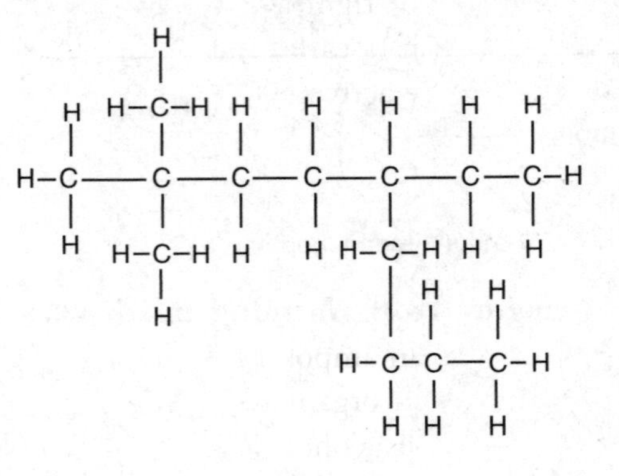

Answer: 5-ethyl-2,2-dimethylnonane
Solution:

First, pick the longest chain. This is bold-faced in the diagram below. The carbons are attached by single bonds, so this is an alkane. Because the longest chain has nine carbons, it is a nonane.

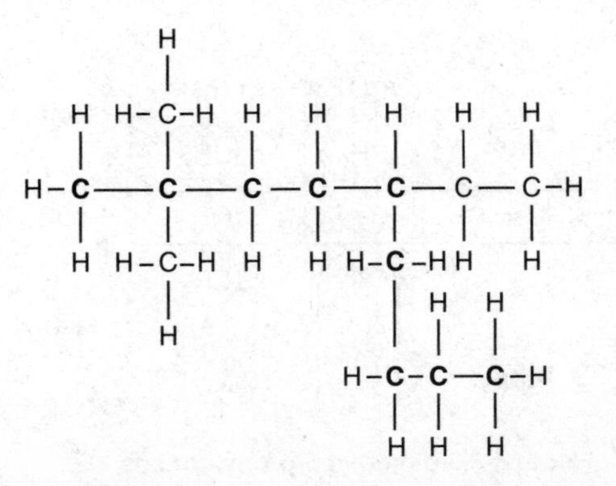

Next, the longest chain should be numbered from one end to the other with the lowest number(s) going to the branches. For the preceding example the numbering of the chain (bold-face carbon atoms) would be:

```
1 2 3 4 5
        6
        7 8 9
```

Once these numbers have been assigned, do not alter them later.

All carbon atoms that are not part of the nine-atom main chain are branches. Branches have *-yl* endings. It may help you to circle the carbon atoms belonging in the branches. In the above example, there are three branches. Two consist of only one carbon and are called methyl groups. The remaining branch has two carbons, so it is an ethyl group. The branches are arranged alphabetically. If there is more than one of a particular type, use a prefix (*di*-, *tri*-, *tetra*-, etc.). The two methyl groups are designated dimethyl. The position of each branch is indicated with a number already determined for the main chain. Each branch must get its own number, even if it is identical to one already used.

In the above example this gives: 5-ethyl-2,2-dimethylnonane

a. ethyl before methyl (alphabetical—prefixes are ignored)
b. two methyl groups = dimethyl
c. three branches = three numbers

Numbers are separated from other numbers by commas, and numbers are separated from letters by a hyphen.

Another type of isomerism is optical isomerism. These molecules are capable of rotating light to either the left or right and are said to be optically active. The presence of an asymmetric or chiral carbon (a carbon atom with four different groups attached to it) will make a compound optically active.

Common Functional Groups

If chemistry students had to learn the properties of each of the millions of organic compounds, they would face an impossible task. Luckily, chemists find that having certain arrangements of atoms in an organic molecule causes those molecules to react in a similar fashion. For example, methyl alcohol, CH_3-OH, and ethyl alcohol, CH_3-CH_2-OH, undergo the same types of reactions. The –OH group is the reactive part of these types of molecule. These reactive groups are called functional groups. Instead of learning the properties of individual molecules, one can simply learn the properties of functional groups.

In our study of the simple hydrocarbons, there are only two functional groups. One is a carbon-to-carbon double bond. Hydrocarbons that contain a carbon-to-carbon double bond are called alkenes. Naming alkenes is very similar to naming alkanes. The major difference is that the carbon base has an *-ene* ending instead of the *-ane* ending. The carbon backbone of the base hydrocarbon is numbered so the position of the double bond has the lowest location number.

The other hydrocarbon functional group is a carbon-to-carbon triple bond. Hydrocarbons that contain a triple bond are called alkynes. Alkynes use the *-yne* ending on the base hydrocarbon. The presence of a double or triple bond make these hydrocarbons unsaturated.

The introduction of other atoms (N, O, Cl, etc.) to organic compounds gives rise to many other functional groups. The major functional groups are shown in Table 18.2, on the next page.

Table 18.2 Common Functional Groups

FUNCTIONAL GROUP	COMPOUND TYPE	SUFFIX OR PREFIX OF NAME	EXAMPLE	SYSTEMATIC NAME (COMMON NAME)
>C=C<	alkene	-ene	$H_2C{=}CH_2$	ethene (ethylene)
—C≡C—	alkyne	-yne	H—C≡C—H	ethyne (acetylene)
—C—O—H	alcohol	-ol	$H_3C{-}O{-}H$	methanol (methyl alcohol)
—C—X: (X=halogen)	haloalkane	halo-	$H_3C{-}Cl$	chloromethane (methyl chloride)
—C—N—	amine	-amine	$H_3C{-}CH_2{-}NH_2$	ethylamine
—C(=O)—H	aldehyde	-al	$H_3C{-}C(=O){-}H$	ethanal (acetaldehyde)
—C—C(=O)—C—	ketone	-one	$H_3C{-}C(=O){-}CH_3$	propanone (acetone)
—C(=O)—O—H	carboxylic acid	-oic acid	$H_3C{-}C(=O){-}O{-}H$	ethanoic acid (acetic acid)
—C(=O)—O—C—	ester	-oate	$H_3C{-}C(=O){-}O{-}CH_3$	methyl ethanoate (methyl acetate)
—C(=O)—N—	amide	-amide	$H_3C{-}C(=O){-}NH_2$	ethanamide (acetamide)

Macromolecules

As we mentioned in the introduction to this chapter, carbon has the ability to bond to itself in long and complex chains. These large molecules, called **macromolecules**, may have molecular masses in the millions. They are large, complex molecules, but most are composed of repeating units called **monomers**. Figure 18.4 shows two macromolecules, cellulose and nylon, and indicates their repeating units.

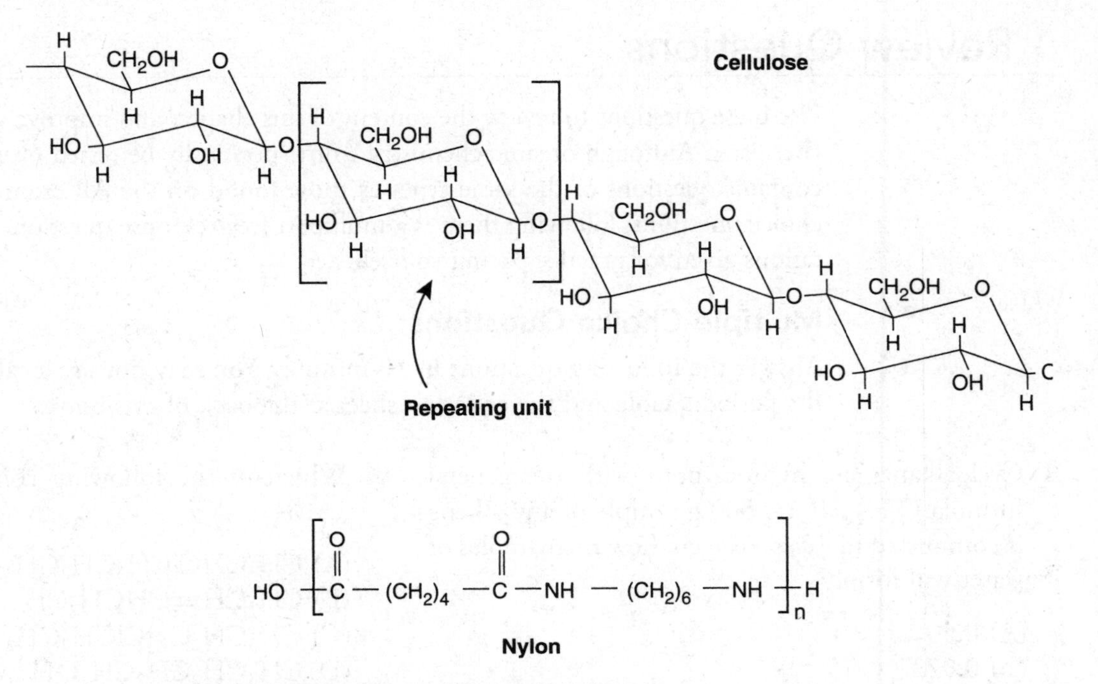

Figure 18.4 Two macromolecules.

Macromolecules are found in nature. Cellulose, wool, starch, and DNA are but a few of the macromolecules that occur naturally. Carbon's ability to form these large, complex molecules is necessary to provide the diversity of compounds needed to make up a tree or a human being. But many of the useful macromolecules that we use every day are created in the lab and industrial complex by chemists. Nylon, rayon, polyethylene, and polyvinyl chloride are all synthetic macromolecules. They differ by which repeating units (monomers) are joined together in the polymerization process. Our society has grown to depend on these plastics, these synthetic fabrics. The complexity of carbon compounds is reflected in the complexity of our modern society.

Experiments

Any experiment would probably apply the concepts of organic chemistry in a synthesis situation.

Common Mistakes to Avoid

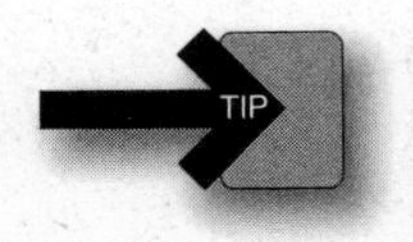

1. When writing organic formulas, make sure that every carbon has four bonds.
2. When naming alkanes, make sure to number the carbon chain so the sum of all location numbers is as small as possible.

3. When naming branched alkanes, be sure to consider the branches when finding the longest carbon chain. The longest chain isn't always the one in which the carbon atoms all lie in a horizontal line.
4. In naming identical substituents on the longest carbon chain, be sure to use repeating location numbers, separated by commas (2,2-dimethyl).
5. **Be sure that every carbon has four bonds!**

› Review Questions

Use these questions to review the content of this chapter and improve your understanding of chemistry. Although organic chemistry won't specifically be tested on the exam, this section contains questions of the same types as those found on the AP exam. First are 5 multiple-choice questions; following them is a multipart free-response question. Follow the time limitations given for practice pacing yourself.

Multiple-Choice Questions

Answer the following questions in 10 minutes. You may not use a calculator. You may use the periodic table and the equation sheet at the back of this book.

1. Cycloalkanes are hydrocarbons with the general formula C_nH_{2n}. If a 0.500 g sample of any alkene is combusted in excess oxygen, how many moles of water will form?

(A) 0.50
(B) 0.072
(C) 0.036
(D) 1.0

2.
```
       OH
       |
CH3–CH–CH2–CH3
```

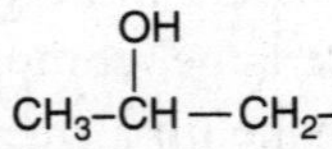

The organic compound shown above would be classified as:

(A) an organic base
(B) an ether
(C) an alcohol
(D) an aldehyde

3.
```
       O
       ||
HO—C—CH2–CH3
```

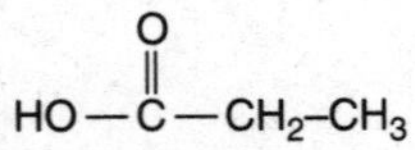

The above compound would be classified as:

(A) an aldehyde
(B) a ketone
(C) an ester
(D) a carboxylic acid

4. Which of the following compounds is optically active?

(A) $CH_3CHClCH_2CH_2CH_3$
(B) $CH_3CH{=}CHCH_2CH_3$
(C) $CH_3CH_2CHClCH_2CH_3$
(D) $CH_3CH_2CH_2CH_2OH$

5. A carboxylic acid may be represented as:

(A) ROH
(B) RCHO
(C) R-O-R′
(D) RCOOH

Answers and Explanations for the Multiple-Choice Questions

1. C—The general formula, C_nH_{2n}, means that 1 mol of H_2O will form per mole of empirical formula unit, regardless of the value of *n*. The moles of water formed are the mass of the alkene divided by the empirical formula mass. (0.500 g alkene)(1 mol alkene/14 g alkene)(1 mol H_2O/mol alkene) = 0.036 mol.

2. C—Organic bases are, in general, amines (contain N). An ether would have an oxygen single-bonded to two carbons (R groups). An aldehyde has oxygen double-bonded to a carbon at the end of a chain. Aldehydes (RCHO) and alcohols (ROH) are often confused because of the similarity in their general formulas. Ketones have oxygen double-bonded to a carbon not at the end of a chain.

3. D—Based on classification of organic compounds.

4. A—Redrawing the structures may help you to recognize the correct answer. An optical isomer must be a carbon atom with four *different* groups attached to it. For A the groups on the second carbon are: $CH_3—$, H, Cl, and $—CH_2CH_2CH_3$. Answer C is misleading. It is similar to A, but two of the groups, the $—CH_2CH_3$ groups, are the same.

5. D—A = alcohol, B = aldehyde, C = ether

Free-Response Question

You have 20 minutes to answer the following question. You may use a calculator and the tables in the back of the book.

Question 1

The alkane hexane, C_6H_{14}, has a molecular mass of 86.17 g mol^{-1}.

(a) Like all hydrocarbons, hexane will burn. Write a balanced chemical equation for the complete combustion of hexane. This reaction produces gaseous carbon dioxide, CO_2, and water vapor, H_2O.

(b) The complete combustion of 10.0 g of hexane produces 487 kJ. What is the molar heat of combustion (ΔH) of hexane?

(c) Determine the pressure exerted by the carbon dioxide formed when 5.00 g of hexane is combusted. Assume the carbon dioxide is dry and stored in a 20.0 L container at 27°C.

(d) Hexane, like most alkanes, may exist in different isomeric forms. The structural formula of one of these isomers is pictured below. Draw the structural formula of any two other isomers of hexane. Make sure all carbon atoms and hydrogen atoms are shown.

```
    H  H  H  H  H  H
    |  |  |  |  |  |
 H—C—C—C—C—C—C—H
    |  |  |  |  |  |
    H  H  H  H  H  H
```

Answer and Explanation for the Free-Response Question

Note that, while organic chemistry is not an AP topic, all of the materials in these questions depend upon basic AP chemistry knowledge, which is why this chapter can be valuable.

(a) $2\ C_6H_{14} + 19\ O_2 \rightarrow 12\ CO_2 + 14\ H_2O$

Give yourself 2 points for the answer shown above, or for the coefficients: 1, 9/2, 6, and 7. Give yourself 1 point if you have one or more, not all, of the elements balanced.

(b) (–487 kJ/10.0 g hexane)(86.17 g hexane/mol hexane) $= -4.20 \times 10^3$ kJ/mol

Give yourself 2 points for the above setup and correct answer (this requires a negative sign in the answer). If the setup is partially correct, give yourself 1 point.

(c) The ideal gas equation should be rearranged to the form $P = nRT/V$.

$$\text{Moles} = n = (5.00\ \text{g}\ C_6H_{14})\left(\frac{1\ \text{mol}\ C_6H_{14}}{86.17\ \text{g}\ C_6H_{14}}\right)\left(\frac{12\ \text{mol}\ CO_2}{2\ \text{mol}\ C_6H_{14}}\right) = 0.3481\ \text{mol}\ CO_2$$

This answer has an extra significant figure. The mole ratio should match the one given in your balanced equation. You will not be penalized again for an incorrectly balanced equation.

You will lose a point if you do not include a hexane-to-CO_2 conversion.

$R = 0.08206$ L atm $\text{mol}^{-1}\ \text{K}^{-1}$ (This value is in your test booklet.)

$$T = 27°\text{C} + 273 = 300.0\ \text{K}$$

In this case, there is no penalty if you forget to use the Kelvin temperature.

$$V = 20.0\ \text{L}$$

$P = (0.3481\ \text{mol}\ CO_2)(0.08206\ \text{L atm mol}^{-1}\ \text{K}^{-1})(300.0\ \text{K})/(20.0\ \text{L}) = 0.429\ \text{atm}$

Give yourself 2 points for the correct setup and answer. Give yourself 1 point if you did everything correctly, except the mole ratio or the Kelvin conversion.

(d) You may need to redraw one or more of your answers to match the answers shown below.

Give yourself 1 point for each correct answer, with a 2-point maximum. There are no bonus points for additional answers.

```
    H H H                      H H H
     \|/                        \|/
  H  C  H  H  H           H  H  C  H  H
  |  |  |  |  |           |  |  |  |  |
H—C——C——C——C——C—H       H—C——C——C——C——C—H
  |  |  |  |  |           |  |  |  |  |
  H  H  H  H  H           H  H  H  H  H

    H H H                         H H H
     \|/                           \|/
  H  C  H  H                  H  H  C  H
  |  |  |  |                  |  |  |  |
H—C——C——C——C—H              H—C——C——C——C—H
  |  |  |  |                  |  |  |  |
  H  C  H  H                  H  C  H  H
    /|\                         /|\
   H H H                       H H H
```

These compounds are 2-methylpentane, 3-methylpentane, 2,2-dimethylbutane, and 2,3-dimethylbutane, respectively. These four, along with the original *n*-hexane, are the only isomers. If you think you have another isomer, you have simply redrawn one of these. Try naming your answer and see if it matches one of these names.

Total your points. The maximum is 8 points. Subtract one point if all your answers do not have the correct number of significant figures.

❯ Rapid Review

- Organic chemistry is the chemistry of carbon and its compounds.
- Hydrocarbons are organic compounds of just carbon and hydrogen atoms.
- Alkanes are hydrocarbons in which there are only single bonds.
- Alkanes are named in a very systematic way. Review the rules for naming alkanes.
- Isomers are compounds that have the same molecular formulas but different structural formulas. Review the writing of the various structural isomers of alkanes. **Make sure that each carbon atom has four bonds.**
- A functional group is a group of atoms that is the reactive part of the molecule. Review the general functional groups.
- Macromolecules are large molecules that may have molecular masses in the millions. Macromolecules are generally composed of repeating units called monomers.

CHAPTER 19

Experimental Investigations

IN THIS CHAPTER

Summary: The free-response portion of the AP exam will contain a question concerning an experiment, and there may also be a few multiple-choice questions on one or more of these experiments. This chapter reviews the basic experiments that the AP Exam Committee believes to be important. You should look over all of the experiments in this chapter and pay particular attention to any experiments you did not perform. In some cases you may find, after reading the description, that you did a similar experiment. Not every AP class does every experiment, but any of these experiments may appear on the AP exam.

The free-response questions on recent exams have been concerned with the equipment, measurements, and calculations required. In some cases, sources of error are considered. To answer the question completely, you will need an understanding of the chemical concepts involved.

To discuss an experiment, you must be familiar with the equipment needed. In the keywords section at the beginning of this chapter is a complete list of equipment for the experiments (see also Figure 19.1). Make sure you are familiar with each item. You may know an item by a different name, or you may need to talk to your teacher to get additional information concerning an item.

In some cases, the exam question will request a list of the equipment needed, while in other cases you will get a list from which to choose the items you need. Certain items appear in many experiments. These include the analytical balance, beakers, support stands, pipets, test tubes, and Erlenmeyer flasks. Burets, graduated cylinders, clamps, desiccators, drying ovens, pH meters, volumetric flasks, and thermometers are also commonly used. If you are not sure what equipment to choose, these serve as good guesses. Most of the remaining equipment appears in three or fewer experiments.

You will need to know the basic measurements required for the experiment. For example, you may need to measure the initial and final temperatures. Do not make the mistake of saying you measure the change in temperature. You ***calculate*** the change in temperature from your measured initial and final temperatures. You do not need to give a lot of detail when listing the required measurements, but you need to be very specific in what you ***measure***. Many students have lost exam points for not clearly distinguishing between measured and calculated values.

The basic calculations fall into two categories. Simple calculations, such as the change in temperature or the change in volume, are the easiest to forget. Simple calculations may also include mass-to-mole conversions. The other calculations normally involve entering values into one of the equations given at the beginning of the previous chapters of this book.

Keywords and Equations

Pay particular attention to the specific keywords and equation in the chapters associated with the individual experiments.

$A = abc$ (A = absorbance; a = molar absorbtivity; b = path length; c = concentration)

analytical balance
barometer
beaker(s)
buret
burner
calorimeter
capillary tubes
centrifuge
clamp
crucible and cover
cuvettes
desiccator
drying oven
electrodes
Erlenmeyer flask
evaporating dish
filter crucibles and adapters
filter flasks
forceps
funnel
graduated cylinder
hot plate
ice
ion exchange resin or silica gel
Meker burner
mortar and pestle
pH meter
pipet
power supply (battery)
Pt or Ni test wire
rubber tubing
spectrophotometer
stirrer
stopwatch
support stand
test-tube rack
test tube(s)
thermometer
tongs
triangle crucible support
voltmeter
volumetric flask
wash bottle
watch glass
water bath
wire gauze

Experiment 1: Spectroscopy

Synopsis

Specific experiments that are performed in this investigation are an introduction to the field of spectroscopy. They are designed to demonstrate the relationship between the amount of light absorbed by some solutions and their concentrations. Light of a specific wavelength is passed through both the solvent and a sample. The amount of light transmitted by the solvent is subtracted from the amount of light transmitted by the sample. If you made a

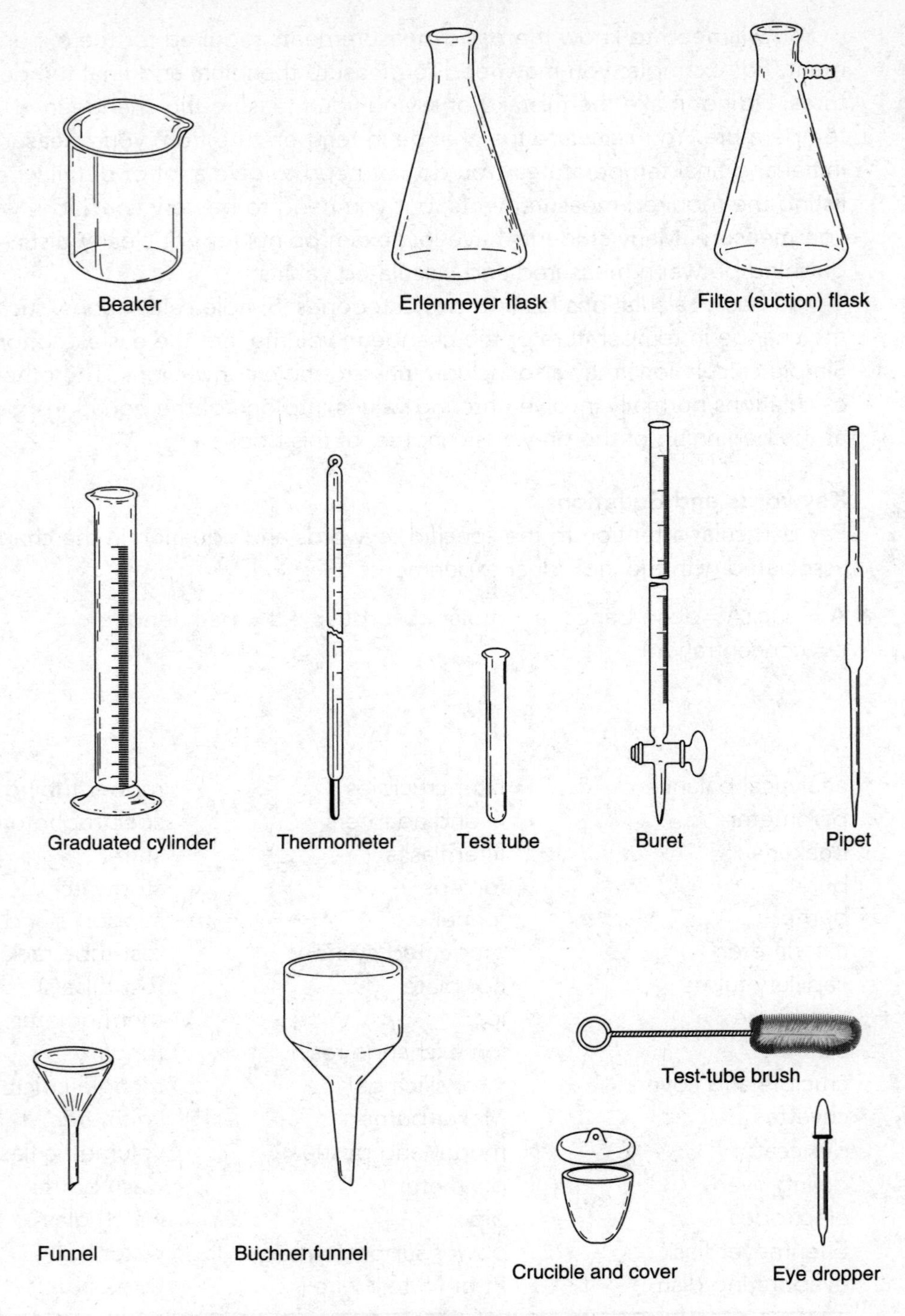

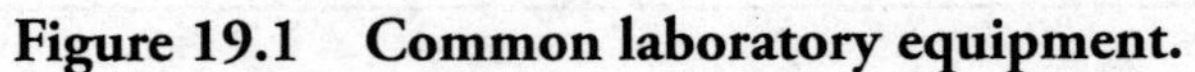
Figure 19.1 Common laboratory equipment.

Crucible tongs

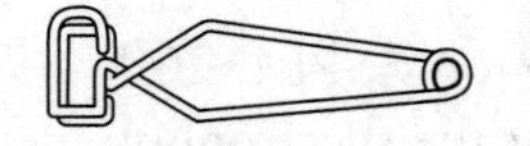

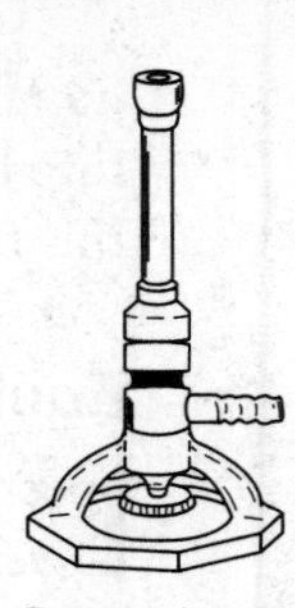

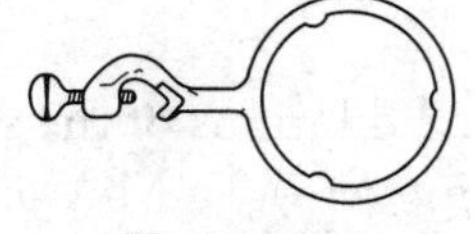

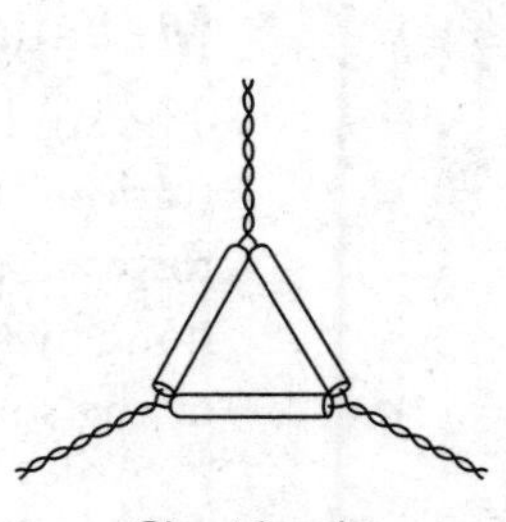

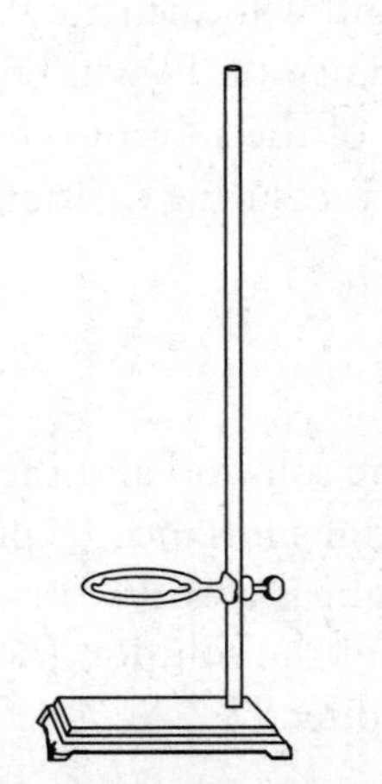

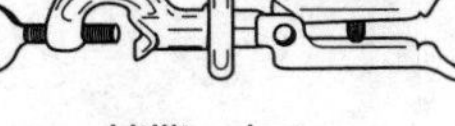

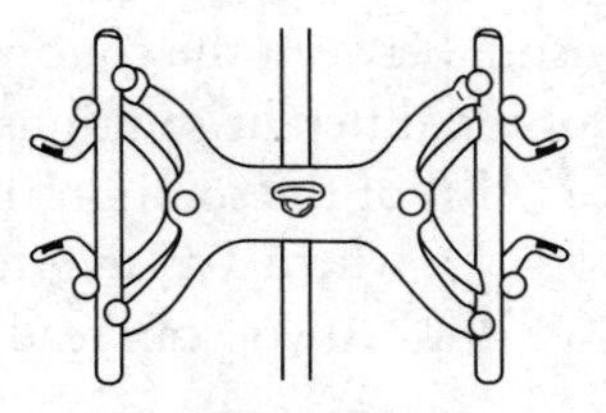

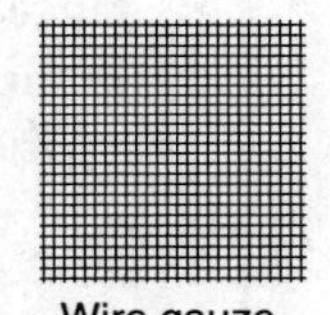

Evaporating dish

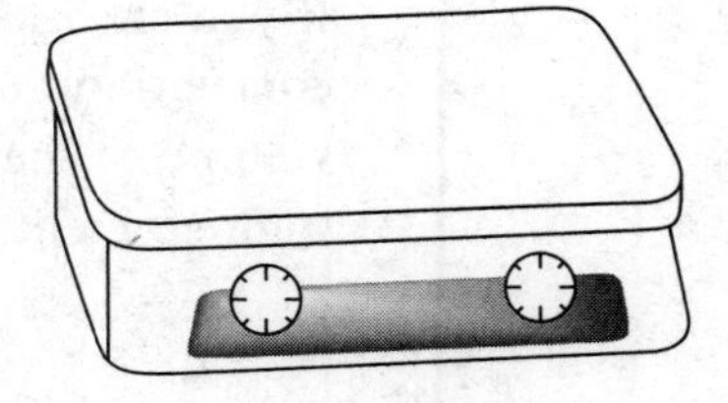

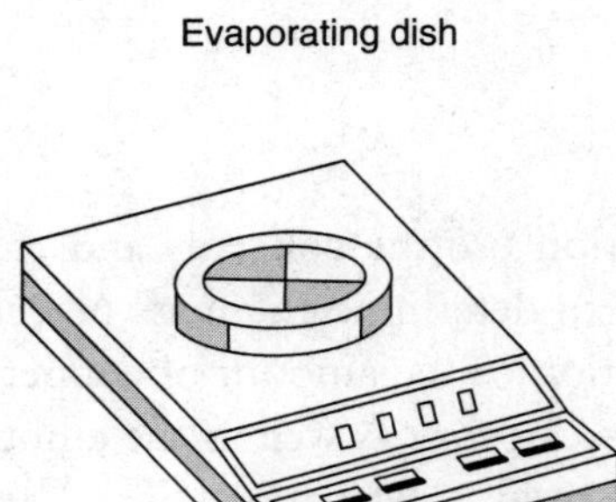

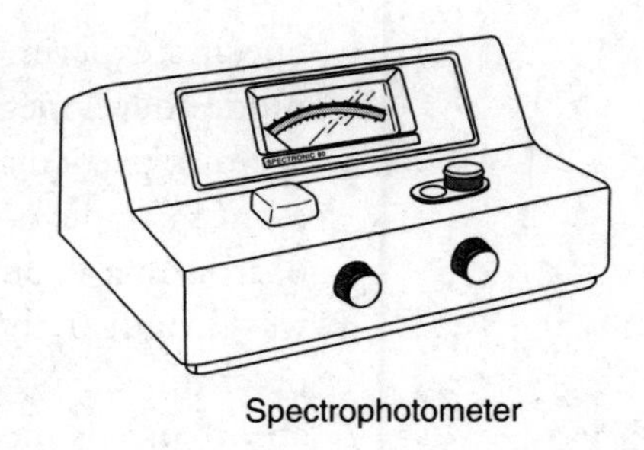

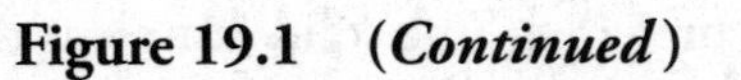

Figure 19.1 ***(Continued)***

number of measurements at different concentrations, you could create a graphical relationship between the amount of light absorbed and the concentration of the solution. By using this relationship, you could determine the concentration of an unknown solution.

Equipment

Spectrophotometer (commonly SPEC 20)
Cuvettes (sample tubes for the spectrophotometer)
Stock solutions (of known concentrations) of the solute (commonly some dye)
Solution of unknown concentration (may be a household substance)
Assorted glassware, including volumetric glassware

Measurements

The student will make several dilutions of the stock solution and will calculate the concentration of each dilution (using $M_1V_1 = M_2V_2$). The transmittance (%T) will be measured for each solution (remembering to subtract the transmittance of the solvent—this may be done by adjusting the spectrophotometer to 100% T and then measuring the transmittance of the solution).

Calculations

To determine the relationship between the concentration of the solution and the transmittance, plot the molarity of the different solutions versus the transmittance (expressed as a decimal). The absorbance (Abs) of the solution (how much light is absorbed) is calculated by the formula Abs = −log (T), where T is the transmittance of the solution (not the percent transmittance). On a SPEC 20 you can read absorbance directly.

Comments

If you are asked for the mass of the solute in the unknown, you first determine its molar concentration using your spectroscopy data. Then, using the molar concentration, the volume of the solution, and the molar mass of the solute, you can calculate the grams of solute present in the sample.

Experiment 2: Spectrophotometry

Synopsis

Specific experiments that are performed in this investigation use the concepts and techniques developed in Experiment 1: Spectroscopy in order to determine the mass percentage of a particular substance in a solid sample. Determination of the amount of copper in a brass sample is a common experiment that is used in this category as well as the amount of iron in a vitamin pill. First, the "best" wavelength to be used is determined. The "best" wavelength is the one that gives the maximum absorbance of the chemical species being determined. Next, solutions of the solute being determined are prepared and their absorbance is measured using a spectrophotometer. A plot of absorbance versus concentration (Beer's law) is prepared. The solid sample is dissolved and diluted to a certain volume. The absorbance of a portion of this sample is measured and its concentration is determined using the graph. From this information the mass of the substance can be found. Using this mass information and the mass of the sample allows you to calculate the mass percentage of the substance in the sample.

Equipment

Spectrophotometer (commonly SPEC 20)
Cuvettes (sample tubes for the spectrophotometer)
Stock solution (known concentration) of the solute
Sample to be analyzed
Assorted glassware, including volumetric glassware

Measurements

The student will make several dilutions of the stock solution (solution of known concentration of the substance being determined) and will calculate the concentration of each dilution (using $M_1V_1 = M_2V_2$). The absorbance of one of the stock solutions is measured at a number of wavelengths (generally 400–700 nm in 10 to 20 nanometer increments) using a spectrophotometer. The data of absorbance versus wavelength is plotted, and the wavelength that gives the maximum absorbance is chosen to be used for the rest of the experiment. The absorbance of each of the dilutions is measured. A plot of absorbance versus concentration (Beer's law plot) is prepared either by hand or using a spreadsheet. The solid sample is dissolved (if it is copper, this will require the use of nitric acid) and diluted to a certain volume. The concentration of that solution is determined using the Beer's law plot. Using the concentration of the solution and the solution's volume, you can calculate the moles and then grams of the substance. Using the initial mass of the sample, you can finally calculate the mass percentage of the substance in the sample.

Calculations

You can determine the concentrations of the diluted stock solution by using the dilution equation ($M_1V_1 = M_2V_2$). The mass percentage is calculated by:

$$(\text{grams substance}/\text{grams sample}) \times 100\%.$$

Comments

If you are doing a brass analysis for percentage of copper, you will dissolve the brass in concentrated nitric acid. Be extremely careful. The nitric acid is corrosive and the NO gas that is produced is toxic. On the AP exam be sure to stress safety if you are describing this process.

Experiment 3: Gravimetric Analysis

Synopsis

Specific experiments that are performed in this investigation use determination of the mass of a specific substance in a sample by precipitation, drying, and weighing. A common experiment done in this category is determination of the hardness of a water sample. The hardness of a water sample is related to the amounts of calcium, magnesium, and iron ions in solution. These ions may be precipitated as the carbonate salts. For simplicity's sake, hard-water samples are commonly prepared with only one of these ions, generally calcium. The carbonate salt is precipitated, separated from the solution by suction filtration, and dried in a drying oven. The mass of the dry salt is determined, and the water sample hardness is calculated as mg calcium carbonate per liter of water sample.

Equipment

Various salt solutions of known concentration
Analytical balance
Drying oven
Suction filtration apparatus
Büchner funnel
Filter paper
Aspirator
Ring stands
Assorted glassware, including volumetric glassware

Measurements

The student will make several measurements in gravimetric analysis, especially mass and volume determinations.

Calculations

If a water hardness analysis is being done, the grams of calcium carbonate per milliliter of water sample are initially calculated. This value is then converted to milligrams of calcium carbonate per liter of water sample (hardness) using appropriate conversions.

Comments

All measurements must be done accurately, especially the mass and volume measurements.

Experiment 4: Titration

Synopsis

In the titration procedure, the concentration of an acid is determined by adding small quantities at a time of a base of known concentration until the point at which the moles of base equal the moles of acid present (the equivalence point). It is possible to do a titration by adding small amounts of a solution of a known concentration of acid to determine the concentration of a base solution. Many times this neutralization point cannot be determined unaided, so an indicator or a pH meter is used. The point at which a color change happens with the indicator or an abrupt change in pH occurs with the pH meter is called the endpoint of the titration. Knowing the volume of the unknown acid, the concentration of the base, and the number of milliliters it took to reach the endpoint allows you to calculate the concentration of the unknown acid. The concentration of an unknown base can be determined by titration with an acid of known concentration (Figure 19.2).

Equipment

Burets
Erlenmeyer flasks
Pipets
Acid-base indicators
pH meter
Base or acid solution of known concentration
Assorted glassware

Measurements

You will be placing a required volume of the unknown acid (or base) solution into the Erlenmeyer flask with a pipet. The buret will be filled with the base (or acid) solution.

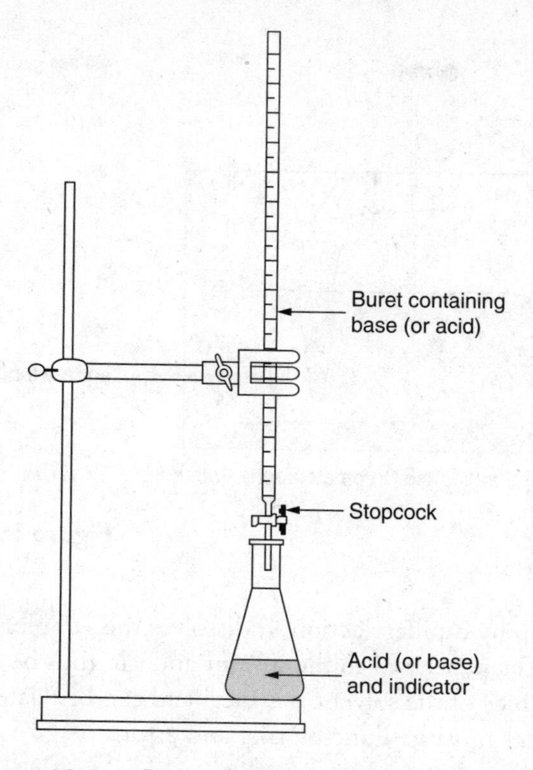

Figure 19.2 General acid–base titration setup

Be sure to record your initial volume. You will add small amounts of base drop by drop until the indicator changes color. Record the final volume. The final volume reading minus the initial volume reading is the volume of base added.

Calculations

The calculation of the concentration of the base is essentially a stoichiometry reaction. Most of the time you will be able to generalize the process using the equation:

$$H^+ + OH^- \rightarrow H_2O.$$

From the molarity of the base and the volume used you can calculate the moles of base (OH^-). Because of the 1:1 stoichiometry that also will be the moles of acid. Dividing that by the liters of acid solution pipetted into the flask gives the acid's molarity.

Comments

This type of titration can be performed with a pH meter without an indicator. The pH readings will be plotted against the volume. The endpoint is the point of inflection of the curve. A titration, either with an indicator or a pH meter, can be used to determine the acid content of household substances such as fruit juices or sodas.

Experiment 5: Chromatography

Synopsis

Many times the components (solutes) in a solution cannot be separated by simple physical means. This is especially true of polar solutes because of their interactions. One method that is commonly used is chromatography. A very small amount of the solution is spotted onto a strip of filter paper or chromatography paper and allowed to dry. The strip is placed vertically into a jar containing a small amount of solvent. As the solvent is drawn up the

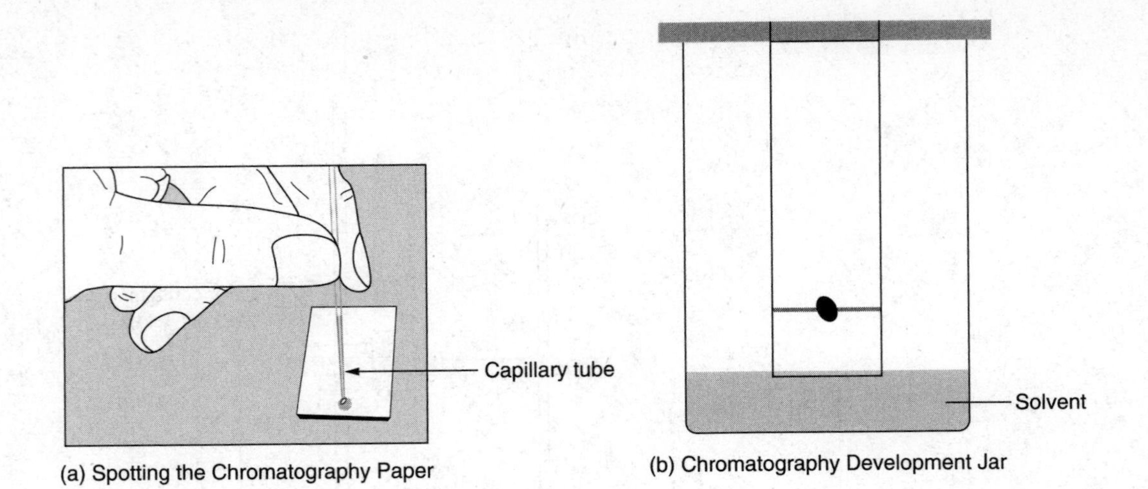

(a) Spotting the Chromatography Paper

(b) Chromatography Development Jar

Figure 19.3

strip by capillary action, it dissolves the sample. The various solutes have different affinities to the paper and to the solvent and can thus be separated as the solvent moves up the strip. Choice of the solvent is critical and can be related to its polarity; however the choice sometimes must be done by trial and error.

Equipment

Filter paper or chromatography paper
Chromatography jar
Various solvents
Metric rules
Sample to be analyzed
Assorted glassware

Measurements

The student will make measurements of the distance that each component travels and the distance that the solvent traveled.

Calculations

The calculations involve determining the R_f value for each component. The R_f value is the distance the component travels divided by the distance the solvent traveled. Substances that interact strongly with the paper do not travel very far (low R_f values), while those that interact strongly with the solvent travel much farther (high R_f values).

Comments

Chromatography is a very powerful separation technique.

Experiment 6: Determination of the Type of Bonding in Solid Samples

Synopsis

In this type of experiment, the student is given a set of bottles that contain solids of various types of bonding—ionic, covalent, or metallic. The student uses various physical and chemical tests to determine the bonding type. These tests might include melting point, conductivity, solubility, etc. along with observations of physical properties such as luster and hardness.

Equipment

Assorted solids—ionic, covalent, metals
Assorted solvents—polar and nonpolar
Conductivity tester
pH paper
Thermometer
Assorted glassware

Measurements

A number of measurements and observations may be made:

Luster: Metals tend to have a metallic luster; solid nonmetals often have a dull luster.
Melting point: Ionic solids and metals have high melting points; covalent compounds have lower melting points.
Solubility: Ionic compounds and polar covalent solids are generally soluble in water; metals and nonpolar covalent solids are generally insoluble or very slightly soluble in water.
Conductivity: Aqueous solutions of ionic compounds are conductors; aqueous solutions of most polar covalent compounds are nonconductors.

Calculations

There are generally no calculations associated with this experiment.

Comments

Many other tests could be used: pH of the aqueous solutions, solubility on organic solvents, and so on.

Experiment 7: Stoichiometry

Synopsis

In this experiment, you are asked to verify the results of an experiment by checking both the stoichiometric calculations and the procedure. You will be asked to determine the percent by mass of substances such as sodium bicarbonate in a mixture. You will do this by making use of the unique properties of the components in this mixture.

Equipment

Bunsen burners and strikers
Digital balances
Ring stands and rings
Ceramic triangles
Crucibles and lids
Assorted glassware, including volumetric glassware

Measurements

A weighed sample mixture of sodium bicarbonate and sodium carbonate is heated to constant mass. The sodium bicarbonate decomposes to sodium carbonate, carbon dioxide (gas), and water vapor: $2\ NaHCO_3(s) \rightarrow Na_2CO_3(s) + H_2O(g) + CO_2(g)$. The loss of mass is the loss in mass of $CO_2 + H_2O$. Examining the equation for the decomposition reaction, you can see that there is a 1:1 ratio of moles of water and carbon dioxide.

Calculations

If you let z = moles CO_2 = moles H_2O, then the total grams of mass lost can be shown as the sum of the moles of each (which will be the same) times the molar mass of each substance:

$$\text{Mass lost (grams)} = (z \times 18.02\text{ g } H_2O/\text{mole}) + (z \times 44.01\text{g } CO_2/\text{mole}).$$

You can then solve for z, the number of moles. As you can see from the balanced equation, the moles of $NaHCO_3$ solid that decomposed is $2z$. The mass of $NaHCO_3$ that decomposed will be:

$$2z \times 84.02\text{g } NaHCO_3/\text{mole}.$$

The percent of $NaHCO_3$ in the mixture will be the mass of the sodium bicarbonate divided by the mass of the mixture sample times 100%:

$$(\text{grams } NaHCO_3/\text{grams mixture}) \times 100\%.$$

Comments

In order to increase the precision (and hopefully the accuracy) of the determination, several runs should be made and an average taken. This same procedure may be applied to many other reactions and mixtures. These samples could also be analyzed by a titration procedure.

Experiment 8: Redox Titration

Synopsis

In this experiment, the concentration of a substance will be determined by using a redox titration. The titrant will need to be standardized before it can be used in the titration. Commonly, the redox titration involves the titration of hydrogen peroxide (H_2O_2) with potassium permanganate ($KMnO_4$), with the goal of analyzing the commercial hydrogen peroxide that can be found in a pharmacy. The $KMnO_4$ solution can be standardized against a $Fe(NH_4)_2(SO_4)_2{\cdot}6H_2O$ solution. You will prepare a standard (known concentration) solution of the $Fe(NH_4)_2(SO_4)_2{\cdot}6H_2O$, a sulfuric acid solution, and a solution of potassium permanganate. The redox half-reactions involved in the standardization are:

$$Fe^{2+}(aq) \rightarrow Fe^{3+}(aq) + 1\ e^- \text{ and}$$

$$MnO_4^-(aq) + 8\ H^+(aq) + 5\ e^- \rightarrow Mn^{2+}(aq) + 4\ H_2O(l),$$

giving an overall redox reaction of:

$$5\ Fe^{2+}(aq) + MnO_4^-(aq) + 8\ H^+(aq) \rightarrow 5\ Fe^{3+}(aq) + Mn^{2+}(aq) + 4\ H_2O(l)$$

The half-reactions involved in the titration of the hydrogen peroxide are:

$$H_2O_2(aq) \rightarrow O_2(g) + 2\ H^+(aq) + 2\ e^- \text{ and}$$

$$MnO_4^-(aq) + 8\ H^+(aq) + 5\ e^- \rightarrow Mn^{2+}(aq) + 4\ H_2O(l),$$

giving the overall redox-reaction:

$$5\ H_2O_2(aq) + 2\ MnO_4^-(aq) + 6\ H^+(aq) \rightarrow 2\ Mn^{2+}(aq) + 8\ H_2O(l) + 5\ O_2(g).$$

Equipment

Buret
Ring stand and clamps
Pipets of assorted volumes
Pipet bulbs
Assorted glassware, including volumetric glassware

Measurements

You will be making mass measurements of the $Fe(NH_4)_2(SO_4)_2{\cdot}6H_2O$ and the $KMnO_4$ and many volume measurements of the pipets, volumetric flasks, and the buret.

Calculations

For the standardization: from the number of grams of $Fe(NH_4)_2(SO_4)_2{\cdot}6H_2O$ used, you can calculate the moles of Fe^{2+} used. Knowing this, you can determine the moles MnO_4^- used from the stoichiometry in the overall reaction (1 MnO_4^- : 5 Fe^{2+}) and then its molarity.

For the peroxide titration: from the buret volume and the molarity of the $KMnO_4$ solution, you can calculate the moles used in the titration, and applying the overall reaction stoichiometry you can get the moles of hydrogen peroxide (5 H_2O_2 : 2 MnO_4^-). From the moles you can get grams and finally mass percent (assuming the mass of the peroxide solution is 1.00 g/mL).

Comments

Be very careful in making your measurements. The same general procedure can be applied to a number of other systems.

Experiment 9: Chemical and Physical Changes

Synopsis

Commonly this experiment involves separating the components of a mixture by using the chemical and physical properties of the mixture components. This is the basis of the analysis of commercially available samples such as over-the-counter acetaminophen- or aspirin-based pain relievers. The binder (many times sucrose), aspirin, and acetaminophen can be separated by the difference in their solubility in water and organic solvents, their acidity, and the difference in the way they react with hydrochloric acid and sodium bicarbonate solutions.

Equipment

Büchner funnels
Vacuum filtration apparatus
Separatory funnel
Hot plate or drying oven
Assorted glassware

Measurements

You will be making mass measurements of the sample, and every time a component is separated as a solid, it is dried and the mass determined.

Calculations

The overall percent recovery for the sample would be the sum of the masses of all the recovered components divided by the initial mass of the sample:

$$(\text{sum of the grams of all recovered components/grams of sample}) \times 100\%.$$

The percentage of each component can be calculated by dividing the mass of a component by the total mass of all recovered components:

$$(\text{grams of component/sum of the grams of all recovered components}) \times 100\%.$$

Comments

Be very careful in making your measurements. The same general procedure can be applied to a number of other systems.

Be especially careful when handling the acid solutions and organic solvents. Be sure to vent the separatory funnel before opening it.

Experiment 10: Kinetics

Synopsis

In this experiment, some of the factors involved in the speed of a chemical reaction will be explored. Commonly this experiment focuses on the decomposition of calcium carbonate—limestone, $CaCO_3(s)$, and hydrochloric acid, HCl(aq). Pieces of calcium carbonate of different sizes (to test how the speed of reaction varies with surface area) and HCl solutions of different concentrations will be available. The temperature of the reaction mixture can be varied by using an ice bath or heating the mixture. In order to measure the speed of the reaction, the carbon dioxide gas product can be collected in a syringe, or a gas pressure probe can be used to monitor the production of the $CO_2(g)$ as a function of time. The mass of sample consumed (or the decrease in the total mass of the reaction flask) versus time can also be used as an indication of the speed of reaction.

Equipment

Balance
Hotplate
Syringes
Stopwatch
Assorted glassware
Magnetic stirrer and stir bar
Gas pressure probe and data collection device

Measurements

Measurements include the initial and final mass of the calcium carbonate sample, the volume of gas evolved, and time measurements.

Calculations

Calculations commonly involve determining the mass of sample consumed (lost) as a function of time. The results of the mass versus time measurements are commonly plotted.

Comments

When plotting the data, the time is commonly the horizontal axis, while the mass lost or mL of gas produced is the vertical axis.

Be especially careful when handling the hydrochloric acid.

Experiment 11: Rate Laws

Synopsis

In this experiment, you will determine the rate law for a specific chemical reaction. Commonly the reaction involved is the reaction of crystal violet (CV) with sodium hydroxide (NaOH). The progress of the reaction is followed with a spectrophotometer or colorimeter. You will initially create a Beer's law calibration curve by measuring the absorbance of solutions of crystal violet of varying concentrations. Then you will use the same spectrophotometer to follow the change in concentration of crystal violet as it reacts with NaOH as a function of time:

$$CV^+(aq) + OH^-(aq) \rightarrow CVOH(aq).$$

The rate expression for this reaction would be:

$$\text{rate} = k\ [CV^+]^x[OH^-]^y.$$

If we use a large stoichiometric excess of NaOH then the rate equation becomes

$$\text{rate} = k^*\ [CV^+]^x$$

since there is so much hydroxide ion present that its concentration essentially becomes constant.

Equipment

Spectrophotometer (commonly SPEC 20)
Cuvettes (sample tubes for the spectrophotometer)
Pipettes and bulbs
Assorted glassware, including volumetric glassware

Measurements

You will be making measurements of absorbance and time. Be sure to use a blank containing only water and NaOH but no crystal violet.

Calculations

You will be making different concentrations of the stock crystal violet solution by dilution, so that you will use the dilution equation: $M_1V_1 = M_2V_2$. You will be making three graphs: (1) concentration versus time (straight line indicates zero order with respect to CV [$x = 0$ in rate expression]); (2) ln(concentration) versus time (straight line indicates first order with respect to CV [$x = 1$ in rate expression]) and (3) 1/concentration versus time (straight line indicates second order with respect to CV [$x = 2$ in rate expression]).

Comments

Be especially careful when handling the sodium hydroxide solution.

Experiment 12: Calorimetry

Synopsis

In this experiment, you will be measuring the heat produced during the dissolving of various ionic substances in water with the goal of determining which of the salts is most efficient (with respect to cost) in generating heat. Substances to test might include anhydrous calcium chloride ($CaCl_2$), anhydrous sodium carbonate (Na_2CO_3), anhydrous ammonium nitrate (NH_4NO_3), anhydrous sodium acetate ($NaC_2H_3O_2$), and similar salts. You will calculate the change in enthalpy of dissolution in kJ/mol (ΔH_{soln}) by using a coffee-cup calorimeter (see Figure 9.1 in Chapter 9 Thermodynamics). You may be using a magnetic stirrer instead of the stirring wire shown in the figure.

Equipment

Thermometers or temperature probes
Polystyrene cups
Magnetic stirrers and stir bars
Assorted glassware

Measurements

You will be making measurements of the initial and final temperatures of the solutions formed by adding a certain mass of solute to a measured amount of water. The ΔT is the final temperature minus the initial temperature. The value of ΔT is a calculated number and not a measured number.

Calculations

You may be given or may have to calculate the calorimeter constant, C, for your calorimeter—the heat absorbed by the calorimeter per degree of temperature change. The energy of solution formation (q_{rxn}) is calculated by multiplying the mass times the specific heat of the solution (given) times the change in temperature ($q_{rxn} = mc\Delta T$) and the energy of solution (q_{soln}) is calculated by : $q_{soln} = -(q_{rxn} + C\Delta T)$. The enthalpy of dissolution (ΔH_{soln}) is calculated by dividing the q_{soln} (in kJ) by the number of moles of salt used.

Comments

Be especially careful with the ammonium nitrate—it is a strong oxidizer.

Experiment 13: Chemical Equilibrium—Le Châtelier's Principle

Synopsis

Experiments that fall into this category examine systems that are at equilibrium and what happens when that equilibrium is disturbed. Many times this involves having a small tray of reagents and testing an equilibrium system by mixing selected reagents and making observations. You may change concentrations (adding more reagent) or change the temperature of the solutions. This may involve an acid–base equilibrium or complex ion equilibriums. Reactions will be given, and you should be able to describe the stress that you imposed and how the system reacted to that stress.

Equipment

Test tubes
Stirring rods
Spatula
Assorted glassware

Measurements

This experiment involves no measurements, only estimations of volumes and masses.

Calculations

This experiment involves no calculations.

Comments

Be very careful when working with concentrated ammonia and hydrochloric acid. Always wear goggles, gloves, and an apron and keep these reagents in the hood.

Experiment 14: Acid–Base Titrations

Synopsis

Experiments that fall into this category are acid–base titrations involving weak acids or weak bases. Many times the course of the titration is followed by a pH meter and the equivalence point is determined graphically. This allows you to determine not only the concentration of the weak acid or base but also its pK_a or pK_b. Both monoprotic and polyprotic acids may be examined. From an examination of the specific reaction involving a weak acid or base, you should be able to determine whether the solution at the equivalence point will be acidic or basic.

Equipment

Stirring rods
pH meters or pH probes
Buret
Assorted glassware

Measurements

You will be making pH measurements and plotting them against volume of titrant added. In many cases, you will titrate various acids (strong and weak) with a NaOH solution of known concentration. The equivalence point for such a titration is the point at which a dramatic increase in pH occurs; this is called the point of inflection of the curve. The pH at the volume corresponding to half the equivalence point volume is the pK_a of the acid. The same is true of bases, except the pH will be decreasing during the titration. A polyprotic acid or base will give you two points of inflection, and two p*K*s and *K*s may be calculated.

Calculations

The K_a of the acid can be calculated by the equation $K_a = 10^{-pKa}$. If the K_b of a weak base is to be determined, use $K_b = 10^{-pKb}$.

Comments

Be extremely careful when working with the acids and bases and wear all of your personal protective equipment, especially your goggles. When making dilutions, always add the acid (or base) to water, NOT water to acid.

Experiment 15: Buffer pH

Synopsis

A buffer is a substance that resists a change in pH when an acid or base is added to it. It is normally a mixture of a weak acid and its conjugate base. Experiments in this category involve examining the properties of buffers and household substances that are buffers. This will involve titrating a substance with an acid or base while following the course of the titration with a pH meter, plotting the pH versus mL of titrant added, and determining the equivalence point graphically. At any point before the equivalence point you have a buffer present. Common household substances may be tested for their buffer ability. The curve of pH versus mL of a substance that has some buffering ability rises sharply initially and then levels off much more than a titration of a substance that is not a buffer. You can use this to determine whether an unknown solution exhibits any buffering capability.

Equipment

pH meter
Burets and clamps
Magnetic stirrer
Assorted glassware

Measurements

You will be making measurements of volume and pH for a wide variety of substances. The point in a titration involving a buffer that corresponds to halfway to the equivalence point is called the point of maximum buffering.

Calculations

The K_a of the acid can be calculated by the equation $K_a = 10^{-pKa}$. If the K_b of a weak base is to be determined, then use $K_b = 10^{-pKb}$.

Comments

Be careful in handling the acid and base solutions.

Experiment 16: The Capacity of a Buffer

Synopsis

Experiments in this category are designed to explore the capacity of a buffer, which is the amount of acid or base that can be neutralized by the buffer. You can determine this by using different amounts of the conjugate acid and base components or by changing the concentration of each by the same amount. Normally, the higher the concentration of the conjugate acid and base in the buffer, the more moles of added base or acid can be neutralized and thus the higher the buffer capacity. You will also be asked to create a buffer of a specific pH.

Equipment

Balance
Burets and clamps
Assorted glassware

Measurements

You will be making measurements of volume and pH for a wide variety of substances. You will be making graphs of pH versus mL of titrant added.

Calculations

You can calculate the initial pH of a conjugate acid/base buffer by using the following equations:

$$[H^+] = K_a\,[\text{weak acid}]/[\text{conjugate base}];\ \text{then pH} = -\log\,[H^+].$$

If you want a buffer of a certain pH, then put in the K_a of the weak acid you want to use and the $[H^+]$ desired and solve for the ratio of acid to base. If you have a choice of several acid/base systems, then choose the one whose pK_a is closest to the desired pH.

Comments

Be extremely careful when working with the acids and bases and wear all of your personal protective equipment, especially your goggles. When making dilutions, always add the acid (or base) to water, NOT water to acid.

› Common Mistakes to Avoid

1. You *measure* initial and final values, but *calculate* the change.
2. You use an analytical balance to weigh the mass (grams), but not the moles.

› Review Questions

Below you will find a multipart free-response question like the ones in Section II of the exam. Use this question to review the content of this chapter and practice for the AP Chemistry exam. To make this an even more authentic practice for the actual exam, time yourself following the instructions provided.

Free-Response Question

You have 15 minutes to answer the following question. You may use a calculator and the tables in the back of the book.

Question 1

A sample of a solid, weak monoprotic acid, HA, is supplied, along with solid sodium hydroxide, NaOH, a phenolphthalein solution, and primary standard potassium hydrogen phthalate (KHP).

(a) Describe how a standardized sodium hydroxide solution may be prepared for the titration.
(b) Sketch a graph of pH versus volume of base added for the titration.
(c) Sketch the titration curve if the unknown acid was really a diprotic acid.
(d) Describe the steps to determine K_a for HA.
(e) What factor determines which indicator should be chosen for this titration?

Answer and Explanation for the Free-Response Question

(a) A sample of sodium hydroxide is weighed and dissolved in deionized water to give a solution of the approximate concentration desired. (Alternatively, a concentrated NaOH solution could be diluted.)

1. Samples of dried KHP are weighed into flasks and dissolved in deionized water.
2. A few drops of the appropriate acid–base indicator (phenolphthalein) are added to each sample.
3. A buret is rinsed with a little of the NaOH solution; then the buret is filled with NaOH solution.
4. Take an initial buret reading.
5. NaOH solution is titrated into the KHP samples until the first permanent pink color.
6. Take the final buret reading.
7. Using the molar mass of KHP, determine the moles of KHP present. This is equal to the moles of NaOH.
8. The difference in the buret readings is the volume of NaOH solution added (convert this to liters).
9. The molarity of the NaOH solution is the moles of NaOH divided by the liters of NaOH solution added.
10. (Repeat the procedure for each sample.)

Give yourself 2 points for this entire list, if the items are in order. If three or more items are in the wrong order or missing, you get only 1 point. You get 0 points for three or fewer items.

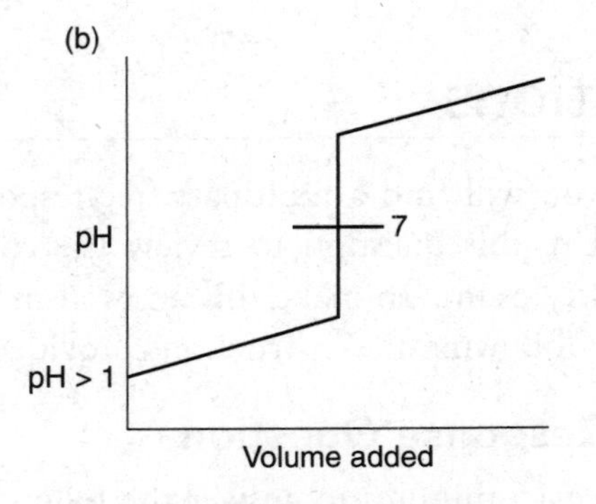

Equivalence point at pH < 7

You get 1 point for this graph. You get an additional point for noting that the equivalence point is greater than 7.

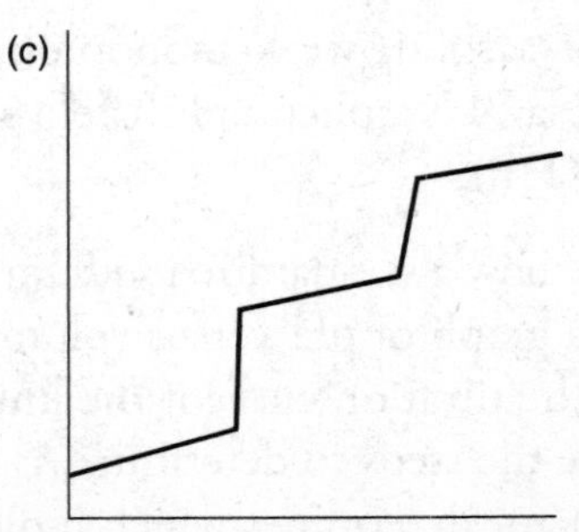

You get 1 point for this graph. You must show two steps.

(d) There are several related ways to do this problem. One method is to split the sample into two portions. Titrate one portion to the equivalence point. Add the titrated sample to the untitrated sample, and add a volume of deionized water equal to the volume of NaOH solution added. The pH of this mixture is equal to the pK_a of the acid (this corresponds to a half-titrated sample).
   You get 1 point for anything concerning a half-titrated sample and an additional point for pH = pK_a.

(e) The pH at the equivalence point must be close to the pK_a of the indicator.
   You get 1 point for this answer.

There are a total of 8 points possible.

❯ Rapid Review

Reviewing the experiments should include looking at the synopsis, apparatus, calculations. and comments as well as the appropriate concept chapters, if needed.

- Pay particular attention to any experiment you did not perform.
- Be familar with the equipment used in each experiment.
- Know the basic measurements required in each experiment.
- Know what values are measured and which are calculated.
- Pay attention to significant figures.
- Balances are used to measure the mass of a substance, not the moles.

Build Your Test-Taking Confidence

AP Chemistry Practice Exam 1
AP Chemistry Practice Exam 2

AP Chemistry Practice Exam 1—Multiple Choice

ANSWER SHEET

1 (A) (B) (C) (D)
2 (A) (B) (C) (D)
3 (A) (B) (C) (D)
4 (A) (B) (C) (D)
5 (A) (B) (C) (D)
6 (A) (B) (C) (D)
7 (A) (B) (C) (D)
8 (A) (B) (C) (D)
9 (A) (B) (C) (D)
10 (A) (B) (C) (D)
11 (A) (B) (C) (D)
12 (A) (B) (C) (D)
13 (A) (B) (C) (D)
14 (A) (B) (C) (D)
15 (A) (B) (C) (D)
16 (A) (B) (C) (D)
17 (A) (B) (C) (D)
18 (A) (B) (C) (D)
19 (A) (B) (C) (D)
20 (A) (B) (C) (D)
21 (A) (B) (C) (D)
22 (A) (B) (C) (D)
23 (A) (B) (C) (D)
24 (A) (B) (C) (D)
25 (A) (B) (C) (D)
26 (A) (B) (C) (D)
27 (A) (B) (C) (D)
28 (A) (B) (C) (D)
29 (A) (B) (C) (D)
30 (A) (B) (C) (D)
31 (A) (B) (C) (D)
32 (A) (B) (C) (D)
33 (A) (B) (C) (D)
34 (A) (B) (C) (D)
35 (A) (B) (C) (D)
36 (A) (B) (C) (D)
37 (A) (B) (C) (D)
38 (A) (B) (C) (D)
39 (A) (B) (C) (D)
40 (A) (B) (C) (D)
41 (A) (B) (C) (D)
42 (A) (B) (C) (D)
43 (A) (B) (C) (D)
44 (A) (B) (C) (D)
45 (A) (B) (C) (D)
46 (A) (B) (C) (D)
47 (A) (B) (C) (D)
48 (A) (B) (C) (D)
49 (A) (B) (C) (D)
50 (A) (B) (C) (D)
51 (A) (B) (C) (D)
52 (A) (B) (C) (D)
53 (A) (B) (C) (D)
54 (A) (B) (C) (D)
55 (A) (B) (C) (D)
56 (A) (B) (C) (D)
57 (A) (B) (C) (D)
58 (A) (B) (C) (D)
59 (A) (B) (C) (D)
60 (A) (B) (C) (D)

The AP exam is a timed exam; keep this in mind as you prepare. When taking the various tests presented in this book, you should follow the AP exam rules as closely as possible. Anyone can improve his or her score by using notes, books, or an unlimited time. You will have none of these on the AP exam, so resist the temptation to use them on practice exams. Carefully time yourself, do not use other materials, and use a calculator only when expressly allowed to do so. After you have finished an exam, you may use other sources to go over questions you missed or skipped. We have seen many students get into trouble because the first time they attempted a test under "test conditions" was on the test itself.

AP Chemistry Practice Exam 1, Section I (Multiple Choice)

Time—1 hour and 30 minutes
NO CALCULATOR MAY BE USED WITH SECTION I

Answer the following questions in the time allowed. You may use the periodic table in the back of the book.

Use the following information to answer questions 1–7.

The reaction of iron metal with hydrochloric acid generates aqueous iron(II) chloride and hydrogen gas. The balanced chemical equation for the reaction is:

$$Fe(s) + 2\ HCl(aq) \rightarrow FeCl_2(aq) + H_2(g)$$

Iron(III) oxide, a component of rust, reacts with hydrochloric acid to generate aqueous iron(III) chloride and water. The balanced chemical equation for the reaction is:

$$Fe_2O_3(s) + 6\ HCl(aq) \rightarrow 2\ FeCl_3(aq) + 3\ H_2O(l)$$

A student weighed a small flask both with and without a sample of rusty iron and recorded the masses. Next, she connected the flask to the system shown below.

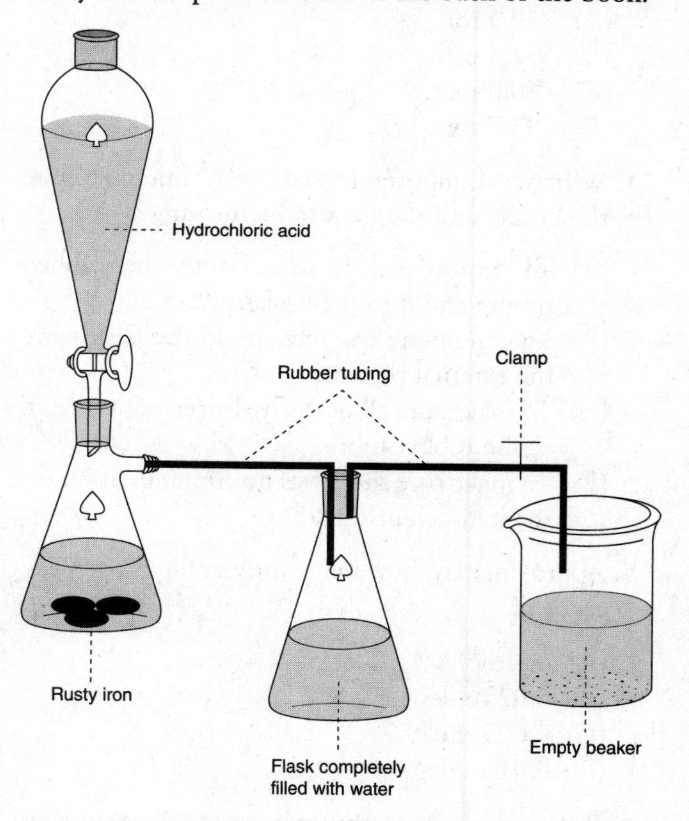

The flask in the middle and the rubber tubing leading to the beaker were completely filled with water, and then the clamp was removed. Some hydrochloric acid was added to the flask containing the rusty iron. The water level in the second flask dropped as the generated hydrogen gas displaced the water into the beaker. More hydrochloric acid was added until the iron completely dissolved and generation of hydrogen gas ceased. After the system returned to room temperature, the beaker was raised until the water in the beaker was at the same level as in the second flask. When the liquid levels were the same, the clamp was replaced to prevent further transfer. The student completed the following data table in her lab book.

Mass of empty flask	175.245 g
Mass of flask plus rusty iron	176.604 g
Volume of displaced water	285 mL
Barometric pressure	752.3 torr
Room temperature	21.0°C
Vapor pressure of water at 21.0°C	18.6 torr

GO ON TO THE NEXT PAGE

1. What type of reaction generated the hydrogen gas?

(A) Single displacement
(B) Combination
(C) Decomposition
(D) Combustion

2. What type of reaction is the reaction of hydrochloric acid with the rust?

(A) Single displacement
(B) Double displacement
(C) Decomposition
(D) Combustion

3. What is the partial pressure of the hydrogen gas in the flask?

(A) 770.9 torr
(B) 752.3 torr
(C) 760.0 torr
(D) 733.7 torr

4. Why was it important to adjust the liquid level in the beaker and the flask to be the same?

(A) To remove excess water from the rubber tubing and into the beaker
(B) To equilibrate the pressure in the flask with the external pressure
(C) To make sure all of the hydrogen gas was out of the rubber tubing
(D) To make sure there was no contamination by the hydrochloric acid

5. Approximately, how many moles of hydrogen gas formed?

(A) 0.1 moles
(B) 0.02 moles
(C) 0.005 moles
(D) 0.01 moles

6. If the sample were pure iron, approximately how many moles of hydrogen gas would form?

(A) 0.04 moles
(B) 0.002 moles
(C) 0.02 moles
(D) 0.2 moles

7. In another experiment, a 2.520 g sample of rusty iron generated 0.0205 moles of hydrogen gas. What was the approximate percentage of pure iron in the sample?

(A) 45%
(B) 100%
(C) 25%
(D) 75%

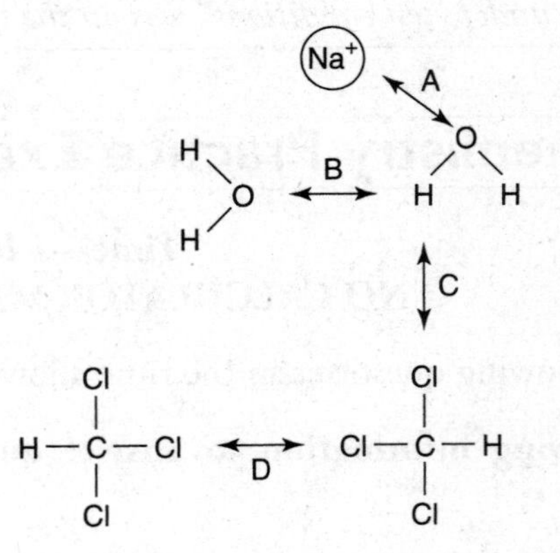

8. Which of the labeled arrows in the diagram above represents the strongest intermolecular force?

(A) Arrow **A**
(B) Arrow **B**
(C) Arrow **C**
(D) Arrow **D**

9. If 75% of a sample of pure ${}^{3}_{1}H$ decays in 24.6 yr, what is the half-life of ${}^{3}_{1}H$?

(A) 24.6 yr
(B) 18.4 yr
(C) 12.3 yr
(D) 6.15 yr

10. Alkenes are hydrocarbons with the general formula C_nH_{2n}. If a 0.420 g sample of any alkene is combusted in excess oxygen, how many moles of water will form?

(A) 0.0300
(B) 0.450
(C) 0.015
(D) 0.300

11.

Acid	**K_a, Acid Dissociation Constant**
H_3PO_4	7.2×10^{-3}
$H_2PO_4^-$	6.3×10^{-8}
HPO_4^{2-}	4.2×10^{-13}

Using the above information, choose the best answer for preparing a pH = 7.9 buffer.

(A) K_2HPO_4
(B) K_3PO_4
(C) $K_2HPO_4 + KH_2PO_4$
(D) $K_2HPO_4 + K_3PO_4$

12. At constant temperature, a change in volume will NOT affect the moles of the substances present in which of the following?

(A) $N_2(g) + 3\ H_2(g) \rightarrow 2\ NH_3(g)$
(B) $CO(g) + Cl_2(g) \rightarrow COCl_2(g)$
(C) $PCl_3(aq) + Cl_2(g) \rightarrow PCl_5(g)$
(D) $CO(g) + H_2O(g) \rightarrow CO_2(g) + H_2(g)$

Use the following information to answer questions 13–17.

pH versus volume of titrant added

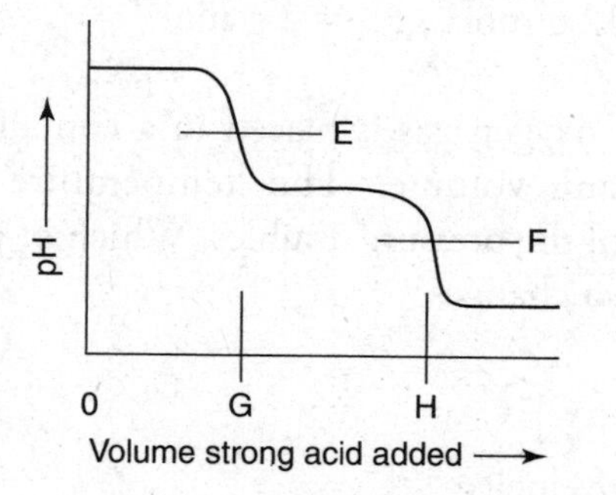

Soda ash is a substance that contains one or two of the following chemicals—sodium bicarbonate ($NaHCO_3$), sodium carbonate (Na_2CO_3), and sodium hydroxide (NaOH). The composition may vary; therefore, it is necessary to titrate samples with a strong acid like HCl to determine the actual composition. The above is a generic graph indicating the possible results for a sample. E and F represent the pH at the endpoints with the possibility that E may shift slightly and one or the other may not be present. G and H will depend on the composition of the sample with the possibility that one may not be present.

13. Soda ash may contain two of the three compounds. Which combination cannot occur?

(A) $NaHCO_3$ and Na_2CO_3
(B) $NaHCO_3$ and NaOH
(C) Na_2CO_3 and NaOH
(D) All combinations are possible.

14. Analysis of one sample found only Na_2CO_3 to be present. What information, derived from the graph, would make this apparent?

(A) The volume added to reach G is equal to the volume added to go from G to H.
(B) Only the E endpoint is present.
(C) The volume added to reach G is greater than the volume added to go from G to H.
(D) Only the F endpoint is present.

15. Analysis of another sample found only $NaHCO_3$ to be present. What information, derived from the graph, would make this apparent?

(A) The volume added to reach G is equal to the volume added to go from G to H.
(B) Only the E endpoint is present.
(C) The volume added to reach G is greater than the volume added to go from G to H.
(D) Only the F endpoint is present.

16. Analysis of another sample found only NaOH to be present. What information, derived from the graph, would make this apparent?

(A) < 7
(B) $= 7$
(C) > 7
(D) Unknown

17. The procedure for analyzing soda ash requires the chemist to heat the sample to boiling for a few minutes near point F. Why is this necessary?

(A) To keep the products in solution
(B) To keep the reactants in solution
(C) To make the reaction go faster
(D) To expel CO_2

18. Which of the following CANNOT behave as both a Brønsted base and a Brønsted acid?

(A) HPO_4^{2-}
(B) $C_2O_4^{2-}$
(C) HSO_4^-
(D) $HC_2O_4^-$

19. A student mixes 50.0 mL of 0.10 M nickel(II) nitrate, $Ni(NO_3)_2$, solution with 50.0 mL of 0.10 M NaOH. A green precipitate of nickel(II) hydroxide forms, and the concentration of the hydroxide ion becomes very small. Which of the following correctly places the concentrations of the remaining ions in order of decreasing concentration?

(A) $[Na^+] > [Ni^{2+}] > [NO_3^-]$
(B) $[Ni^{2+}] > [NO_3^-] > [Na^+]$
(C) $[Na^+] > [NO_3^-] > [Ni^{2+}]$
(D) $[NO_3^-] > [Na^+] > [Ni^{2+}]$

20. An experiment to determine the molecular mass of a gas begins by heating a solid to produce a gaseous product. The gas passes through a tube and displaces water in an inverted, water-filled bottle. The mass of the solid is measured, as is the volume and the temperature of the displaced water. Once the barometric pressure has been recorded, what other information is needed to finish the experiment?

(A) The heat of formation of the gas
(B) The density of the water
(C) The mass of the displaced water
(D) The vapor pressure of the water

$$2\ ClF(g) + O_2(g) \rightarrow Cl_2O(g) + OF_2(g) \qquad \Delta H° = 167.5\ kJ$$

$$2\ F_2(g) + O_2(g) \rightarrow 2\ OF_2(g) \qquad \Delta H° = -43.5\ kJ$$

$$2\ ClF_3(l) + 2\ O_2(g) \rightarrow Cl_2O(g) + 3\ OF_2(g) \qquad \Delta H° = 394.1\ kJ$$

21. Using the information given above, calculate the enthalpy change for the following reaction:

$$ClF(g) + F_2(g) \rightarrow ClF_3(l)$$

(A) −135.1 kJ
(B) +135.1 kJ
(C) 270.2 kJ
(D) −270.2 kJ

22. When lithium sulfate, Li_2SO_4, is dissolved in water, the temperature increases. Which of the following conclusions may be related to this?

(A) Lithium sulfate is less soluble in hot water.
(B) The hydration energies of lithium ions and sulfate ions are very low.
(C) The heat of solution for lithium sulfate is endothermic.
(D) The solution is not an ideal solution.

Use the information on the containers in the following diagram to answer questions 23–25.

A	B	C	D
Kr	CH_4	O_2	H_2
120 K	120 K	120 K	120 K
1.0 mole	1.0 mole	1.0 mole	1.0 mole
1.0 L	1.0 L	1.0 L	1.0 L

Approximate molar masses:

$Kr = 84\ g\ mol^{-1}$, $CH_4 = 16\ g\ mol^{-1}$,
$O_2 = 32\ g\ mol^{-1}$, $H_2 = 2\ g\ mol^{-1}$

23. A sample of oxygen gas is placed in a container with constant volume. The temperature is changed until the pressure doubles. Which of the following also changes?

(A) Density
(B) Moles
(C) Average velocity
(D) Number of molecules

24. From the following, choose the gas that probably shows the least deviation from ideal gas behavior.

(A) Kr
(B) CH_4
(C) O_2
(D) H_2

GO ON TO THE NEXT PAGE

25. If each of the containers sprang a small leak, the gases would effuse out through the hole. Which of the gases would take the longest time to effuse out of the container?

(A) Kr
(B) CH_4
(C) O_2
(D) H_2

26. The specific rate constant, *k*, for radioactive lawrencium-256 is 86 h^{-1}. What mass of a 0.0500 ng sample of lawrencium-256 remains after 58 s?

(A) 0.0500 ng
(B) 0.0250 ng
(C) 0.0125 ng
(D) 0.00625 ng

Use the information on the containers in the following diagram to answer questions 27–31 concerning the following equilibrium:

$$CO(g) + 2H_2(g) \rightarrow CH_3OH(g)$$

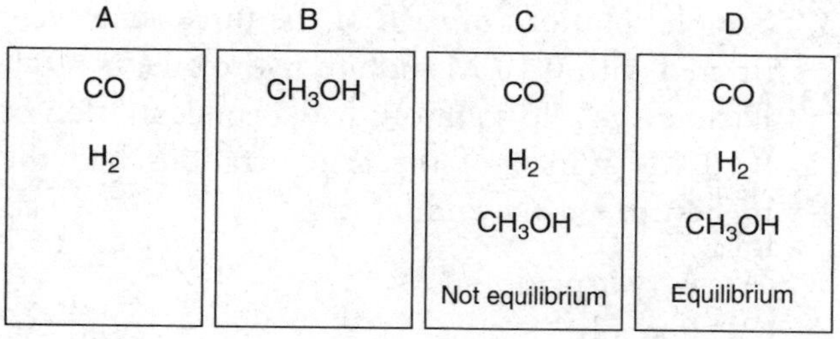

A	B	C	D
CO	CH_3OH	CO	CO
H_2		H_2	H_2
		CH_3OH	CH_3OH
		Not equilibrium	Equilibrium

27. A 1.00 L flask (container C) is filled with 0.70 mole of H_2 and 0.60 mole of CO and allowed to come to equilibrium. At equilibrium, there are 0.40 mole of CO in the flask. What is the value of K_c, the equilibrium constant, for the reaction?

(A) 0.74
(B) 3.2
(C) 0.0050
(D) 5.6

28. Which of the following applies when the reaction in container B approaches equilibrium?

(A) $H_2(g)$ forms twice as fast as CO(g).
(B) $H_2(g)$ forms at the same rate as CO(g).
(C) $H_2(g)$ forms at half the rate of CO(g).
(D) It is not possible to predict the relative rates.

29. The mixture in container D is in equilibrium. Which of the following is true?

(A) The rate of the forward and reverse reactions is equal to zero.
(B) The rate of the forward reaction is equal to the rate of the reverse reaction.
(C) The pressure in the system is increasing.
(D) The pressure in the system is decreasing.

30. The mixture in container A goes to equilibrium. If the initial moles of $H_2(g)$ is twice the initial moles of CO(g), which of the following is true?

(A) Both reactants are limiting; therefore, the reaction will continue until there are zero moles of each remaining.
(B) The reaction continues until the rate of the forward reaction equals that of the reverse reaction.
(C) The pressure of the system increases until the system equals equilibrium.
(D) No reaction occurs until a catalyst is added.

31. As the mixture in container B approaches equilibrium, the partial pressure of CH_3OH gas decreases by 1.0 atm. What is the net change in the total pressure of the system?

(A) 0.0 atm
(B) +2.0 atm
(C) −1.0 atm
(D) +3.0 atm

Use the information on standard reduction potentials in the following table to answer questions 32–36.

	E° (V)
$Cl_2(g) + 2\ e^- \rightarrow 2\ Cl^-(aq)$	+1.36
$S_2O_6^{2-}(aq) + 4\ H^+(aq) + 2\ e^- \rightarrow 2\ SO_2(g) + 2\ H_2O(l)$	+0.56
$Cu^{2+}(aq) + 2\ e^- \rightarrow Cu(s)$	+0.34
$Sn^{2+}(aq) + 2\ e^- \rightarrow Sn(s)$	−0.14
$2\ H_2O(l) + 2\ e^- \rightarrow H_2(g) + 2\ OH^-(aq)$	−0.83
$Ca^{2+}(aq) + 2\ e^- \rightarrow Ca(s)$	−2.87
$K^+(aq) + e^- \rightarrow K(s)$	−2.92

GO ON TO THE NEXT PAGE

32. An electrolysis cell was constructed with two platinum electrodes in a 1.00 M aqueous solution of KCl. An odorless gas evolved from one electrode and a gas with a distinctive odor evolved from the other electrode. Choose the correct statement from the following list.

(A) The odorless gas was oxygen.
(B) The odorless gas was the result of oxidation.
(C) The gas with the distinctive odor was the result of oxidation.
(D) The odorless gas was evolved at the positive electrode.

33. There is a galvanic cell involving a copper, Cu, electrode in a 1.0 M copper(II) sulfate, $CuSO_4$, solution and a tin, Sn, electrode in a 1.0 M tin(II) sulfate, $SnSO_4$, solution. What is the cell potential?

(A) +0.48 V
(B) +0.20 V
(C) −0.20 V
(D) 0.00 V

34. A student attempts to prepare an electrolysis cell to produce calcium metal (Ca) from an aqueous solution of calcium chloride, $CaCl_2$, using a 6.0 V battery. The student was unsuccessful. Why was the student unable to produce calcium metal?

(A) The voltage from the battery was insufficient to force the reaction to occur.
(B) Reduction of chloride ion occurred in preference to reduction of calcium ion.
(C) Calcium chloride solutions do not conduct electricity.
(D) Reduction of water occurred in preference to reduction of calcium ion.

35. A student mixed an acidic 1.0 M potassium dithionate, $K_2S_2O_6$, solution with a 1.0 M tin(II) chloride, $SnCl_2$, solution containing some elemental tin, Sn, and observed a gas with a distinctive odor. What was the gas?

(A) $H_2(g)$
(B) $SO_2(g)$
(C) $Cl_2(g)$
(D) $Sn(g)$

36. A galvanic cell has a tin, Sn, electrode in a compartment containing a 1.0 M tin(II) sulfate, $SnSO_4$, solution and a platinum, Pt, electrode in a compartment containing 1.0 M copper(II) chloride, $CuCl_2$, solution. A salt bridge containing a 1.0 M potassium sulfate, K_2SO_4, solution connects the two compartments. What is the net ionic equation for the cell reaction?

(A) $CuCl_2(aq) + Sn(s) \rightarrow SnSO_4(aq) + Cu(s)$
(B) $Sn^{2+}(aq) + Cu(s) \rightarrow Cu^{2+}(aq) + Sn(s)$
(C) $Cu^{2+}(aq) + Sn(s) \rightarrow Sn^{2+}(aq) + Cu(s)$
(D) $2Cl^-(aq) + 2H_2O(l) \rightarrow H_2(g) + 2OH^-(aq) + Cl_2(g)$

Use the information on the acids in the following diagram to answer questions 37–40.

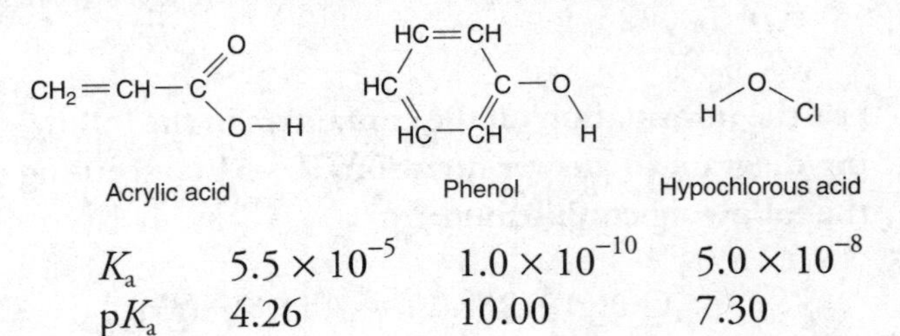

	Acrylic acid	Phenol	Hypochlorous acid
K_a	5.5×10^{-5}	1.0×10^{-10}	5.0×10^{-8}
pK_a	4.26	10.00	7.30

37. Sample solutions of each of the three acids were titrated with 0.10 M sodium hydroxide, NaOH. Each of the acid solutions had a concentration of 0.10 M. Which of the acid titrations had the highest pH at the endpoint?

(A) Acrylic acid
(B) Phenol
(C) Hypochlorous acid
(D) They all had a pH of 7 at the endpoint.

38. A solution is prepared by mixing 500.0 mL of 0.20 M acrylic acid with 500.0 mL of 0.10 M potassium hydroxide, KOH. What is the pH of the solution?

(A) 4.26
(B) 7.00
(C) 0.70
(D) 13.00

GO ON TO THE NEXT PAGE

39. Hypochlorous acid is an unstable compound, and one of the decomposition products is chlorine gas, Cl_2. The decomposition of the acid lowers its concentration over time. What effect will the decomposition of one-fourth of the acid have on the pH at the endpoint of the titration with sodium hydroxide?

(A) The pH will shift to a lower pH.
(B) The pH will shift to a higher pH.
(C) There is very little change.
(D) It is impossible to determine.

40. Three 25.00 mL samples of approximately 0.10 M phenol were removed from a container and placed in 250 mL beakers. The samples were titrated with standard potassium hydroxide, KOH, solution. Cresol red was the indicator. The samples required 31.75, 42.38, and 41.75 mL to reach the endpoint. Which of the following might explain why one of the samples required less base to titrate?

(A) The indicator was added too late.
(B) The wrong indicator was used.
(C) There was an acid contaminating the unclean beaker.
(D) There was a base contaminating the unclean beaker.

41. The plot of ln [A] versus time gives a straight line. This implies the rate law is

(A) Rate = $k[A]^2$
(B) Rate = $k[A]$
(C) Rate = $k[A]^0$
(D) Rate = $k[A]^{-1}$

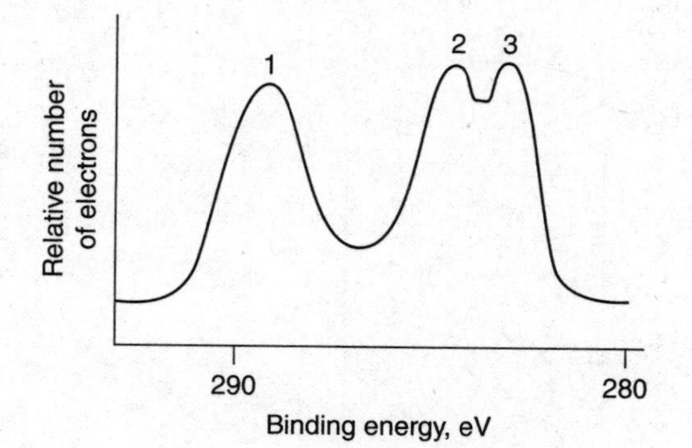

The photoelectron spectrum above shows the energy required to remove a 1s electron from a carbon atom in ethyl chloroformate, shown below.

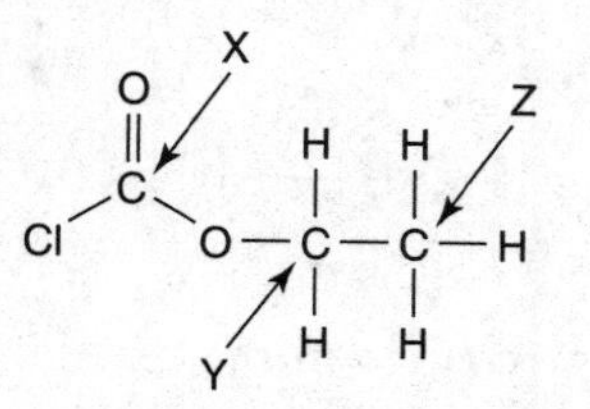

42. What is the best explanation of the position of peak 1 in the spectrum?

(A) The peak is due to atom Z because of the strong interaction of the 1s orbitals on three hydrogen atoms with the 1s orbital on this carbon atom.
(B) The peak is due to atom X because it is the only sp^2 hybridized carbon atom in the compound.
(C) The peak is due to atom X because it is attached to three very electronegative atoms.
(D) The peak is due to atom Z because it is the farthest from the very electronegative chlorine and oxygen atoms.

COMPOUND	FORMULA	MOLAR MASS G mol^{-1}	MELTING POINT °C
Diethyl ether	C_2H_5-O-C_2H_5	74.12	−116.2
Butyl alcohol	C_4H_9OH	74.12	−88.6
Butylamine	C_4H_9-NH_2	73.14	−49.1
Diethyl amine	C_2H_5-NH-C_2H_5	73.14	−49.8

43. According to the data in the table above, which of the compounds has the weakest intermolecular forces?

(A) Diethyl ether
(B) Butyl alcohol
(C) Butylamine
(D) Diethyl amine

Ion	Ionic Radius (pm) (cubic environment)
Na^+	132
Cd^{2+}	124
La^{3+}	130

$$M^{n+}(g) + x\,H_2O(l) \rightarrow M(H_2O)_x^{\,n+}(aq)$$
$$\Delta H = \text{Heat of hydration}$$

44. The energy change in the reaction given above is the heat of hydration. Assuming the value of x is the same in all cases, which of the following correctly predicts the relative order of the heats of hydration and gives a correct explanation?

(A) $Na^+ > Cd^{2+} > La^{3+}$ because ions of elements lower on the periodic table have lower hydration energies.
(B) $Cd^{2+} > La^{3+} > Na^+$ because smaller radii lead to higher hydration energies.
(C) $La^{3+} > Cd^{2+} > Na^+$ because higher charges lead to higher hydration energies.
(D) All are about the same because their ionic radii are similar.

45. Which of the following may serve as both a Brønsted-Lowry base and as a Brønsted-Lowry acid?

(A) F^-
(B) H_2CO_3
(C) $CO_3^{\,2-}$
(D) $HCO_3^{\,-}$

46. A certain reaction is nonspontaneous at 1 atm and 298 K. Which of the following combinations must apply for this reaction to become spontaneous at lower temperatures?

(A) $\Delta H < 0$, $\Delta S < 0$, and $\Delta G = 0$
(B) $\Delta H > 0$, $\Delta S < 0$, and $\Delta G > 0$
(C) $\Delta H < 0$, $\Delta S < 0$, and $\Delta G > 0$
(D) $\Delta H > 0$, $\Delta S > 0$, and $\Delta G > 0$

47. What is the ionization constant, K_a, for a weak monoprotic acid if a 0.6 molar solution has a pH of 2.0?

(A) 1.7×10^{-4}
(B) 1.7×10^{-2}
(C) 6.0×10^{-6}
(D) 2.7×10^{-3}

$[:\ddot{O}-\ddot{N}=\ddot{O}]^-$ NO_2^- $[:\ddot{O}-N(=\ddot{O})(-:\ddot{O}:)]^-$ NO_3^-

48. The Lewis electron-dot diagrams for two nitrogen-oxygen species are in the above diagram. Which of these has the smaller bond angle and why?

(A) NO_2^- because there are two resonance structures instead of three
(B) NO_3^- because the three oxygen atoms push away from each other more than two
(C) NO_2^- because the pair of nonbonding electrons on the nitrogen atom "pushes" the bonds closer together
(D) NO_3^- because there are three resonance structures instead of two

49. Which of the following best represents the result for the reaction of 50.0 mL of a 0.10 M barium hydroxide, $Ba(OH)_2$, solution with 25.0 mL of a 0.10 M potassium sulfate, K_2SO_4, solution to form a precipitate of $BaSO_4$?

(A)

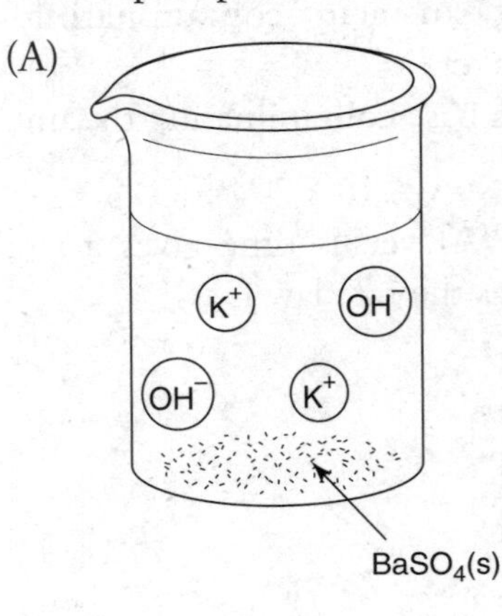

(B)

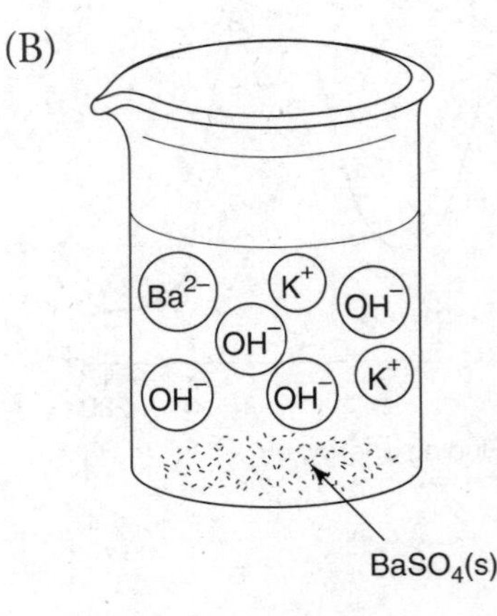

GO ON TO THE NEXT PAGE

(C)

(D)

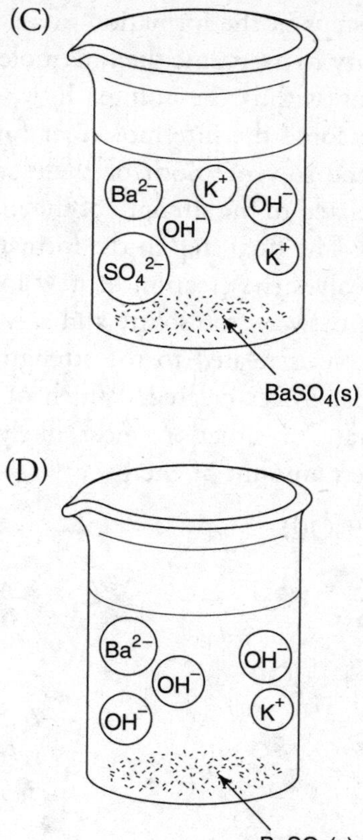

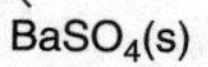

50. A saturated aqueous solution of sodium chloride, NaCl, is heated to the boiling point. Which of the following best represents this system?

(A)

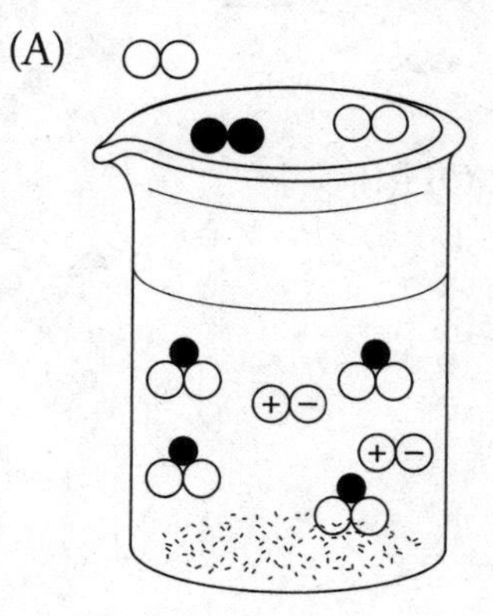

(B)

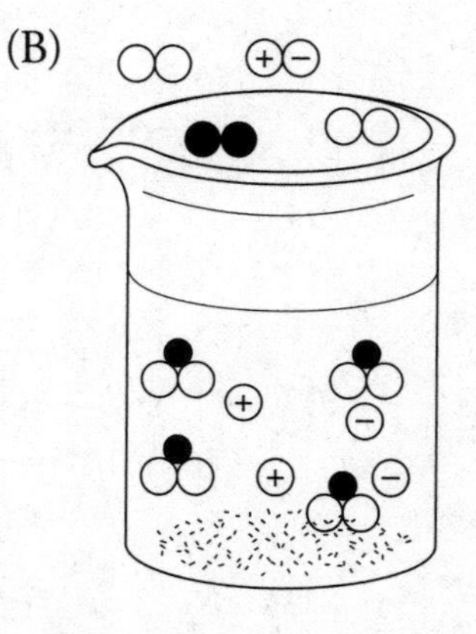

(C)

(D)

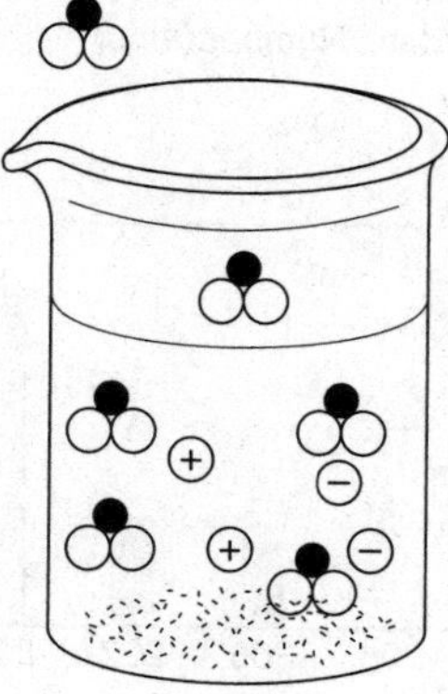

The average distribution of lead isotopes on Earth is in the following table:

ISOTOPE	PERCENT ABUNDANCE
^{204}Pb	1.4 %
^{206}Pb	24.1 %
^{207}Pb	22.1 %
^{208}Pb	52.4 %

However, the distribution of lead isotopes in uranium and thorium ores is different because the radioactive decay of these two isotopes generates specific lead isotopes as illustrated in the following table. This will alter the isotopic composition of lead depending upon the concentration of radioisotope(s) and age of the sample.

PARENT ISOTOPE	LEAD ISOTOPE FORMED
^{238}U	^{206}Pb
^{235}U	^{207}Pb
^{232}Th	^{208}Pb

51. Which of the following mass spectra best illustrates the lead isotope distribution of a lead ore contaminated with thorium?

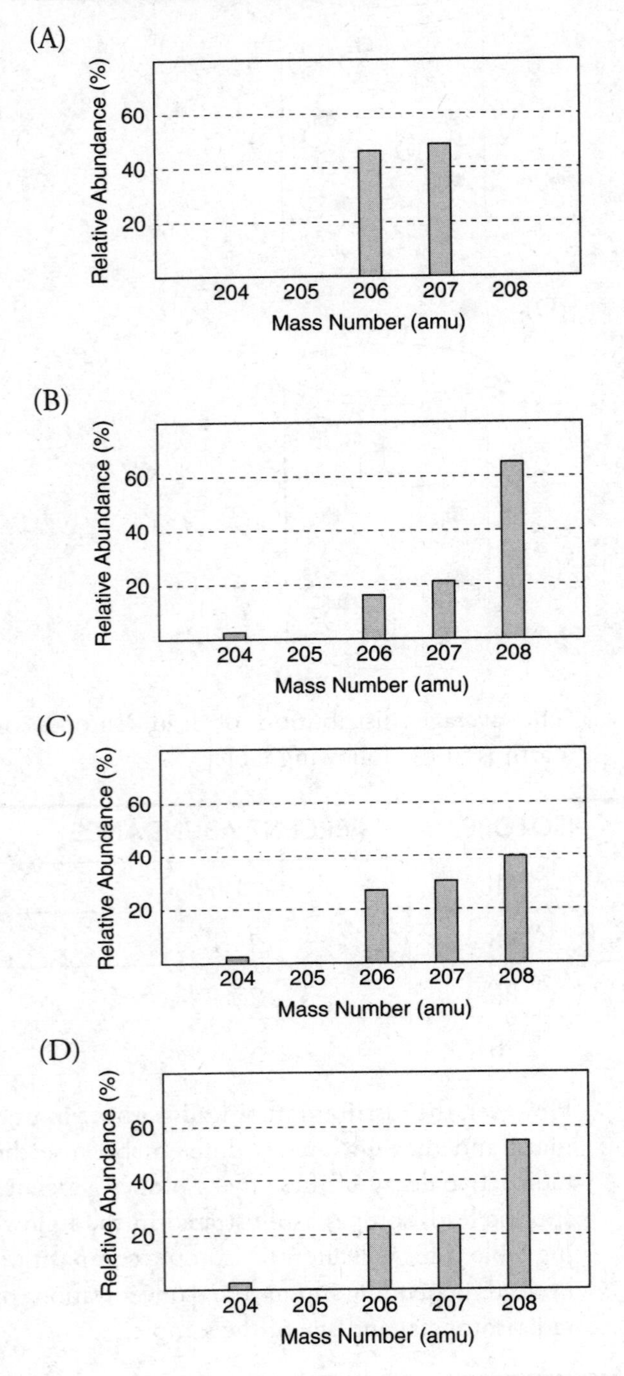

52. There are three steps in the formation of a solution. It is necessary to overcome the intermolecular forces present within the solute. It is also necessary to overcome the intermolecular forces present within the solvent. Both of these steps require energy related to the strength of the intermolecular forces. The final step in the formation of a solution involves the creation of new intermolecular forces between the solute and solvent. This energy release is related to the strength of the intermolecular forces created. Which of the following illustrates a situation most likely to release the greatest amount of energy?

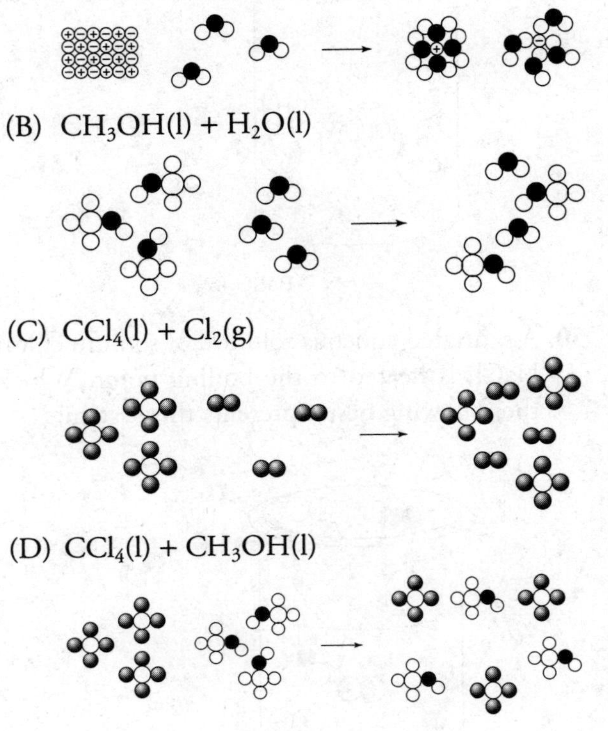

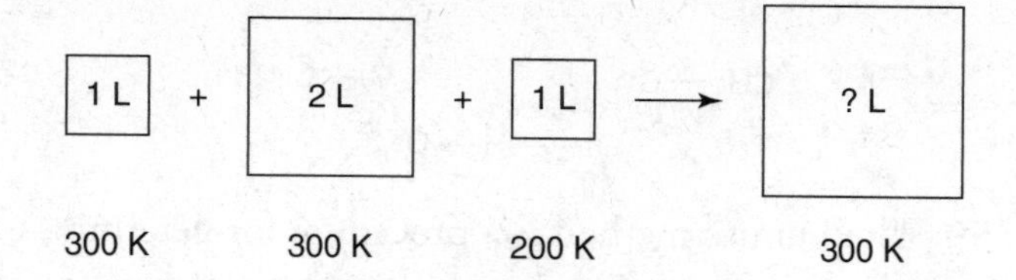

53. The contents in the three containers on the left in the diagram above are transferred to the container on the right. The volumes of the original containers are exactly the values indicated. The pressure in all four containers is 1.0 atm. What is the volume of the container on the right?

(A) 4.00 L
(B) 3.50 L
(C) 4.50 L
(D) 5.00 L

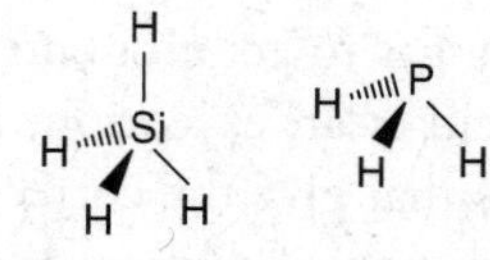

54. The diagram above shows the structure of molecules of SiH_4 and PH_3. The boiling point of SiH_4 is 161 K, and the boiling point of PH_3 is 185 K. Which of the following is the best explanation of why the boiling point of PH_3 is higher?

(A) The molar mass of PH_3 is greater.
(B) SiH_4 has weaker covalent bonds than PH_3.
(C) Only PH_3 can form hydrogen bonds.
(D) PH_3 has stronger intermolecular forces because it is polar and SiH_4 is not.

$CO(g) + Cl_2(g) \rightarrow COCl_2(g) \qquad \Delta H = -109\ \text{kJ}$

55. The above equilibrium is established in a closed system. A laser quickly increases the temperature of the system by 10°C. Which of the following graphs best illustrates the rate of the reverse reaction?

(A)

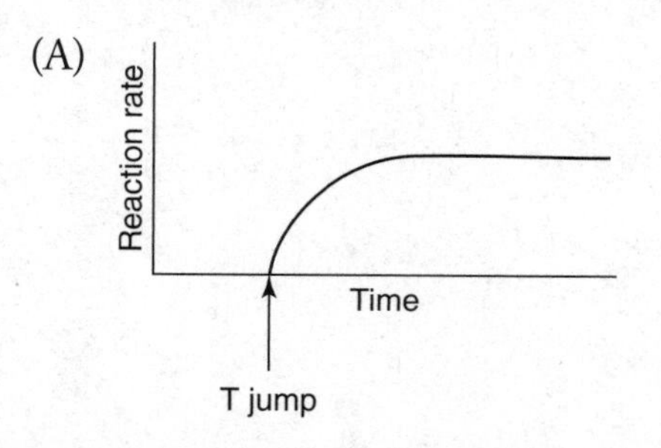

(B)

(C)

(D)

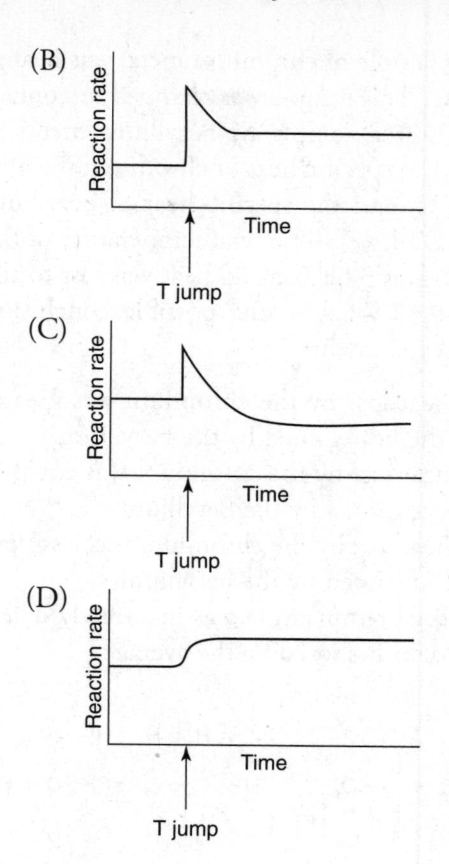

Step 1: $Cl_2(g) \rightarrow 2\ Cl(g)$ (fast) $\qquad k_1$ and k_{-1}
Step 2: $Cl(g) + CH_4(g) \rightarrow HCl(g) + CH_3(g)$ (slow) $\qquad k_2$
Step 3: $Cl(g) + CH_3(g) \rightarrow CH_3Cl(g)$ (fast) $\qquad k_3$

56. What is the rate law expression for the above reaction?

(A) Rate = $k[CH_4]\,[Cl_2]^{1/2}$
(B) Rate = $k[CH_4][Cl_2]$
(C) Rate = $k[Cl_2]$
(D) Rate = $\dfrac{[HCl][CH_3]}{[Cl][CH_4]}$.

57. A 25.00 g sample of chromium metal was heated to 75.00°C. This sample was clamped in contact with a 47.00 g sample of beryllium metal at 25.00°C. The specific heat of chromium metal is 0.450 J/g°C, and the specific heat of beryllium metal is 1.82 J/g°C. The final temperature of the two metals was 30.8°C as no heat was lost to the surroundings. What is one possible conclusion from this experiment?

(A) The heat lost by the chromium was greater than the heat gained by the beryllium.
(B) The heat lost by the chromium was equal to the heat gained by the beryllium.
(C) The heat lost by the chromium was less than the heat gained by the beryllium.
(D) The final temperature was incorrectly determined, as it should be the average.

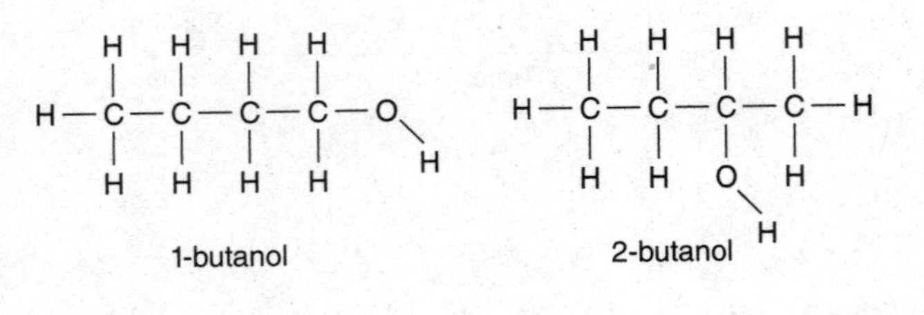

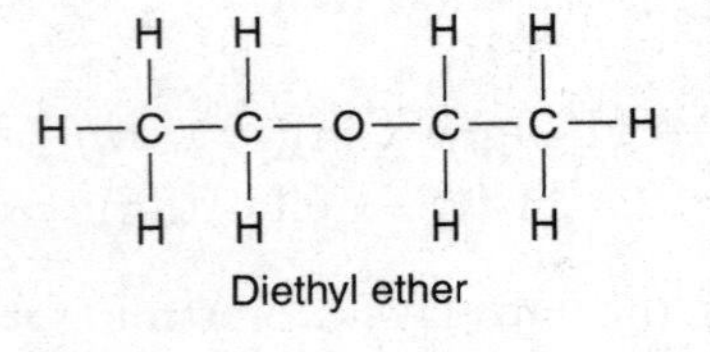

Compound	Melting Point (°C)
1-butanol	−88.6
2-butanol	−88.5
diethyl ether	−116.2

58. Which of the following best explains why the melting point of diethyl ether is lower than the other two compounds in the diagram and tables above?

(A) The oxygen in the center distorts the shape so the molecules do not tangle.
(B) It has a lower molar mass.
(C) It is the only one that cannot hydrogen bond.
(D) It is a more symmetrical molecule.

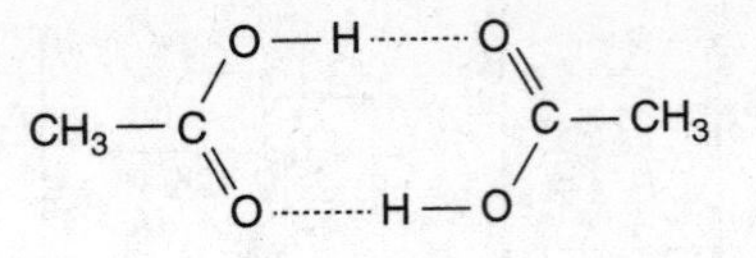

59. The Dumas method is a procedure for determining the molar mass of a gas. In this procedure the mass of a gas is divided by the moles of gas determined from the ideal gas equation ($n = PV/RT$). The molar masses of some compounds, such as acetic acid illustrated above, show significant deviations from the "correct" values. Why does the molar mass of acids, such as acetic acid, show a deviation?

(A) Acetic acid loses a hydrogen ion, so the molar mass is that of the acetate ion, which is less than that of acetic acid.
(B) Acetic acid is a liquid, and this method requires a gas to get accurate results.
(C) Acetic acid is an exception; it is one of the few acids that give the correct molar mass.
(D) The formation of dimers, as seen in the diagram, gives a molar mass double the true molar mass.

60. In which of the following groups are the species listed correctly in order of increasing radius?

(A) Sr, Ca, Mg
(B) Se^{2-}, S^{2-}, O^{2-}
(C) Mn^{3+}, Mn^{2+}, Mn
(D) I^-, Br^-, Cl^-

STOP: End of AP Chemistry Practice Exam 1, Section I (Multiple Choice).

› Answers and Explanations for Exam 1, Section I (Multiple Choice)

1. A—The reaction is a single displacement reaction as the iron metal displaces hydrogen gas from the hydrochloric acid.

2. B—This is a double displacement reaction where two compounds combine to form the two products.

3. D—The pressure inside the flask is the sum of the partial pressures. Therefore, the pressure of hydrogen gas is the total pressure (752.3 torr) minus the vapor pressure of water (18.6 torr). The leveling of the water in the beaker and flask adjusted the pressure in the flask to the external (barometric) pressure.

4. B—If the two liquid levels are the same, the pressures must be the same.

5. D—The ideal gas equation ($PV = nRT$) gives the moles of hydrogen gas formed.

$$\text{Moles } H_2 = n = \frac{PV}{RT}$$
$$= \frac{(733.7 \text{ torr}/760 \text{ torr})(0.285 \text{ L})}{(0.0821 \text{ L atm mol}^{-1}\text{K}^{-1})(294.2 \text{ K})}$$
$$= 0.0114 \text{ moles } H_2.$$

It is easier to calculate the answer by simple rounding as:

$$\frac{(1 \text{ atm})(0.3 \text{ L})}{(0.08 \text{ L atm mol}^{-1}\text{K}^{-1})(300 \text{ K})} \approx \frac{(1)}{(0.1)}\left(\frac{0.3}{300}\right)$$
$$= \left(\frac{0.3}{30.0}\right)$$
$$= 0.01 \text{ moles}$$

6. C—The moles of hydrogen gas formed equal the moles of iron reacting (see the balanced chemical equation). If the sample were pure iron, the mass of iron would be (176.604 − 175.245) g = 1.359 g sample (= Fe). The moles of iron are the mass of iron divided by its molar mass (55.84 g mol^{-1}).

$$\text{Moles } H_2 = \text{mole Fe} = \frac{1.359 \text{ g}}{55.84 \text{ g mol}^{-1}}$$
$$\approx \frac{1.359}{52} \approx \frac{1.3}{4(13)} \approx \frac{0.13}{4} \approx 0.0325$$

Due to rounding, the actual answer must be a little smaller (0.02434 moles).

7. A—According to the equation for the formation of hydrogen gas from iron, the moles of iron equals the moles of hydrogen. Therefore, 0.0205 moles of iron reacted. The mass of reacting iron is (0.0205 mole Fe) (55.84 g mol^{-1}) = 1.14 grams Fe. The percentage of pure iron in the sample is the mass of the iron divided by the mass of the sample then multiplied by 100%.

$$\text{Mass percentage} = \frac{1.14 \text{ g Fe}}{2.520 \text{ g}}(100\%) \approx 50\%$$
$$= 45.2\%$$

8. A—This represents an ion-dipole force, which is stronger than a hydrogen bond (B), a dipole-dipole force (C), or a London dispersion force (D).

9. C—After one half-life, 50% would remain. Another half-life would reduce this by one-half to 25%. The total amount decayed is 75%. Thus, 24.6 years must be two half-lives of 12.3 years each.

10. A—One mole of alkene, C_nH_{2n}, will form n moles of water. It is possible to determine the value of n by dividing the mass of the alkene by the empirical formula mass (CH_2 = 14 g/mol). This gives $\frac{0.420 \text{ g}}{14 \text{ g/mol}} = \frac{3\ (0.140) \text{ g}}{14 \text{ g/mol}} = 0.030$ mol.

11. C—The pK_a for $H_2PO_4^-$ is nearest to the pH value needed. Thus, the simplest buffer would involve this ion.

12. D—If there are equal numbers of moles of gas on each side of the equilibrium arrow, then volume or pressure changes will not affect the equilibrium.

13. B—The strong base (NaOH) would react with the hydrogen in the sodium bicarbonate to yield water and sodium carbonate. This reaction would continue until either all the NaOH or all the $NaHCO_3$ reacted.

14. A—To reach G, it is necessary to convert all the carbonate ion to bicarbonate ion, which means the moles of bicarbonate ion formed must equal the moles of carbonate ion originally present. The moles of bicarbonate ion formed would require the same number of moles of acid as that required to convert the carbonate to bicarbonate. Equal moles would mean equal volumes added.

15. D—F is the endpoint for the reaction of bicarbonate ion with acid, since only bicarbonate is present; this is the only observable endpoint.

16. B—If only NaOH were present, then the titration with HCl would be an example of a strong acid–strong base reaction, which has an endpoint of 7.

17. D—It is necessary to heat the solution to expel CO_2. Any CO_2 remaining in the solution would produce carbonic acid, which would lower the pH leading to a premature end of the analysis.

18. B—All can behave as Brønsted bases (accept a hydrogen ion). Only B cannot behave as an acid (donate a hydrogen ion).

19. D—Initially, doubling the volume will result in halving the concentrations. Next, consider the reaction. Some of the nickel remains (the remainder precipitated), the sodium does not change (soluble), and two nitrate ions (soluble) are formed per nickel(II) nitrate.

20. D—Water, whenever present, will contribute its vapor pressure.

$$2\ ClF(g) + O_2(g) \rightarrow Cl_2O(g) + OF_2(g) \qquad \Delta H° = 167.5\ kJ$$

$$2\ F_2(g) + O_2(g) \rightarrow 2\ OF_2(g) \qquad \Delta H° = -43.5\ kJ$$

$$2\ ClF_3(l) + 2\ O_2(g) \rightarrow Cl_2O(g) + 3\ OF_2(g) \qquad \Delta H° = 394.1\ kJ$$

21. A—

$$\tfrac{1}{2}[2\ ClF(g) + O_2(g) \rightarrow Cl_2O(g) + OF_2(g)] \qquad \tfrac{1}{2}(167.5\ kJ)$$

$$\tfrac{1}{2}[2\ F_2(g) + O_2(g) \rightarrow 2\ OF_2(g)] \qquad \tfrac{1}{2}(-43.5\ kJ)$$

$$\tfrac{1}{2}[Cl_2O(g) + 3\ OF_2(g) \rightarrow 2\ ClF_3(l) + 2\ O_2(g)] \qquad -\tfrac{1}{2}(394.1\ kJ)$$

$$ClF(g) + F_2(g) \rightarrow ClF_3(l) \qquad -135.1\ kJ$$

22. A—The increase in temperature indicates that this is an exothermic process. Exothermic processes shift toward the starting materials when heated.

23. C—The average velocity is related to temperature.

24. D—The smaller the molecule and the less polar (more nonpolar) the gas is, the smaller the deviation from ideal gas behavior.

25. A—The gases are all at the same temperature; therefore, the gas with the greatest molar mass will have the lowest average velocity and, for this reason, will take the longest time to effuse.

26. C—It is necessary to use the half-life relationship for first-order kinetics. This relationship is $t_{1/2} = 0.693/k$, and $t_{1/2} = (0.693/86\ h^{-1})(3{,}600\ s/h) = 29\ s$. To save time on the exam you can approximate this equation as $t_{1/2} = (0.7/90)(3{,}600)$. Dividing 3,600 by 90 gives 40, and 40 times 0.7 is equal to 28. If the half-life is ≈ 28 s, then the time (58 s) is equivalent to about two half-lives, so one-fourth of the sample should remain.

27. D—The loss of 0.20 mol of CO means that 0.40 mol of H_2 reacted (leaving 0.30 mol) and 0.20 mol of CH_3OH formed. Dividing all the moles by the volume gives the molarity, and:

$$K_c = (0.20)/(0.40)(0.30)^2 = (0.10)/(0.20)(0.30)^2 = 0.10/(0.20 \times 0.090) = 0.10/0.018 = 5.6$$

28. A—Based upon the stoichiometry of the reaction, $H_2(g)$ will form twice as fast as CO(g).

29. B—At equilibrium, there is no net change because the forward and reverse reactions are going at the same rate.

30. B—At equilibrium, the rates of the forward and reverse reactions are always the same. At equilibrium, the amount of none of the species can be zero. The pressure of the system will decrease until the system reaches equilibrium.

31. B—The loss of 1.0 atm of $CH_3OH(g)$ leads to the formation of 1.0 atm of $CO(g)$ and 2.0 atm of $H_2(g)$; therefore, the net change is $(-1.0 + 1.0 + 2.0)$ atm = 2.0 atm.

32. C—Hydrogen (odorless) evolves at the cathode (negative), and chlorine (distinctive odor) evolves at the anode (positive).

33. A—The copper electrode is the cathode (+0.34 V), and the tin electrode is the anode (reverse sign, +0.14 V). Cell voltage = (+0.34 + 0.14) V = 0.48 V.

34. D—It is possible to reduce both water and the calcium ion; however, the reduction potential for water is lower, so water will reduce in preference to calcium ion.

35. B—The tin is a sufficiently strong reducing agent to reduce the dithionate ion to sulfur dioxide, a gas with a distinctive odor. None of the other substances will form. Hydrogen is odorless, chlorine has a distinctive odor, and tin is a solid (metal) at room temperature.

36. C—Using the cell potentials, the copper(II) ions will undergo reduction (use the half-reaction from the table) and the tin metal will undergo oxidation (reverse the half-reaction from the table). Combine the resultant half-reactions and cancel electrons to get this answer. A is not net ionic. B is the reverse reaction. D is nonspontaneous.

37. B—The weakest acid (smallest K_a) will have the highest pH at the endpoint.

38. A—The KOH will neutralize one-half of the acid. The solution will contain equal concentrations of the conjugate base and the conjugate acid. A solution containing equal concentrations of the conjugate acid and the conjugate base of a weak acid is a buffer solution with a $pH = pK_a$.

39. C—Less acid would require less base to reach the endpoint. Therefore, even though there is less conjugate base at the endpoint, the volume is less, which means that the concentration of conjugate base remains about the same. The same concentration of conjugate base means the same pH.

40. D—A base contaminant would react with some of the phenol leaving less to titrate.

41. B—This plot gives a straight line only for a first-order reaction.

42. C—The very electronegative chlorine and oxygen atoms attached to atom X pull electrons away from the carbon atom. This causes the carbon atom to hold on to the remaining electrons more strongly, which means that it requires more energy to remove a 1s electron from this atom.

43. A—The compound with the lowest melting point has the weakest intermolecular forces.

44. C—The ionic radii are too similar to make a significant difference, so only the charge differences are important. The greater the charge, the greater the attraction for the polar water molecules is. The greater the attraction, the greater the lattice energy is.

45. D—The bicarbonate ion, HCO_3^-, may behave as a Brønsted-Lowry base by accepting a hydrogen ion to form H_2CO_3 and as a Brønsted-Lowry acid by donating a hydrogen ion to become CO_3^{2-}.

46. C—Nonspontaneous at 1 atm and 298 K means $\Delta G > 0$. To become spontaneous at a lower temperature means entropy impeded the reaction (entropy was negative). The enthalpy must be negative or the reaction could never be spontaneous.

47. A—Enter the information into the K_a expression. A pH of 2.0 means that $[H^+] = 10^{-2.0}$ M or 1.0×10^{-2}. $K_a = [H^+][A^-]/[HA]$; $[H^+] = [A^-] = 1.0 \times 10^{-2}$ $[HA] = 0.6$.

48. C—These are both trigonal planar species, which ideally have a 120° bond angle. However, the presence of the lone pair on the nitrogen reduces the angle slightly.

49. B—The reaction is:

$$Ba(OH)_2(aq) + K_2SO_4(aq) \rightarrow BaSO_4(s) + 2\,KOH(aq)$$

The potassium sulfate is the limiting reagent, and all of the sulfate will precipitate along with half of the barium. All the hydroxide and potassium remain in solution. For every barium ion precipitated, there will be two potassium ions remaining in solution.

50. D—This shows the sodium ions and chloride ions as separate ions, which is a property of strong electrolytes. The water molecules are present in both the liquid and gas states. Other diagrams incorrectly show water dissociating, sodium chloride ion pairs, and sodium chloride vaporizing.

51. B—Thorium produces ^{208}Pb; therefore, the peak for this isotope should be enhanced as it is here.

52. A—This creates ion-dipole forces, which are the strongest intermolecular force generated by any of the solutions in this problem. Creating the strongest intermolecular force releases the greatest amount of energy.

53. C—There are several ways of solving this problem. One way is to determine the moles present in the original containers, which must be the same as in the final container. In each case, moles = n = PV/RT. Numbering the containers from left to right as 1, 2, 3, and 4 gives:

$$n_4 = n_1 + n_2 + n_3$$

$$= \left(\frac{(1.0\text{ atm})(1\text{ L})}{\left(0.0821\frac{\text{L}\cdot\text{atm}}{\text{mol}\cdot\text{K}}\right)(300\text{ K})}\right) + \left(\frac{(1.0\text{ atm})(2\text{ L})}{\left(0.0821\frac{\text{L}\cdot\text{atm}}{\text{mol}\cdot\text{K}}\right)(300\text{ K})}\right) + \left(\frac{(1.0\text{ atm})(1\text{ L})}{\left(0.0821\frac{\text{L}\cdot\text{atm}}{\text{mol}\cdot\text{K}}\right)(200\text{ K})}\right) n_4$$

$$= \left(\frac{1}{(0.0821)(300)}\right) + \left(\frac{2}{(0.0821)(300)}\right) + \left(\frac{1}{(0.0821)(200)}\right)$$

$$= \left(\frac{3}{(0.0821)(300)}\right) + \left(\frac{1}{(0.0821)(200)}\right)$$

$$= \left(\frac{1}{(0.0821)(100)}\right) + \left(\frac{1}{(0.0821)(200)}\right)$$

$$= \left(\frac{1}{0.0821}\right)\left(\frac{1}{100} + \frac{1}{200}\right)\text{mol}$$

$$V_4 = n_4RT/P =$$

$$= \frac{\left(\left(\frac{1}{0.0821}\right)\left(\frac{1}{100} + \frac{1}{200}\right)\text{mol}\right)\left(0.0821\frac{\text{L}\cdot\text{atm}}{\text{mol}\cdot\text{K}}\right)(300\text{ K})}{1.0\text{ atm}}$$

$$= \frac{\left(\left(\frac{1}{100} + \frac{1}{200}\right)\right)(\text{L})(300)}{1.0}$$

$$= (0.010 + 0.050)\text{L}(300) = 4.50\text{ L}.$$

On the exam, it is not necessary to write out all these steps. Take shortcuts.

54. D—Stronger intermolecular forces lead to higher boiling points. PH_3 has dipole–dipole forces, which are stronger than the London dispersion forces present in SiH_4.

55. B—The reaction is exothermic; therefore, a temperature increase will increase the rate of the reverse reaction (eliminating C). The reverse reaction began with a nonzero rate (eliminating A). The sudden change in temperature will cause a sudden increase in the rate with a decrease to a constant rate (eliminating D).

56. A—The reactants are Cl_2 and CH_4, so only these substances can appear in the rate law. The second step is the slow step, so it is the rate-determining step and gives the rate law. The slow step involves one CH_4 and one-half a Cl_2. The one and the one-half are the exponents.

57. B—This must be true according to the Law of Conservation of Energy (First Law of Thermodynamics).

58. C—Both 1-butanol and 2-butanol are capable of hydrogen bonding. The intermolecular forces in diethyl ether are dipole–dipole forces. Dipole–dipole forces are weaker than hydrogen bonding; therefore, the melting point is lower.

59. D—Strong hydrogen bonds hold two molecules of acetic acid together to give a molar mass that is double the "true" molar mass.

60. C—Increasing sizes indicate decreasing charge, lower position in a column on the periodic table, or position to the left in a period on the periodic table. Note, this type of explanation is unacceptable on the free-response questions of the AP Exam.

AP Chemistry Practice Exam 1, Section II (Free Response)

Time—1 hour and 30 minutes

Answer the following questions in the time allowed. You may use a calculator and the resources at the back of the book. Write the answers on separate sheets of paper.

Question 1

Compound	K_{sp}
$Cr(OH)_2$	1.0×10^{-17}
$Cr(OH)_3$	6.3×10^{-31}
$Fe(OH)_2$	7.9×10^{-16}
$Pb(OH)_2$	1.1×10^{-20}
$Mg(OH)_2$	6.0×10^{-10}
$Mn(OH)_2$	1.9×10^{-13}
$Sn(OH)_2$	6.3×10^{-27}

Use the K_{sp} data given above to answer the following questions.

(a) Excess manganese(II) hydroxide, $Mn(OH)_2$, is added to 100.0 mL of deionized water. What is the pH of the solution?
(b) A solution that is 0.10 M in Mg^{2+} and 0.10 M in Fe^{2+} is slowly made basic. What is the concentration of Fe^{2+} when Mg^{2+} begins to precipitate?
(c) Two beakers are filled with water. Excess chromium(III) hydroxide is added to one and excess tin(II) hydroxide is added to the other. Which beaker has the higher concentration of metal ions? Calculate the concentration of metal ion in each beaker to support your prediction.
(d) Chromium(III) hydroxide, $Cr(OH)_3$, is less soluble than chromium(II) hydroxide, $Cr(OH)_2$. Explain.
(e) Calculate the grams of lead(II) hydroxide, $Pb(OH)_2$, that will dissolve in 1.00 L of water.

Question 2

$$2\ PCl_3(g) + O_2(g) \rightarrow 2\ POCl_3(g)$$

Thermodynamic values related to the above reaction are given in the table below.

SUBSTANCE	ΔH_f° (kJ/mol)	S° (J/mol K)	BONDS	BOND ENERGIES (kJ/mol)
$PCl_3(g)$	−287	312	P–Cl	331
$O_2(g)$	0	205.0	O=O	498
$POCl_3(g)$	−542.2	325	O–O	204

(a) Determine the enthalpy change for the above reaction.
(b) Estimate the PO bond energy.
(c) Is the PO bond a single or a double bond? Justify your answer.
(d) Calculate the entropy change for the reaction.
(e) Is this reaction spontaneous or nonspontaneous at 25°C? Justify your prediction.

Question 3

A sample of a solid, weak monoprotic acid, HA, is available along with a standard sodium hydroxide solution. The sodium hydroxide solution was standardized with potassium hydrogen phthalate (KHP).

(a) List the apparatus required to titrate an HA solution.
(b) Sketch a pH versus volume-of-base-added curve for the titration.
(c) Sketch the titration curve if the unknown acid was really a diprotic acid.
(d) Describe the steps required to determine the molar mass of HA.
(e) How would the molar mass of HA change if the KHP contained an inert impurity?

Question 4

The following equipment is available for the determination of the molar mass of a volatile solid.

analytical balance	thermometer	beaker(s)	support stand and clamp
stoppers	glass tubing	buret	pipet(s)
test tube(s)	stopwatch	hot plate	flask(s)
crucible and lid	clay triangle	wire gauze	graduated cylinder

(a) Which of the above equipment is necessary for this experiment?
(b) In addition to information that you can determine by using the above equipment, what information is necessary to determine the molar mass of the unknown solid?
(c) Describe how the above equipment may be used to calculate the molar mass of the unknown solid.

Question 5

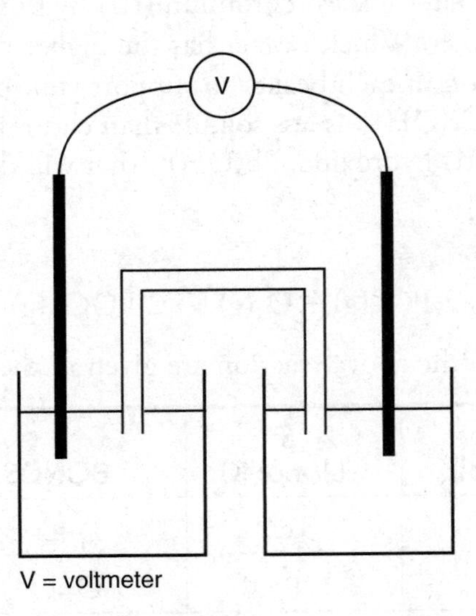

The above galvanic cell is constructed with a cadmium electrode in a 1.0 M $Cd(NO_3)_2$ solution in the left compartment and a silver electrode in a 1.0 M $AgNO_3$ solution in the right compartment. The salt bridge contains a KNO_3 solution. The cell voltage is positive.

$$Cd^{2+} + 2\,e^- \rightarrow Cd \qquad E° = -0.40\ V$$
$$Ag^+ + 1\,e^- \rightarrow Ag \qquad E° = +0.80\ V$$

(a) What is the balanced net ionic equation for the reaction, and what is the cell potential?
(b) Write the expression for the reaction quotient, *Q*, of the cell. Explain why any substances from the net ionic equation do not appear in *Q*.

Question 6

Relate each of the following to atomic properties and the principles of bonding.

(a) Draw the Lewis electron-dot structures for CO_2 and CO. Explain the polarity of these compounds.
(b) The compound C_2H_3F is polar, but compounds with the general formula $C_2H_2F_2$ are sometimes polar and sometimes nonpolar. Show the structures and explain.
(c) There are two isomers with the formula C_2H_6O. One of the isomers is more soluble in water than the other isomer. Use the structures of these two compounds to explain the difference in solubility.

Question 7

A flask, filled with water, is inverted into a water bath. Hydrogen gas was introduced to the flask through a tube until it displaced all of the water in the flask. The volume of the sample was 490.0 mL at 26°C, and the pressure in the room was 754 mm Hg. The vapor pressure of water at 26°C is 25.2 mm Hg.

(a) Calculate the number of grams of hydrogen in the flask.
(b) Calculate how many grams of water vapor are present in the flask.
(c) Determine the density (in $g\ L^{-1}$) of the gas mixture in the flask.

STOP. End of AP Chemistry Practice Exam 1.

› Answers and Explanations for Exam 1, Section II (Free Response)

Question 1

(a) The volume of the solution is irrelevant. The equilibrium

$$Mn(OH)_2(s) \rightleftharpoons Mn^{2+}(aq) + 2\ OH^-(aq) \text{ is important.}$$

The mass-action expression for this equilibrium is:

$$K_{sp} = [Mn^{2+}][OH^-]^2 = 1.9 \times 10^{-13}.$$

Setting $[Mn^{2+}] = x$ and $[OH^-] = 2x$, and inserting into the mass-action expression gives:

$$(x)(2x)^2 = 4x^3 = 1.9 \times 10^{-13}.$$

Solving for x gives $x = 3.6 \times 10^{-5}$, and $[OH^-] = 2x = 7.2 \times 10^{-5}$.

You get 1 point for the correct $[OH^-]$.

There are two common ways to finish the problem. You do not need to show both.

(i) $pOH = -\log [OH^-] = -\log 7.5 \times 10^{-5} = 4.14$
$pH = 14.00 - pOH = 14.00 - 4.14 = 9.86$

(ii) $[H^+] = K_w/[OH^-] = 1.0 \times 10^{-14}/7.5 \times 10^{-5} = 1.4 \times 10^{-10}$
$pH = -\log [H^+] = -\log 1.4 \times 10^{-10} = 9.86$

You get 1 point for the correct pH. If you got the wrong $[OH^-]$ value, but use it correctly, you still get 1 point.

(b) The important equilibria are: $M(OH)_2(s) \rightleftharpoons M^{2+}(aq) + 2\ OH^-(aq)$, where M = Mg or Fe. It is necessary to determine the hydroxide ion concentration when the magnesium begins to precipitate.

$K_{sp} = [Mg^{2+}][OH^-]^2 = 6.0 \times 10^{-10}$
$[OH^-]^2 = K_{sp}/[Mg^{2+}] = 6.0 \times 10^{-10}/0.10 = 6.0 \times 10^{-9}$
$[OH^-] = 7.7 \times 10^{-5}$

Using this value with the iron equilibrium gives:

$K_{sp} = [Fe^{2+}][OH^-]^2 = 7.9 \times 10^{-16}$
$[Fe^{2+}] = K_{sp}/[OH^-]^2 = (7.9 \times 10^{-16})/(7.7 \times 10^{-5})^2 = 1.3 \times 10^{-7}$ M

You get 1 point for the correct $[OH^-]$ and 1 point for the correct $[Fe^{2+}]$. Alternately, you get 1 point if you did only part of the procedure correctly. There is a maximum of 2 points for this part.

(c) Using the appropriate mass-action expressions:

$K_{sp} = [Sn^{2+}][OH^-]^2 = 6.3 \times 10^{-27}$
$[Sn^{2+}] = x$ and $[OH^-]^2 = 2x$
$(x)(2x)^2 = 4x^3 = 6.3 \times 10^{-27}$
$x = 1.2 \times 10^{-9}\ M = [Sn^{2+}]$

$K_{sp} = [Cr^{3+}][OH^-]^3 = 6.3 \times 10^{-31}$
$[Cr^{3+}] = x$ and $[OH^-] = 3x$
$(x)(3x)^3 = 27x^4 = 6.3 \times 10^{-31}$
$x = 1.2 \times 10^{-8}\ M = [Cr^{3+}]$

The tin(II) hydroxide beaker has the lower metal ion concentration.

You get 1 point for the correct beaker. You also get 1 point for each metal ion concentration you got correct. Your answers do not need to match exactly, but they should round to the same value.

(d) The higher the charge on the cation, the less soluble a substance is, because there is a greater attraction between the ions (higher lattice energy).

You get 1 point for this answer.

(e) The mass-action expression is:

$K_{sp} = [Pb^{2+}][OH^-]^2 = 1.1 \times 10^{-20}$
$[Pb^{2+}] = x$ and $[OH^-] = 2x$
$(x)(2x)^2 = 4x^3 = 1.1 \times 10^{-20}$
$x = 1.4 \times 10^{-7}$ M
$(1.4 \times 10^{-7}$ mol/L$)(1.00$ L$)(241.2$ g/mol$) = 3.4 \times 10^{-5}$ g

You get 1 point for the correct answer (or an answer that rounds to this answer). You get 1 point for the setup.

Total your points for the different parts. There is a maximum of 10 points possible. Subtract 1 point if all answers did not have the correct number of significant figures.

Question 2

(a) $\Delta H_{rxn}° = [2(-542.2)] - [2(-287) + 1(0)] = -510$ kJ

The setup (products – reactants) is worth 1 point, and the answer is worth 1 point. You do not need to get the exact answer, but your answer should round to this one.

(b) The answer from part **a** equals the bonds broken minus the bonds formed. Both phosphorus molecules have three P–Cl bonds, and O_2 has an O=O bond.

$[(2 \times 3 \times 331) + (498)] - [(2\ PO) + (2 \times 3 \times 331)] = -510$ kJ
PO = 504 kJ

The setup (broken – formed) is worth 1 point, and the answer is worth 1 point. You do not need to get the exact answer, but your answer should round to this one. If your answer from **a** was wrong, but you use it correctly in this part, you still get your answer point.

(c) It is a double bond. The value from part **b** is much higher than the single-bond values from the table.

You get 1 point for the correct prediction, and 1 point for the explanation. If you got the wrong answer for part **b**, you can still get 1 or 2 points if you used the answer correctly on this part.

(d) $\Delta S_{rxn}° = [2(325)] - [2(312) + 1(205.0)] = -179$ J/K

The setup (products – reactants) is worth 1 point, and the answer is worth 1 point. You do not need to get the exact answer, but your answer should round to this one.

(e) It is necessary to calculate the free-energy change.

$\Delta G_{rxn}° = \Delta H_{rxn}° - T\Delta S_{rxn}° = -510.$kJ $- (298$ K$)(1$ kJ/1000 J$)(-179$ J/K$) = -457$ kJ

The negative value means the reaction is spontaneous.

You get 1 point for the prediction that the reaction is spontaneous. The setup (plugging into the equation) is worth 1 point if you remember to change the temperature to kelvin and the joule to kilojoule conversion. An additional 1 point comes from the answer. If you got the wrong value in either part **a** or **b**, but used it correctly, you will still get the point for the answer. The free-energy equation is part of the material supplied in the exam booklet.

Total your points for the different parts. There is a maximum of 11 points possible. Subtract 1 point if all your answers do not have the correct number of significant figures.

Question 3

(a) *analytical balance   *buret   clamp
desiccator   drying oven   *Erlenmeyer flask
pH meter   pipet   support stand
wash bottle

You get 1 point if you have ALL the starred items. You get 1 point for the other items. There is a maximum of 2 points. If you have only some of the starred items, your maximum is 1 point.

(b)

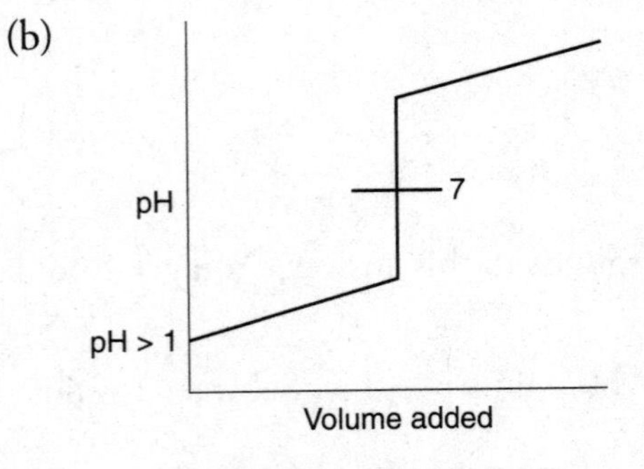

Equivalence point at pH < 7

You get 1 point for this graph. You get 1 point for noting that the equivalence point is greater than 7.

(c)

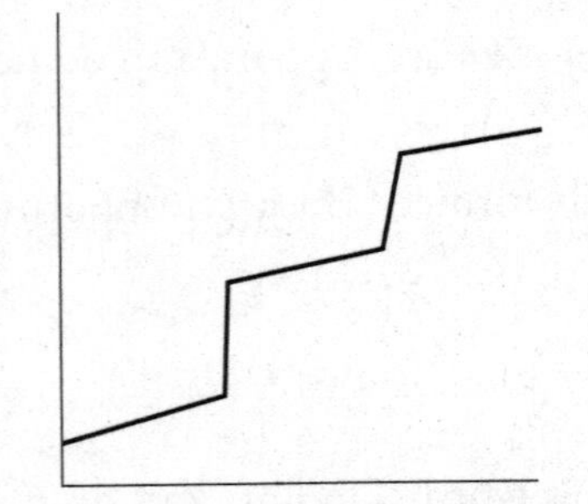

You get 1 point for this graph. You must show two regions.

(d) Weigh a sample of HA.

Titrate HA versus standard NaOH to find the volume of NaOH solution required to neutralize the acid.

Multiply the concentration of the NaOH solution times the volume used to get the moles of NaOH.

The moles of HA is the same as the moles of NaOH. Divide the mass of HA by the moles of HA.

You get 2 points if you list all five steps. If you miss one or more steps, you get only 1 point. You get 0 points if you get none of the steps correct. There are no bonus points for more steps or more details.

(e) If the KHP contained an inert impurity, the concentration of the NaOH solution would be too low. If the concentration of the NaOH solution were too low, then more solution would be needed for the titration of HA. This would yield a lower number of moles of HA, giving a higher molar mass.

You get 1 point for the NaOH concentration being low. You get 1 point for predicting a higher molar mass. If you incorrectly predicted the NaOH concentration to be too high, you can get 1 point if you predicted a lower molar mass.

Total your points. There is a maximum of 9 possible points.

Question 4

(a) analytical balance, thermometer, support stand and clamp, stoppers, glass tubing, hot plate, flask(s), clay triangle, wire gauze, beaker(s), graduated cylinder

You must have analytical balance, thermometer, glass tubing, hot plate, and flask(s) to get 1 point. Having the remaining items will get you a second point.

(b) You need the atmospheric pressure (from a barometer).

You get 1 point for this answer.

(c)

1. Determine the mass of the flask with a stopper and a small piece of glass tubing.
2. Add a small quantity of the unknown solid.
3. Place the stoppered flask (vapor may exit through the glass tube) in a hot water bath consisting of a large beaker partially filled with water and sitting on a hot plate.
4. Continue heating until the solid has completely vaporized. Use the thermometer to measure the temperature (T) of the water, and remove the flask from the hot water bath to cool. Dry the flask.
5. Weigh the flask after it cools to room temperature.
6. Fill the flask with water and use the graduated cylinder to determine the volume (V) of the flask. (As an alternative, it is possible to determine the volume using a pipet.)
7. Record the atmospheric pressure (P).
8. Use the ideal gas equation to determine the moles of solid ($n = PV/RT$). Use the gas constant and the values of T, V, and P previously determined.
9. Determine the mass (m) of the solid by subtracting the mass of the empty flask from that of the cooled flask.
10. Divide the mass of the sample (m) by the moles (n) determined from the ideal gas equation.

There are 10 steps; you must have each of the steps to get one of the variables (T, V, P, n, and m) to get 1 point. The other steps will get you 1 more point.

Total your points. There are 5 possible points.

Question 5

(a) Use the two half-reactions provided.

$$Cd^{2+} + 2\,e^- \rightarrow Cd \qquad E° = -0.40\ V$$
$$Ag^+ + 1\,e^- \rightarrow Ag \qquad E° = +0.80\ V$$

It is necessary to reverse the first reaction, the silver half-reaction needs to be doubled, and the reactions and voltages added:

$$Cd \rightarrow Cd^{2+} + 2\,e^- \qquad E° = +0.40\ V$$
$$2(Ag^+ + 1\,e^- \rightarrow Ag) \qquad E° = +0.80\ V$$
$$Cd(s) + 2\,Ag^+(aq) \rightarrow Cd^{2+}(aq) + 2\,Ag(s) \qquad E° = +1.20\ V$$

You get 1 point for the correct equation. You also get 1 point for the correct cell voltage.

(b) $Q = [Cd^{2+}]/[Ag^+]^2$

This answer is worth 1 point.

The remaining substances (Cd and Ag) are solids. Solids do not appear in Q expressions.

You get 1 point for the explanation.

Total your points; there are 4 points possible. Subtract 1 point if all answers do not have the correct number of significant figures.

Question 6

(a) :Ö=C=Ö: :C≡O:

CO_2 is linear and nonpolar. The different electronegativities of C and O make CO polar.

You get 1 point for each correct Lewis structure and 1 point if you explain both polarities correctly. There is a maximum of 3 points.

(b) There is one compound with the formula C_2H_3F, and there are three compounds with the formula $C_2H_2F_2$. The structures are:

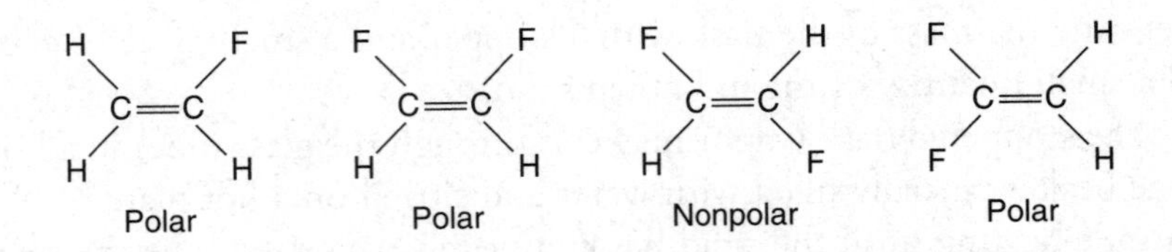

The fluorine atoms are the most electronegative atoms present, and the bonds to them are polar covalent. The only nonpolar compound is the result of the polar C—F bonds pulling equally in opposite directions.

You get 1 point for ALL the structures, and 1 point for a correct explanation. There is a maximum of 2 points.

(c) The structures are CH_3—O—CH_3 and CH_3CH_2OH.

The first compound (dimethyl ether) is polar, but not as soluble in water as the second compound (ethanol), which is capable of hydrogen bonding to water.

You get 1 point if you show both structures, and you get 1 point for a correct explanation. The names shown in parentheses are not required.

Total your points for the problem. There is a maximum of 7 possible points.

Question 7

(a) $P_{total} = 754$ mm Hg $\quad\quad P_{hydrogen} = 754 - 25.2 = 729$ mm Hg

$P = 729$ mm Hg/760 mm Hg $= 0.959$ atm

$V = 490.0$ mL $= 0.4900$ L $\quad\quad T = 26°C = 299$ K

$R = 0.08206$ L atm mol^{-1} K^{-1}

$n = PV/RT$

$= (0.959 \text{ atm} \times 0.4900 \text{ L})/(0.08206 \text{ L atm mol}^{-1} \text{ K}^{-1} \times 299 \text{ K})$

$= 0.0192 \text{ mol } H_2 \ (2.02 \text{ g } H_2/\text{mol } H_2) = 0.0388 \text{ g } H_2$

Give yourself 1 point for the correct answer (no deduction for rounding differently). You must include ALL parts of the calculation (including "=").

Give yourself 1 point for the correct equation, or for any other correct calculation. Do not give yourself more than 2 points total for this part.

(b) $P_{water} = 25.2$ mm Hg/760 mm Hg $= 0.0332$ atm. T and V are the same as in part **a**.

$n = PV/RT = (0.0332 \text{ atm} \times 0.4900 \text{ L})/(0.08206 \text{ L atm mol}^{-1} \text{ K}^{-1} \times 299 \text{ K})$

$= 6.63 \times 10^{-4} \text{ mol } H_2O$

Mass $= (6.63 \times 10^{-4} \text{ mol } H_2O) \times (18.015 \text{ g } H_2O/\text{mol } H_2O) = 0.0119 \text{ g } H_2O$

Give yourself 1 point for the correct answer (no deduction for rounding differently). You must include ALL parts of the calculation (including "="). Give yourself 1 point for the correct equation, or for any other correct calculation. Do not give yourself more than 2 points total for this part.

(c) Total mass = 0.0388 g H_2 + 0.0119 g H_2O = 0.0507 g
Density = 0.0507 g/0.4900 L = 0.103 g/L

Give yourself 1 point for the correct answer (no deduction for rounding differently). You must include ALL parts of the calculation (including "="). Give yourself 1 point for the correct equation, or for any other correct calculation. Do not give yourself more than 2 points total for this part.

Total score = sum of parts **a–c**. If any of the final answers has the incorrect number of significant figures, subtract 1 point.

Scoring Sheet for Chemistry Practice Exam 1

Multiple-Choice Section

No. correct (60 max) ____ = ____ Subtotal A

Free-Response Section

Question 1 (10 points max): __ number correct = Subtotal Q1
Question 2 (11 points max): __ number correct = Subtotal Q2
Question 3 (9 points max): __ number correct = Subtotal Q3
Question 4 (5 points max): __ number correct = Subtotal Q4
Question 5 (4 points max): __ number correct = Subtotal Q5
Question 6 (7 points max): __ number correct = Subtotal Q6
Question 7 (6 points max): __ number correct = Subtotal Q7

Subtotal Q1 + Q2 + Q3 + Q4 + Q5 + Q6 + Q7 = ____ Subtotal B

Total Raw Score = Subtotal A + [1.11 × Subtotal B]

______ + ______ = ______

Approximate Conversion Scale:

Total Raw Score	Approximate AP Grade
68–120	5
52–67	4
40–51	3
24–39	2
0–23	1

AP Chemistry Practice Exam 2—Multiple Choice

ANSWER SHEET

1 Ⓐ Ⓑ Ⓒ Ⓓ
2 Ⓐ Ⓑ Ⓒ Ⓓ
3 Ⓐ Ⓑ Ⓒ Ⓓ
4 Ⓐ Ⓑ Ⓒ Ⓓ
5 Ⓐ Ⓑ Ⓒ Ⓓ
6 Ⓐ Ⓑ Ⓒ Ⓓ
7 Ⓐ Ⓑ Ⓒ Ⓓ
8 Ⓐ Ⓑ Ⓒ Ⓓ
9 Ⓐ Ⓑ Ⓒ Ⓓ
10 Ⓐ Ⓑ Ⓒ Ⓓ
11 Ⓐ Ⓑ Ⓒ Ⓓ
12 Ⓐ Ⓑ Ⓒ Ⓓ
13 Ⓐ Ⓑ Ⓒ Ⓓ
14 Ⓐ Ⓑ Ⓒ Ⓓ
15 Ⓐ Ⓑ Ⓒ Ⓓ
16 Ⓐ Ⓑ Ⓒ Ⓓ
17 Ⓐ Ⓑ Ⓒ Ⓓ
18 Ⓐ Ⓑ Ⓒ Ⓓ
19 Ⓐ Ⓑ Ⓒ Ⓓ
20 Ⓐ Ⓑ Ⓒ Ⓓ

21 Ⓐ Ⓑ Ⓒ Ⓓ
22 Ⓐ Ⓑ Ⓒ Ⓓ
23 Ⓐ Ⓑ Ⓒ Ⓓ
24 Ⓐ Ⓑ Ⓒ Ⓓ
25 Ⓐ Ⓑ Ⓒ Ⓓ
26 Ⓐ Ⓑ Ⓒ Ⓓ
27 Ⓐ Ⓑ Ⓒ Ⓓ
28 Ⓐ Ⓑ Ⓒ Ⓓ
29 Ⓐ Ⓑ Ⓒ Ⓓ
30 Ⓐ Ⓑ Ⓒ Ⓓ
31 Ⓐ Ⓑ Ⓒ Ⓓ
32 Ⓐ Ⓑ Ⓒ Ⓓ
33 Ⓐ Ⓑ Ⓒ Ⓓ
34 Ⓐ Ⓑ Ⓒ Ⓓ
35 Ⓐ Ⓑ Ⓒ Ⓓ
36 Ⓐ Ⓑ Ⓒ Ⓓ
37 Ⓐ Ⓑ Ⓒ Ⓓ
38 Ⓐ Ⓑ Ⓒ Ⓓ
39 Ⓐ Ⓑ Ⓒ Ⓓ
40 Ⓐ Ⓑ Ⓒ Ⓓ

41 Ⓐ Ⓑ Ⓒ Ⓓ
42 Ⓐ Ⓑ Ⓒ Ⓓ
43 Ⓐ Ⓑ Ⓒ Ⓓ
44 Ⓐ Ⓑ Ⓒ Ⓓ
45 Ⓐ Ⓑ Ⓒ Ⓓ
46 Ⓐ Ⓑ Ⓒ Ⓓ
47 Ⓐ Ⓑ Ⓒ Ⓓ
48 Ⓐ Ⓑ Ⓒ Ⓓ
49 Ⓐ Ⓑ Ⓒ Ⓓ
50 Ⓐ Ⓑ Ⓒ Ⓓ
51 Ⓐ Ⓑ Ⓒ Ⓓ
52 Ⓐ Ⓑ Ⓒ Ⓓ
53 Ⓐ Ⓑ Ⓒ Ⓓ
54 Ⓐ Ⓑ Ⓒ Ⓓ
55 Ⓐ Ⓑ Ⓒ Ⓓ
56 Ⓐ Ⓑ Ⓒ Ⓓ
57 Ⓐ Ⓑ Ⓒ Ⓓ
58 Ⓐ Ⓑ Ⓒ Ⓓ
59 Ⓐ Ⓑ Ⓒ Ⓓ
60 Ⓐ Ⓑ Ⓒ Ⓓ

The AP exam is a timed exam; keep this in mind as you prepare. When taking the various tests presented in this book, you should follow the AP exam rules as closely as possible. Anyone can improve his or her score by using notes, books, or an unlimited time. You will have none of these on the AP exam, so resist the temptation to use them on practice exams. Carefully time yourself, do not use other materials, and use a calculator only when expressly allowed to do so. After you have finished an exam, you may use other sources to go over questions you missed or skipped. We have seen many students get into trouble because the first time they attempted a test under "test conditions" was on the test itself.

AP Chemistry Practice Exam 2, Section I (Multiple Choice)

Time—1 hour and 30 minutes
NO CALCULATOR MAY BE USED WITH SECTION I

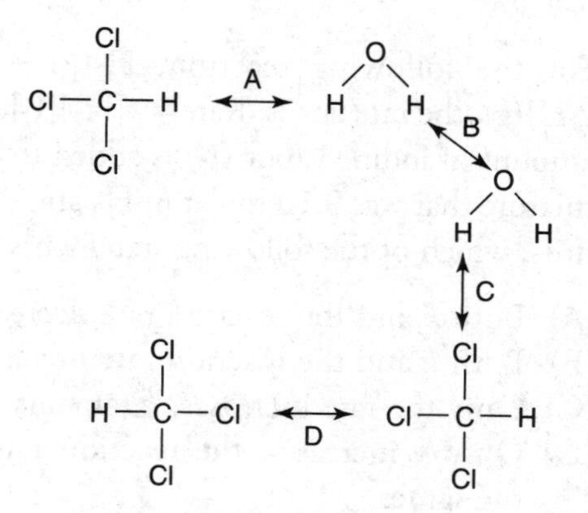

1. Which of the labeled arrows in the diagram above represents the strongest intermolecular force?

(A) A
(B) B
(C) C
(D) D

The following information applies to questions 2–5.

The following reaction takes place in a voltaic cell:

$Zn(s) + Cu^{2+}(1\ M) \rightarrow Cu(s) + Zn^{2+}(1\ M)$

The cell's voltage is measured and found to be +1.10 volts.

2. What happens to the voltage when deionized water is added to the zinc compartment?

(A) There is no change in the voltage.
(B) The voltage becomes zero.
(C) The voltage increases.
(D) The voltage decreases, but stays positive.

3. What happens to the cell voltage when the copper electrode is made larger?

(A) There is no change in the voltage.
(B) The voltage becomes zero.
(C) The voltage increases.
(D) The voltage decreases, but stays positive.

4. What happens to the cell voltage when the salt bridge is replaced with a zinc wire?

(A) There is no change in the voltage.
(B) The voltage becomes zero.
(C) The voltage increases.
(D) The voltage decreases, but stays positive.

5. What happens to the cell voltage after the cell has operated for 15 min?

(A) There is no change in the voltage.
(B) The voltage becomes zero.
(C) The voltage increases.
(D) The voltage decreases, but stays positive.

GO ON TO THE NEXT PAGE

Use the following information to answer questions 6–10.

When heated, most metals will react with oxygen in the air to form oxides. A few metals, when heated in air, will form an oxide contaminated with the metal nitride. It is often possible to remove the contaminating nitride by the addition of water or dilute acid and reheating to give the pure oxide. A student conducts an experiment to determine the formula of a metal oxide by collecting the following data:

Mass of crucible	53.120 g
Mass of crucible plus metal	53.660 g
Mass of crucible plus metal oxide	53.855 g
Atomic mass of metal	58.93 g/mole

The metal is finely powdered to ensure complete reaction. After an initial heating to form an oxide-nitride mixture, the cooled sample is treated with water to react with the nitride, which produces a gas that turns red litmus paper blue. A final careful reheating removes the unreacted water to leave the pure metal oxide.

6. What type of reaction occurs when heating the metal with air to form an oxide-nitride mixture?

(A) Decomposition
(B) Neutralization
(C) Precipitation
(D) Combination

7. What is the approximate percent oxygen in the final metal oxide?

(A) 26%
(B) 74%
(C) 50%
(D) 15%

8. The treatment of the metal nitride released a gas. What was the gas?

(A) N_2
(B) NO
(C) NH_3
(D) H_2O

9. In another experiment on the same metal a different oxide formed. This oxide was 21% oxygen. What was the formula of the second oxide?

(A) M_2O_3
(B) MO
(C) MO_2
(D) M_2O

10. What would be the approximate percent oxygen if the oxide were Fe_3O_4? (Molar masses: O = 16.00 g mol^{-1}, Fe = 55.85 g mol^{-1}, and Fe_3O_4 = 231.55 g mol^{-1}.)

(A) 28%
(B) 7%
(C) 72%
(D) 57%

11. For the following reaction, $H_2(g) + I_2(g) \rightarrow 2\ HI(g)$, the rate law is: Rate = $k[H_2][I_2]$. If a small amount of iodine vapor (I_2) is added to a reaction mixture that was 0.10 molar in H_2 and 0.20 molar in I_2, which of the following statements is true?

(A) Both k and the reaction rate decrease.
(B) Both k and the reaction rate increase.
(C) Only the rate increases, k remains the same.
(D) Only k increases, the reaction rate remains the same.

EXPERIMENT	INITIAL $[H_2]$ (mol L^{-1})	INITIAL [NO] (mol L^{-1})	INITIAL RATE OF FORMATION OF N_2O (mol L^{-1} S^{-1})
1	0.100	0.100	2.80×10^5
2	0.200	0.100	5.60×10^5
3	0.200	0.200	2.24×10^6

12. The table above gives the initial concentrations and rate for three experiments. The reaction is $H_2(g) + 2\ NO(g) \rightarrow N_2O(g) + H_2O(g)$. What is the rate law for this reaction?

(A) Rate = k[NO]
(B) Rate = $k[NO]^2[H_2]^2$
(C) Rate = $k[H_2]$
(D) Rate = $k[NO]^2[H_2]$

13. A student prepared five vinegar samples by pipetting 10.00 mL samples of vinegar into five separate beakers. Each of the samples was diluted with deionized water, and phenolphthalein was added as an indicator. The samples were then titrated with standard sodium hydroxide until the appearance of a permanent pink color indicated the endpoint of the titration. The following volumes were obtained.

	Volumes of Standard NaOH
Sample 1:	43.28 mL
Sample 2:	43.27 mL
Sample 3:	50.00 mL no color change
Sample 4:	43.26 mL
Sample 5:	43.24 mL

Which of the following is the most likely cause for the variation in the results?

(A) Too much deionized water was added to the third sample.
(B) The student forgot to add phenolphthalein to the third sample.
(C) More acetic acid was present in the third vinegar sample.
(D) The student did not properly rinse the buret with sodium hydroxide solution.

14. $Co^{2+} + 2\,e^- \rightarrow Co \quad E° = -0.28\ V$
$Zn^{2+} + 2\,e^- \rightarrow Zn \quad E° = -0.76\ V$

Given the above standard reduction potential, estimate the approximate value of the equilibrium constant for the following reaction:

$$Zn + Co^{2+} \rightarrow Zn^{2+} + Co$$

(A) 10^4
(B) 10^{16}
(C) 10^{-16}
(D) 10^{-8}

Use the following information to answer questions 15–19.

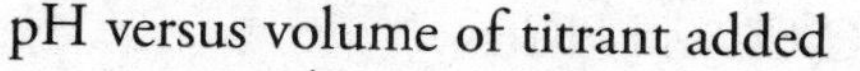

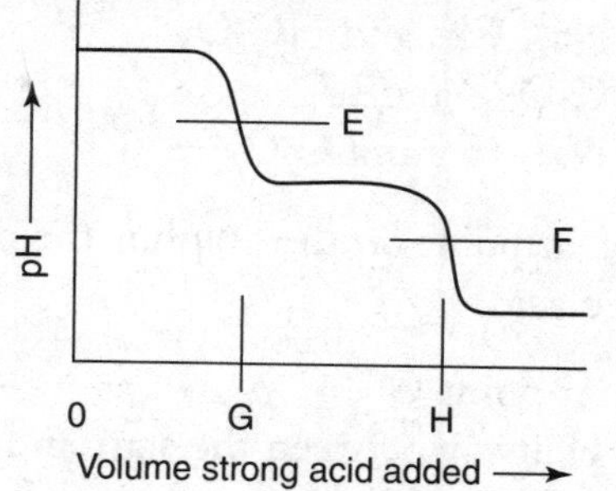

The diagram above represents the idealized titration curve for the reaction of sodium oxalate ($Na_2C_2O_4$) with a strong acid like hydrochloric acid (HCl). E and F represent the pH at the endpoints. G and H will depend on the composition of the sample with the possibility that one may not be present.

15. A trial run used a sample of pure sodium oxalate. How does the volume of acid necessary to reach G compare to the volume of acid necessary to get from G to H?

(A) They are the same.
(B) It takes more to reach point G.
(C) It takes more to get from G to H.
(D) It is impossible to determine.

16. The analysis of a sample contaminated with $NaHC_2O_4$ gave slightly different results. How does the volume of acid necessary to reach G compare to the volume of acid necessary to get from G to H for the second sample?

(A) It takes more to get from G to H.
(B) It takes more to reach point G.
(C) They are the same.
(D) It is impossible to determine.

17. The analysis of a sample contaminated with $NaHC_2O_4$ and NaCl gave slightly different results. How does the volume of acid necessary to reach G compare to the volume of acid necessary to get from G to H for the second sample?

(A) It takes more to get from G to H.
(B) It takes more to reach point G.
(C) They are the same.
(D) It is impossible to determine.

GO ON TO THE NEXT PAGE

18. In addition to water, what are the predominate species in solution at E?

(A) $Na_2C_2O_4$ and HCl
(B) Na^+, Cl^-, and $HC_2O_4^-$
(C) $C_2O_4^{2-}$ and H^+
(D) Na^+, H^+, and $C_2O_4^{2-}$

19. At what point on the graph is the pH = pK_{a2} for oxalic acid?

(A) At point G
(B) Halfway between the start and point G
(C) At point H
(D) Halfway between points G and H

20. Three flexible containers (balloons) hold gas samples. The containers are at the same temperature and pressure. One container has 2.0 g of hydrogen, another has 32.0 g of oxygen, and the third has 44.0 g of carbon dioxide. Pick the FALSE statement from the following list:

(A) The densities increase in the order hydrogen < oxygen < carbon dioxide.
(B) The number of molecules in all the containers is the same.
(C) The volume of all three containers is the same.
(D) The average speed of all the molecules is the same.

21. When cerium(III) acetate, $Ce(C_2H_3O_2)_3$, is dissolved in water, the temperature increases. Which of the following conclusions may be related to this?

(A) The hydration energies of cerium(III) ions and acetate ions are very low.
(B) Cerium(III) acetate is less soluble in hot water.
(C) The solution is not an ideal solution.
(D) The heat of solution for cerium(III) acetate is endothermic.

22. The unusually high melting point of hydrogen fluoride is due to which one of the following?

(A) Ionic bonds
(B) Hybrid orbitals
(C) Resonance structures
(D) Hydrogen bonding

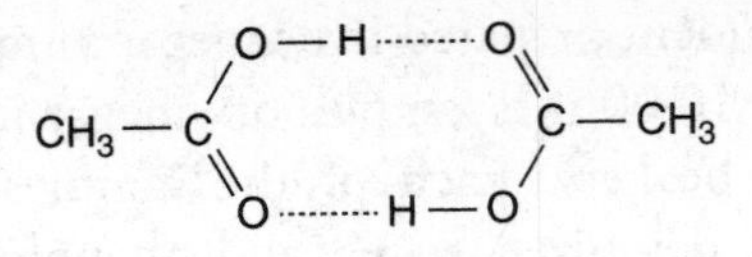

23. The diagram shows the structure of acetic acid dimers that are observed in the gas phase. This is why acetic acid molecules exist as dimers in the gaseous phase:

(A) London dispersion forces
(B) Covalent bonding
(C) Hydrogen bonding
(D) Metallic bonding

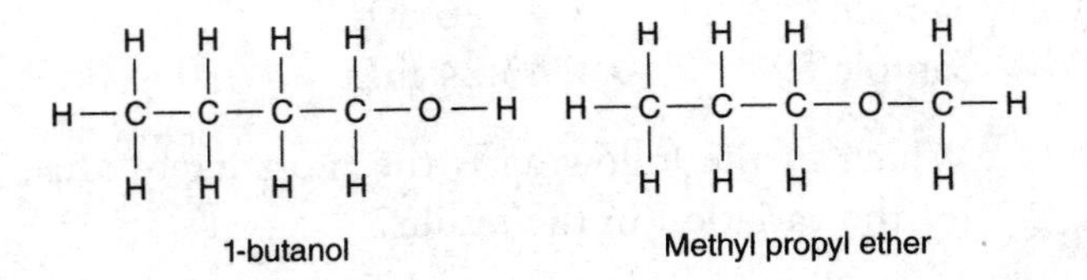

24. The above diagram shows the structures of 1-butanol and methyl propyl ether. Which of the following best explains why 1-butanol has a higher boiling point (117°C) than its isomer, methyl propyl ether (39°C)?

(A) The presence of hydrogen bonding in 1-butanol
(B) The lack of hydrogen bonding in 1-butanol
(C) The higher molecular mass of 1-butanol
(D) The lower specific heat of 1-butanol

Step 1: $(CH_3)_3CBr(aq) \rightarrow (CH_3)_3C^+(aq) + Br^-(aq)$
Step 2: $(CH_3)_3C^+(aq) + H_2O(l) \rightarrow (CH_3)_3COH_2^+(aq)$
Step 3: $(CH_3)_3COH_2^+(aq) \rightarrow H^+(aq) + (CH_3)_3COH(aq)$

25. The above represents a proposed mechanism for the hydrolysis of $(CH_3)_3CBr$. What are the overall products of the reaction?

(A) $(CH_3)_3C^+$ and Br^-
(B) $(CH_3)_3COH_2^+$ and H^+
(C) $(CH_3)_3COH$ and H^+
(D) $(CH_3)_3COH$, H^+, and Br^-

Use the information on the containers in the following diagram to answer questions 26–28.

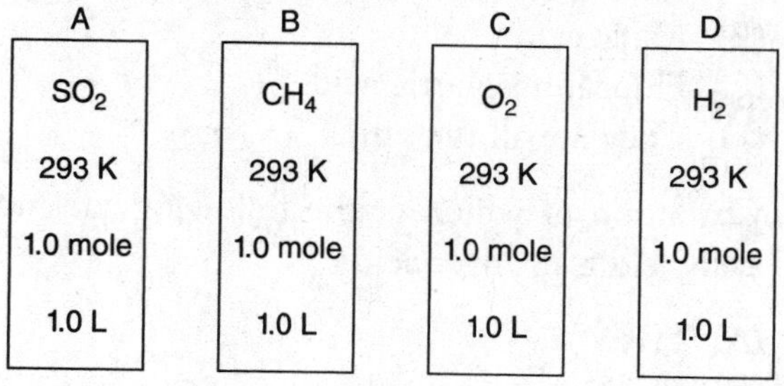

A	B	C	D
SO_2	CH_4	O_2	H_2
293 K	293 K	293 K	293 K
1.0 mole	1.0 mole	1.0 mole	1.0 mole
1.0 L	1.0 L	1.0 L	1.0 L

Approximate molar masses:

$SO_2 = 64\ g\ mol^{-1}$, $CH_4 = 16\ g\ mol^{-1}$,
$O_2 = 32\ g\ mol^{-1}$, $H_2 = 2\ g\ mol^{-1}$

26. If a sample of SO_2 effuses at a rate of 0.0035 mole per hour at 20°C, which of the gases below will effuse at approximately double the rate under the same conditions?

(A) H_2
(B) O_2
(C) CH_4
(D) Impossible to determine

27. Assuming all four gases are behaving ideally, which of the following is the same for all the gas samples?

(A) Kinetic energy
(B) Electron affinity
(C) Free energy
(D) Ionization energy

28. Which of the gases will show the greatest deviation from ideal behavior?

(A) H_2
(B) O_2
(C) CH_4
(D) SO_2

Use the information on the containers in the following diagram to answer questions 29–32 concerning the following equilibrium.

$C_2H_2(g) + H_2O(g) \rightarrow CH_3CHO(g)$ exothermic

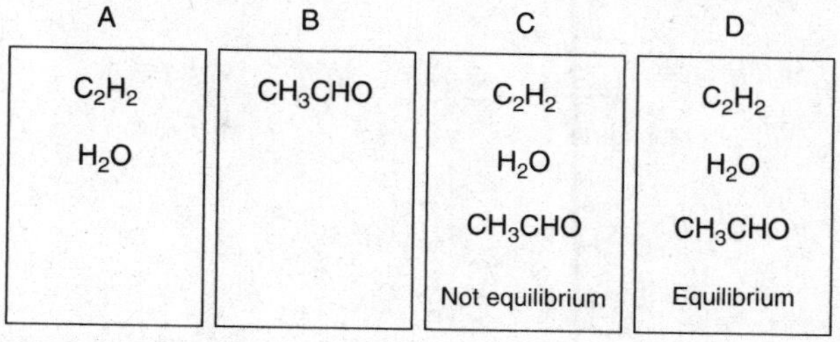

A	B	C	D
C_2H_2	CH_3CHO	C_2H_2	C_2H_2
H_2O		H_2O	H_2O
		CH_3CHO	CH_3CHO
		Not equilibrium	Equilibrium

29. An equilibrium mixture of the reactants is placed in a sealed container (container D) at 150°C. Which of the following changes may increase the amount of the product?

(A) Decreasing the volume of the container
(B) Increasing the volume of the container
(C) Raising the temperature of the container
(D) Adding 1 mole of Ar(g) to the container

30. The mixture in container C has less than the equilibrium amount of CH_3CHO. If this system is allowed to approach equilibrium, what can be said about the rates of the forward and reverse reactions?

(A) Both rates equal zero at equilibrium.
(B) Initially, the reverse rate is greater than the forward rate.
(C) Initially, the forward rate is greater than the reverse rate.
(D) Both rates are initially the same.

31. If the initial pressure in container B is 1.0 atm, what will be the pressure at equilibrium?

(A) > 1.0 atm
(B) = 1.0 atm
(C) < 1.0 atm
(D) Impossible to determine

32. If the moles of H_2O in container A are one-half the moles of C_2H_2, what can be said about the moles of H_2O present at equilibrium?

(A) It will be zero because it is the limiting reagent.
(B) It will remain the same.
(C) It will be higher.
(D) It is impossible to determine.

GO ON TO THE NEXT PAGE

Use the information on the acids in the following diagram to answer questions 33–35.

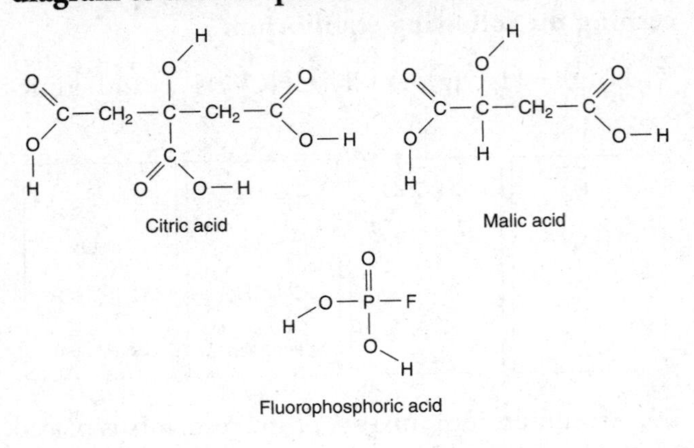

	pH of a 1.0 M solution
Citric acid	1.56
Malic acid	2.23
Fluorophosphoric acid	0.28

33. Solutions of each of the acids in the above diagram are titrated. The acid concentrations were 0.1000 M, and 0.1000 M sodium hydroxide, NaOH, was used for the titrations. Two of the acids required the same quantity of NaOH. Which two acids required the same volume of base for the titration?

(A) Citric acid and malic acid
(B) Malic acid and fluorophosphoric acid
(C) Fluorophosphoric acid and citric acid
(D) The concentrations of all the acids were the same; therefore, all the acids required the same quantity of base.

34. In general, acid dissociation calculations use the generic equation $K_a = \frac{[H^+][A^-]}{[HA]}$, where K_a is the acid equilibrium constant, $[H^+]$ is the equilibrium hydrogen ion concentration, $[A^-]$ is the equilibrium concentration of the conjugate base, and [HA] is the equilibrium concentration of the conjugate acid. The simple application of this equation will not work for these acids. Why will the simple application of this equation not work?

(A) These are polyprotic acids.
(B) The presence of –OH groups makes these bases.
(C) The equation only works for common acids, and these are not common acids.
(D) These are strong acids.

35. Which of the acids is the strongest?

(A) Citric acid
(B) Malic acid
(C) Fluorophosphoric acid
(D) They are all the same.

36. Oxidation of which of the following substances will yield a stronger acid?

(A) HIO
(B) HNO_3
(C) H_2CO_3
(D) HIO_4

37. The approximate boiling points for hydrogen compounds of some of the elements in the nitrogen family are: (SbH_3 15°C), (AsH_3 −62°C), (PH_3 −87°C), and (NH_3, −33°C). The best explanation for the fact that NH_3 does not follow the trend of the other hydrogen compounds is

(A) NH_3 is the only one that is nearly ideal in the gas phase.
(B) NH_3 is the only one that is a base.
(C) NH_3 is the only one that is water-soluble.
(D) NH_3 is the only one to exhibit hydrogen bonding.

38. Choose the reaction expected to have the greatest increase in entropy.

(A) $2\ H_2(g) + O_2(g) \rightarrow 2\ H_2O(g)$
(B) $2\ Mn_2O_7(l) \rightarrow 4\ MnO_2(s) + 3\ O_2(g)$
(C) $C(s) + O_2(g) \rightarrow CO_2(g)$
(D) $2\ Ca(s) + O_2(g) \rightarrow 2\ CaO(s)$

39. A certain reaction is nonspontaneous under standard conditions, but becomes spontaneous at lower temperatures. What conclusions may be drawn under standard conditions?

(A) $\Delta H < 0$, $\Delta S < 0$ and $\Delta G = 0$
(B) $\Delta H > 0$, $\Delta S < 0$ and $\Delta G > 0$
(C) $\Delta H < 0$, $\Delta S < 0$ and $\Delta G > 0$
(D) $\Delta H > 0$, $\Delta S > 0$ and $\Delta G > 0$

GO ON TO THE NEXT PAGE

40. Chlorine forms a number of oxyacids. Which of the following is the correct order of increasing acid strength?

(A) $HClO_3 < HClO_4 < HClO < HClO_2$
(B) $HClO_4 < HClO_3 < HClO_2 < HClO$
(C) $HClO < HClO_2 < HClO_3 < HClO_4$
(D) $HClO_4 < HClO_3 = HClO_2 < HClO$

41. Some properties for the elements silicon (Si), and tin (Sn) are listed below. Mendeleev used the properties of these two elements to predict the properties of eka-silicon (later discovered and named germanium). What are the predicted values for germanium?

PROPERTY	SILICON	GERMANIUM PREDICTED	TIN
Atomic weight	28.08 amu	_____	118.71 amu
Density of element	2.32 g/cm^3	_____	5.75 g/cm^3
Melting point of element	1410°C	_____	232°C

(A) 72.64 amu, 5.35 g/cm^3, and 937°C
(B) 73.40 amu, 5.35 g/cm^3, and 821°C
(C) 72.64 amu, 4.04 g/cm^3, and 937°C
(D) 73.40 amu, 4.04 g/cm^3, and 821°C

42. When heated, silver oxide, Ag_2O, decomposes to silver metal, Ag, plus oxygen gas, O_2. How many moles of oxygen gas will form when 4.64 g of solid silver oxide decomposes? The formula mass of silver oxide is 232.

(A) 0.100 mol
(B) 0.0100 mol
(C) 0.0200 mol
(D) 0.0150 mol

COMPOUND	FORMULA	MOLAR MASS G mol^{-1}	BOILING POINT °C
Ethyl methyl ether	C_2H_5-O-CH_3	60.10	7.4
Isopropyl alcohol	$CH_3CH(OH)CH_3$	60.10	82.3
Propylamine	C_3H_7-NH_2	59.11	47.2
Ethyl methylamine	CH_3-NH-C_2H_5	59.11	36.7

43. According to the data in the table above, which of the compounds has the strongest intermolecular forces?

(A) Isopropyl alcohol
(B) Ethyl methyl ether
(C) Propylamine
(D) Ethyl methylamine

Ion	**Ionic Radius (pm) (cubic environment)**
Na^+	132
Cd^{2+}	124
La^{3+}	130

44. The identities of both the cation and the anion are important factors in determining the lattice energies of an ionic solid. Which of the following correctly predicts and explains the order of the cations listed in the above table?

(A) $Na^+ > Cd^{2+} > La^{3+}$ because ions of elements lower on the periodic table have lower lattice energies.
(B) $Cd^{2+} > La^{3+} > Na^+$ because smaller radii lead to higher lattice energies.
(C) $La^{3+} > Cd^{2+} > Na^+$ because higher charges lead to higher lattice energies.
(D) All are about the same because their ionic radii are similar.

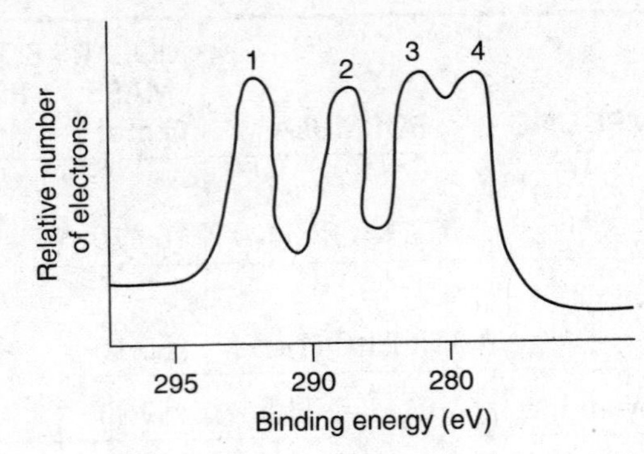

The photoelectron spectrum above shows the energy required to remove a 1s electron from a carbon atom in ethyl trifluoroacetate, shown below.

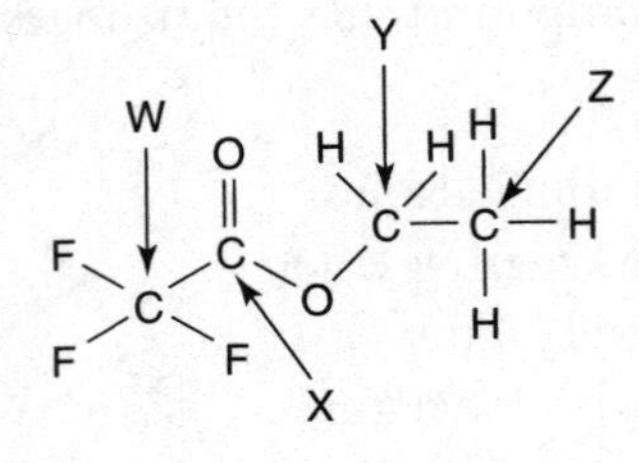

45. What is the best explanation of the position of peak 4 in the spectrum?

(A) The peak is due to atom W because of the strong interaction of the three fluorine atoms with this carbon atom.
(B) The peak is due to atom X because it is the only sp^2 hybridized carbon atom in the compound.
(C) The peak is due to atom Y because it is attached to an oxygen atom and two hydrogen atoms.
(D) The peak is due to atom Z because it is the farthest from the very electronegative fluorine and oxygen atoms.

46. Determine the $OH^-(aq)$ concentration in 0.10 M pyridine (C_5H_5N) solution. (The K_b for pyridine is 9×10^{-9}.)

(A) 9×10^{-9} M
(B) 5×10^{-6} M
(C) 7×10^{-3} M
(D) 3×10^{-5} M

$[:\ddot{O}-\ddot{N}=\ddot{O}]^-$ $\quad$ $:\ddot{O}-\dot{N}=\ddot{O}$ $\quad$ $[\ddot{O}=N=\ddot{O}]^+$

NO_2^- $\quad$ NO_2 $\quad$ NO_2^+

47. The Lewis electron-dot diagrams for three nitrogen-oxygen species are in the above diagram. Which of these has the largest bond angle?

(A) NO_2^-
(B) NO_2
(C) NO_2^+
(D) They are all the same.

$$S_2(g) + C(s) \rightarrow CS_2(g)$$

48. The above equilibrium was established at a certain temperature. The initial pressure of sulfur vapor was 0.42 atm. After establishing equilibrium, the S_2 pressure was 0.40 atm. Determine the value of K_p at this temperature.

(A) 0.02
(B) 200
(C) 0.9
(D) 0.005

49. A solution of a weak base is titrated with a solution of a standard strong acid. The progress of the titration is followed with a pH meter. Which of the following observations would occur?

(A) At the equivalence point, the pH is below 7.
(B) The pH of the solution gradually decreases throughout the experiment.
(C) At the equivalence point, the pH is 7.
(D) The pOH at the equivalence point equals the pK_b of the base.

GO ON TO THE NEXT PAGE

50. A student mixes 100.0 mL of a 0.050 M sulfuric acid, H_2SO_4, solution with 150.0 mL of a 0.10 M potassium hydroxide, KOH, solution. Which of the diagrams below best represents the ions in the solution after the reaction?

(A)

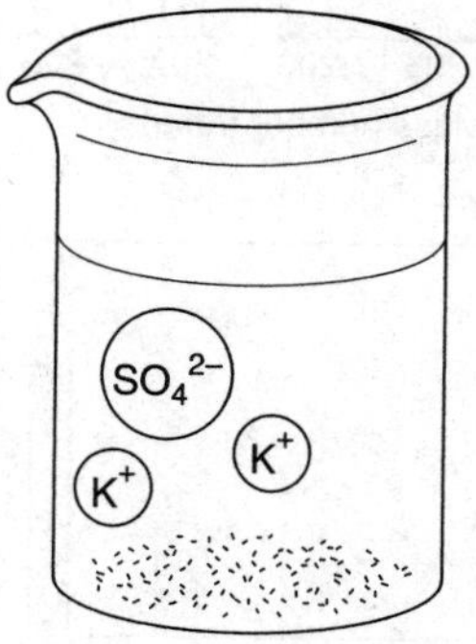

(B)

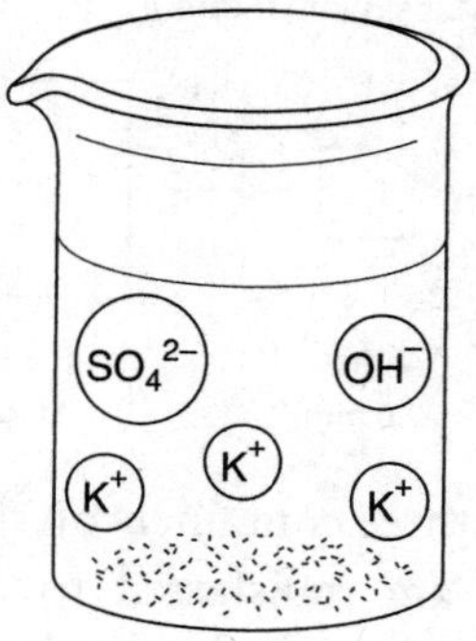

(C)

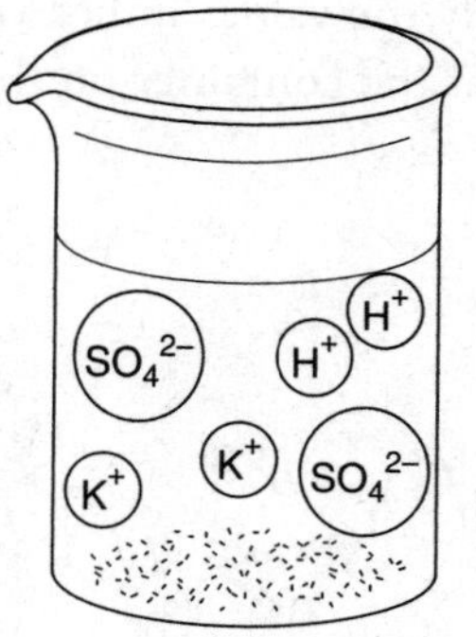

(D)

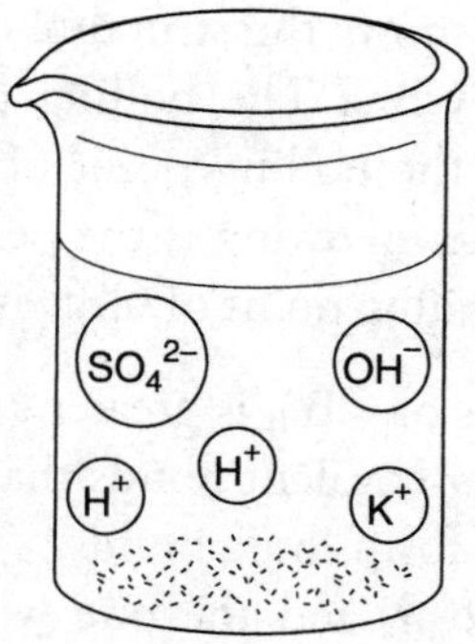

51. Which of the following best represents the result for the reaction of 50.0 mL of a 0.10 M barium hydroxide, $Ba(OH)_2$, solution with 50.0 mL of a 0.10 M sulfuric acid, H_2SO_4, solution to form a precipitate of $BaSO_4$?

(A)

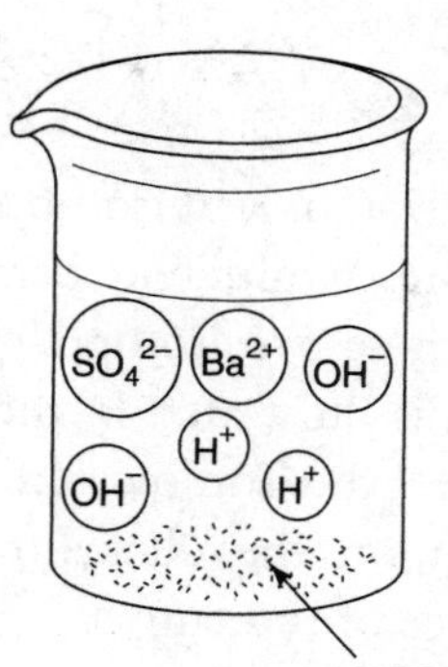

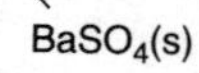

(B)

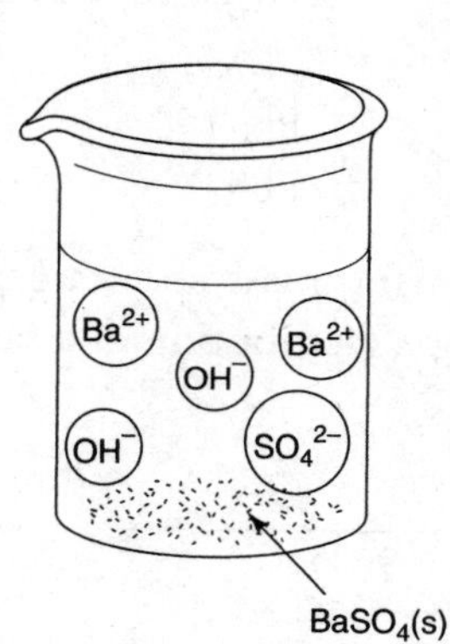

(C)

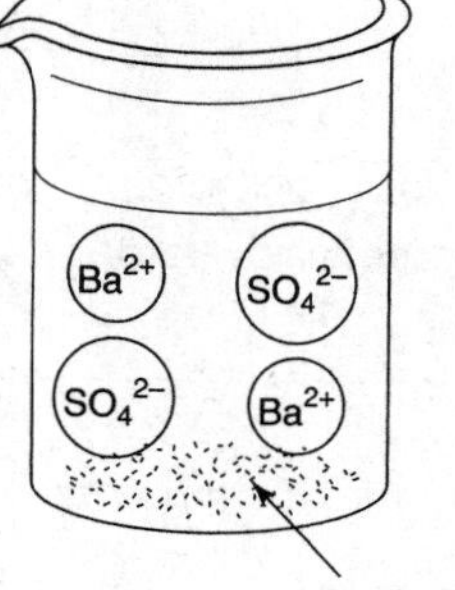

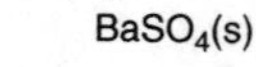

(D)

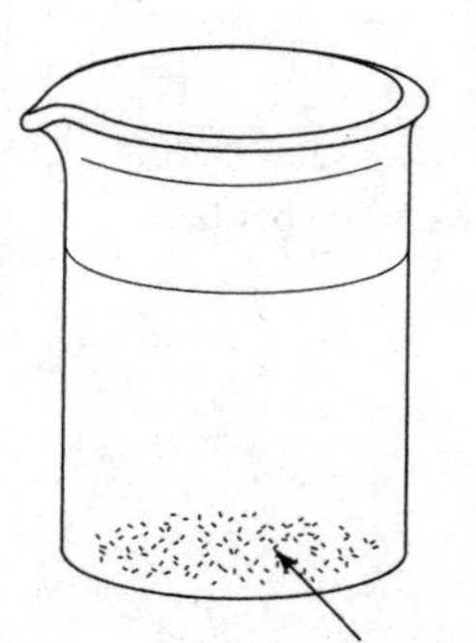

GO ON TO THE NEXT PAGE

The distribution of lead isotopes on Earth is shown in the following table.

Isotope	Percent Abundance
^{204}Pb	1.4%
^{206}Pb	24.1%
^{207}Pb	22.1%
^{208}Pb	52.4%

However, the distribution of lead isotopes in uranium and thorium ores is different because the radioactive decay of these two isotopes generates specific lead isotopes as illustrated in the following table. This will alter the isotopic composition of lead depending upon the concentration of radioisotope(s) and age of the sample.

Parent Isotope	Lead Isotope Formed
^{238}U	^{206}Pb
^{235}U	^{207}Pb
^{232}Th	^{208}Pb

52. Which of the following mass spectra best illustrates the lead isotope distribution of a sample of uranium ore?

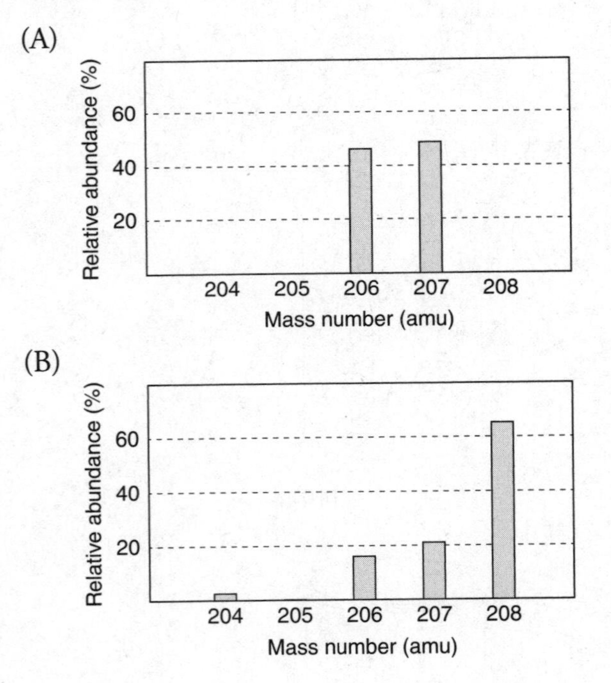

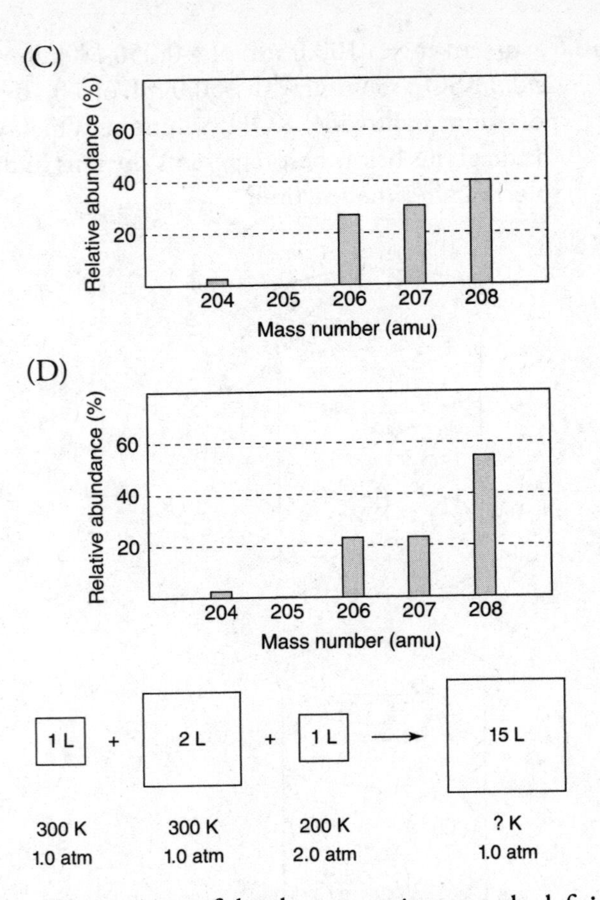

53. The contents of the three containers on the left in the diagram above are transferred to the container on the right. The volumes of the original containers are exactly the values indicated. What is the temperature of the container on the right?

(A) 370 K
(B) 250 K
(C) 1,000 K
(D) 270 K

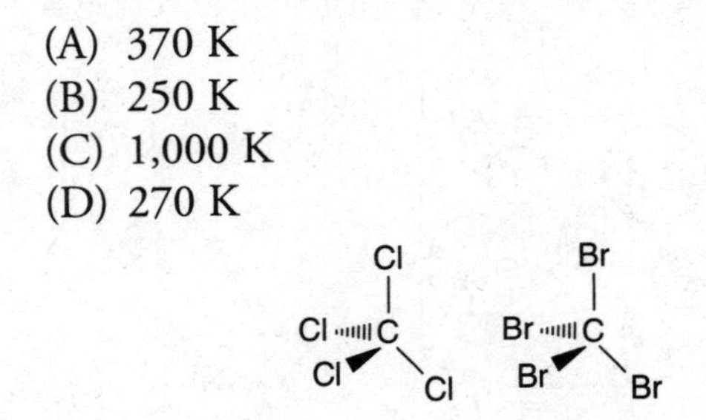

54. The diagram above shows the structure of molecules of CCl_4 and CBr_4. The boiling point of CCl_4 is 350 K and the boiling point of CBr_4 is 463 K. Which of the following is the best explanation of why the boiling point of CBr_4 is higher?

(A) The molar mass of CBr_4 is greater.
(B) CCl_4 has weaker covalent bonds than CBr_4.
(C) Only CBr_4 can form ionic bonds.
(D) Both are nonpolar, and the one with more electrons will have the stronger London dispersion forces.

GO ON TO THE NEXT PAGE

$CO(g) + Cl_2(g) \rightarrow COCl_2(g) \qquad \Delta H = -109\ kJ$

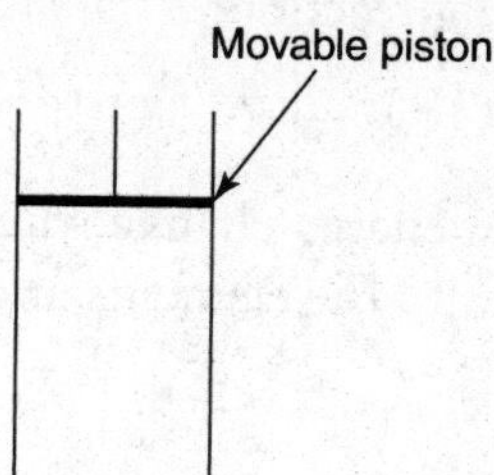

55. The above equilibrium is established in the system illustrated in the figure below the reaction. The piston quickly moves up to increase the volume of the system. Which of the following graphs best illustrates the rate of the reverse reaction?

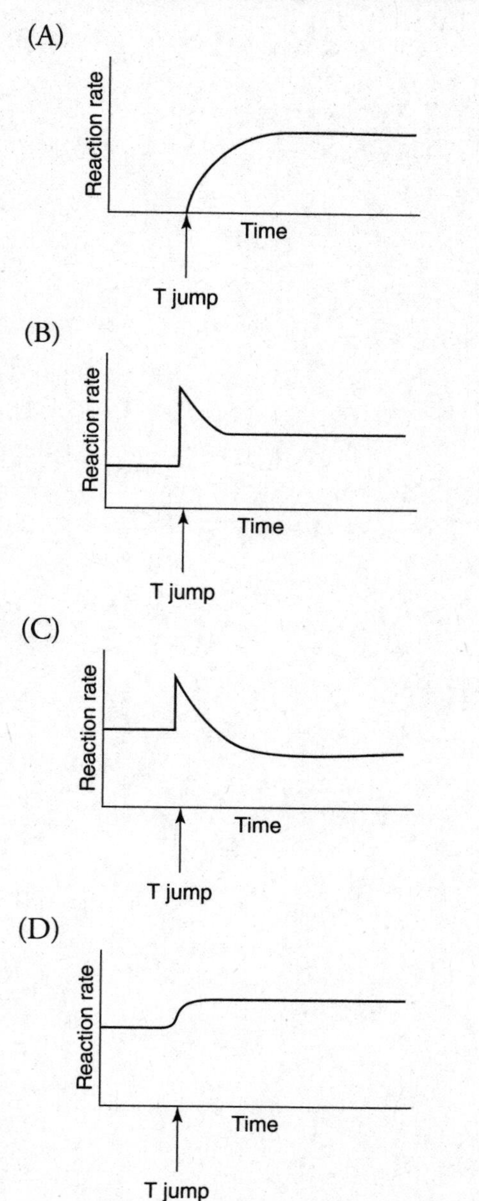

A reaction has the following suggested mechanism:

1. $2\ NO(g) \rightarrow N_2O_2(g)$
2. $N_2O_2(g) + H_2(g) \rightarrow N_2O(g) + H_2O(g)$
3. $N_2O(g) + H_2(g) \rightarrow N_2(g) + H_2O(g)$

56. Referring to the above mechanism, which of the following would support the suggested mechanism?

(A) Heating increases the rate of the reaction.
(B) N_2O_2 is detected by infrared spectroscopy.
(C) The rate constant does not change with temperature.
(D) The process does not reach equilibrium.

57. Determine the final temperature of a sample of hydrogen gas. The sample initially occupied a volume of 7.50 L at 227°C and 875 mm Hg. The sample was heated, at constant pressure, until it occupied a volume of 15.00 L.

(A) 454°C
(B) 727°C
(C) 45°C
(D) 181°C

58. A solution contains 1.00 mole of ammonium chloride, NH_4Cl, and 2.00 mole of ammonia, NH_3. The addition of a small amount of strong acid or strong base has little effect on the pH. Larger quantities of strong acid or strong base will cause a significant change in pH. How many moles of barium hydroxide, $Ba(OH)_2$, may be added before the pH begins to change significantly?

(A) 0.500 mole
(B) 1.00 mole
(C) 2.00 mole
(D) 3.00 mole

59. $4\ NO_2(g) + O_2(g) \rightarrow 2\ N_2O_5(g)$ $\Delta H = -111$ kJ

Determine ΔH for the above reaction if $N_2O_5(s)$ were formed in the above reaction instead of $N_2O_5(g)$. The ΔH of sublimation for N_2O_5 is 54 kJ/mol.

(A) +54 kJ
(B) +219 kJ
(C) +165 kJ
(D) −219 kJ

60. Hydrogen gas burns in oxygen gas according to the following reaction:

$2\ H_2(g) + O_2(g) \rightarrow 2\ H_2O(l)$ $\Delta H = -572$ kJ

What is the energy change when 18.0 g of water decomposes to the elements at constant pressure?

(A) −286 kJ
(B) −572 kJ
(C) +286 kJ
(D) +572 kJ

STOP: End of AP Chemistry Practice Exam 2, Section I (Multiple Choice).

› Answers and Explanations for Exam 2, Section I (Multiple Choice)

1. B—This is a hydrogen bond, which is stronger than a dipole–dipole force (C) or a London dispersion force (A and D).

2. C—This will lower the concentration of Zn^{2+}, causing a shift to the right.

3. A—The size of the electrode is irrelevant.

4. B—A source of cations and anions is necessary to keep the charges in each compartment neutral. If there is no salt bridge, there is no ion source.

5. D—As the cell begins to run, the voltage decreases.

6. D—This is a combination reaction where the metal combines with either oxygen or nitrogen to form either the metal oxide or metal nitride.

7. A—The percent oxygen is 100% times the grams of oxygen divided by the mass of metal oxide. The mass of oxygen is:

(mass of crucible plus metal oxide) − (mass of crucible plus metal) = (53.855 − 53.660) g = 0.195 g oxygen.

The mass of the metal oxide is:

(mass of crucible plus metal oxide) − (mass of crucible) = (53.855 − 53.120) g = 0.735 g metal oxide.

Finally,

$$\text{Percent oxygen} = \frac{0.195\text{ g O}}{0.735\text{ g}}(100\%) = 26.5\%$$

The simplified calculation would be:

$$\text{Percent oxygen} \approx \frac{0.2\text{ g O}}{0.8\text{ g}}(100\%) = 25\%$$

8. C—This is the only one of the gases that is basic (turns red litmus paper blue).

9. B—It is necessary to determine the empirical formula of the compound. If the sample is 21% oxygen, then it is 79% metal (M). Assuming 100 grams of compound, the masses of O and M are 21 g and 79 g, respectively. Converting each of these to moles gives:

$$\text{Moles M} = (79\text{ g M})\left(\frac{1\text{ mole M}}{58.93\text{ g M}}\right) = 1.3\text{ mole M}$$

$$\text{Moles O} = (21\text{ g O})\left(\frac{1\text{ mole O}}{16.00\text{ g O}}\right) = 1.3\text{ mole O}$$

Since the moles are the same, the formula must be MO.

10. A—The general equation to determine the percent oxygen in the sample is:

$$\text{Percent oxygen} = \frac{(\text{mass of oxygen})}{(\text{mass of sample})} \times 100\% =$$

$$\frac{(4 \times 16.00\text{ g O mol}^{-1})}{(231.55\text{ g mol}^{-1})} \times 100\% \approx \frac{(64) \times 100}{(232)} \approx$$

$$\frac{(64) \times 100}{(240)} \approx \frac{(64) \times 10}{(24)} \approx \frac{(32) \times 10}{(12)} \approx \frac{(8) \times 10}{(3)} \approx$$

$$\frac{75}{(3)} \approx 25\%.$$

Due to rounding the actual answer must be greater, which it is, 27.64% oxygen.

11. C—Placing a larger number into the rate law will give a larger rate. The rate constant, *k*, is constant (unless the temperature changes or a catalyst is added).

12. D—The reaction is first order in H_2 and second order in NO. The doubling of the hydrogen concentration in experiments 1 and 2 led to a doubling of the rate, which is why the hydrogen is first order. The doubling of the NO concentration in experiments 2 and 3 led to a quadrupling (2^2) of the rate, which is why the nitrogen oxide is second order.

13. B—The phenolphthalein is the source of the color change. If it is not present, there will be no color change.

14. B—The relationship is:

$\log K = nE°/0.0592 = [2 \times (0.76 - 0.28)]/0.0592 \approx (2 \times 0.5) / 0.06 \approx 17$.

If $\log K \approx 17$, then $K \approx 10^{17}$.

15. A—At point G, all the $C_2O_4^{2-}$ has been converted to $HC_2O_4^-$ and the moles of $HC_2O_4^-$ will equal the moles of $C_2O_4^{2-}$ originally present. It will require an equal volume of acid to titrate an equal number of moles of $HC_2O_4^-$ as required for the $C_2O_4^{2-}$.

16. A—At point G, all the $C_2O_4^{2-}$ has been converted to $HC_2O_4^-$ and the moles of $HC_2O_4^-$ will equal the moles of $C_2O_4^{2-}$ originally present plus the quantity of $HC_2O_4^-$ originally present. It will require a greater volume of acid to titrate a greater number of moles of $HC_2O_4^-$ as required for the $C_2O_4^{2-}$.

17. C—At point G, all the $C_2O_4^{2-}$ has been converted to $HC_2O_4^-$ and the moles of $HC_2O_4^-$ will equal the moles of $C_2O_4^{2-}$ originally present. It will require an equal volume of acid to titrate an equal number of moles of $HC_2O_4^-$ as required for the $C_2O_4^{2-}$. The NaCl will not react; therefore, it will not change the relative amounts.

18. B—All the $C_2O_4^{2-}$ is now $HC_2O_4^-$. The Na^+ did not react, so it is still present. The Cl^- is from the HCl; the H reacted with the $C_2O_4^{2-}$ to form $HC_2O_4^-$ and is no longer present. The Cl^- remained unreacted. Other than water, all species are strong electrolytes and exist as ions in solution.

19. D—The pH will equal the pK_{a2} when the concentration of $HC_2O_4^-$ equals the concentration of $H_2C_2O_4$. This occurs when one-half of the $HC_2O_4^-$ has been converted to $H_2C_2O_4$.

20. D—The average kinetic energy, not the average speed, is the same. Each has one mole of gas, which means that at the same pressure and temperature they will have the same volume. The greater the molar mass, divided by a constant volume, the greater the density. One mole of gas will have Avogadro's number of molecules.

21. B—This is an exothermic process, and exothermic processes do not proceed as well at higher temperatures.

22. D—Hydrogen bonding may occur when hydrogen is attached directly to N, O, or F.

23. C—The two molecules are hydrogen bonded together. Hydrogen bonding is a relatively strong intermolecular force.

24. A—It is the presence of the –OH group, which makes hydrogen bonding possible. Hydrogen bonding is a stronger intermolecular force than the dipole–dipole force present in methyl propyl ether.

25. D—Add the equations together and cancel any species that appear on both sides.

26. C—Use Graham's law; a molecule with one-fourth the molar mass will diffuse at double the rate.

27. A—This is a consequence of kinetic molecular theory. The average kinetic energy depends only on the absolute temperature.

28. D—Heavier molecules, especially polar species, show the greatest deviation under a given set of conditions.

29. A—This will force a shift to the right (less gas). D will yield no change because argon is not part of the equilibrium, B is the opposite of A, and C will decrease the amount of product because heating an exothermic equilibrium will cause a shift toward the reactants.

30. C—The amount of CH_3CHO is too low; therefore, to reach equilibrium, it must form (forward reaction) faster than it decomposes (reverse reaction) to reach equilibrium.

31. A—Two moles of gas come from each mole of gas decomposing. Thus, there is a net increase in the moles of gas, which leads to an increase in pressure.

32. D—To determine the amount, it would be necessary to know the value of the equilibrium constant. B and C cannot be correct because the absence of CH_3CHO would force the reaction toward the right, leading to a decrease in the reactants. A is wrong because none of the equilibrium species can ever be zero at equilibrium.

33. B—Malic and fluorophosphoric acids are both diprotic acids; therefore, they will require the same amount of base. Citric acid is a triprotic acid, which will require more base.

34. A—Each of these is a polyprotic acid; therefore, there is more than one K_a value to consider.

35. C—A 1.0 M solution of fluorophosphoric acid has the lowest pH; therefore, it is the strongest.

36. A—Only A can undergo oxidation, as there are higher oxidation states of iodine. For example, HIO_4 is a possible oxidation product.

37. D—Hydrogen bonding is possible when hydrogen is attached to N, O, and F. Only ammonia has hydrogen attached to one of these elements.

38. B—This is the reaction that has the greatest increase in the number of moles of gas.

39. C—To become spontaneous at a lower temperature means entropy impeded the reaction (entropy was negative). The enthalpy must be negative. Nonspontaneous under standard conditions means: $\Delta G > 0$.

40. C—In general, the more oxygen atoms present, the stronger the oxyacid.

41. D—These are the averages of the silicon and tin values.

42. B—The calculation is:

$$(4.64 \text{ g } Ag_2O)\left(\frac{1 \text{ mole } Ag_2O}{232 \text{ g } Ag_2O}\right)\left(\frac{1 \text{ mole } O_2}{2 \text{ mole } Ag_2O}\right) =$$

$$(4.64)\left(\frac{1}{232}\right)\left(\frac{1 \text{ mole } O_2}{2}\right) =$$

$$(2.32)\left(\frac{1}{232}\right)\left(\frac{1 \text{ mole } O_2}{1}\right) = 0.0100 \text{ mole } O_2$$

43. A—The substance with the highest boiling point has the strongest intermolecular forces.

44. C—The greater the charge, the greater the lattice energy is. The ionic radii are too similar to make a significant difference.

45. C—The very electronegative fluorine and oxygen atoms pull electrons away from the carbon atoms to which they are bonded. This causes these carbon atoms to hold on to the remaining electrons more strongly, which means that it requires more energy to remove a 1s electron from this atom. Atom Z does not have any fluorine or oxygen atoms attached; therefore, it does not hold onto electrons as strongly, which is why it takes the least amount of energy to excite electrons on this atom.

46. D—Use the K_b expression:

$$K_b = \frac{[OH^-][CA]}{[CB]} = 9 \times 10^{-9} = \frac{[x][x]}{[0.10]}$$

which leads to:

$[OH^-] = (0.10 \times 9 \times 10^{-9})^{1/2} = (9 \times 10^{-10})^{1/2} = 3 \times 10^{-5}$ M

47. C—This is a linear species (bond angle 180°), and the other two are bent species (bond angle < 180°) due to the lone electron or pair of electrons on the nitrogen.

48. D—The carbon is a solid and does not enter into the calculation.

	S_2(G)	C(S)	CS_2(G)
Initial	0.42	—	0
Change	$-x$	—	$+x$
Equilibrium	$0.42 - x$	—	$+x$

$0.42 - x = 0.40 \rightarrow x = 0.02$

$$K_p = \frac{P_{CS_2}}{P_{S_2}} = \frac{x}{0.42 - x} \approx \frac{0.02}{0.40} = 0.005$$

49. A—Strong acids and strong bases have pH = 7 at the equivalence point. The presence of a weak base with a strong acid lowers this value.

50. B—The reaction is:

$H_2SO_4(aq) + 2\ KOH(aq) \rightarrow K_2SO_4(aq) + 2\ H_2O(l)$

The sulfuric acid is the limiting reagent, and all of the hydrogen ions from the acid combine with two-thirds of the hydroxide ions added. The sulfate ions, potassium ions, and the remaining hydroxide ion are still in the solution.

51. D—The reaction is $Ba(OH)_2(aq) + H_2SO_4(aq) \rightarrow BaSO_4(s) + 2\ H_2O(l)$

Both the barium hydroxide and the sulfuric acid are limiting. All the barium ions combine with all the sulfate ions to form the $BaSO_4$ precipitate. All the hydroxide ions from the base react with all the hydrogen ions from the acid to produce water. No ions remain in solution in any significant concentration.

52. A—Uranium produces ^{206}Pb and ^{207}Pb; therefore, these peaks should be the only peaks present.

53. C—There are several ways of solving this problem. One way is to determine the moles present in the original containers, which must be the same as in the final container. In each case, moles = $n = PV/RT$. Numbering the containers from left to right as 1, 2, 3, and 4 gives:

$$n_4 = n_1 + n_2 + n_3 =$$

$$\left(\frac{(1.0\text{ atm})(1\text{ L})}{\left(0.0821\frac{\text{L}\cdot\text{atm}}{\text{mol}\cdot\text{K}}\right)(300\text{ K})}\right)+$$

$$\left(\frac{(1.0\text{ atm})(2\text{ L})}{\left(0.0821\frac{\text{L}\cdot\text{atm}}{\text{mol}\cdot\text{K}}\right)(300\text{ K})}\right)+$$

$$\left(\frac{(1.0\text{ atm})(1\text{ L})}{\left(0.0821\frac{\text{L}\cdot\text{atm}}{\text{mol}\cdot\text{K}}\right)(200\text{ K})}\right)$$

$$n_4 = \left(\frac{1}{(0.0821)(300)}\right)$$

$$+\left(\frac{2}{(0.0821)(300)}\right)+\left(\frac{1}{(0.0821)(200)}\right)$$

$$=\left(\frac{3}{(0.0821)(300)}\right)+\left(\frac{1}{(0.0821)(200)}\right)$$

$$=\left(\frac{1}{(0.0821)(100)}\right)+\left(\frac{1}{(0.0821)(200)}\right)$$

$$=\left(\frac{1}{0.0821}\right)\left(\frac{1}{100}+\frac{1}{200}\right)\text{mol}$$

$$T_4 = PV/nR =$$

$$\frac{(1.0\text{ atm})(15\text{ L})}{\left(\left(\frac{1}{0.0821}\right)\left(\frac{1}{100}+\frac{1}{200}\right)\text{mol}\right)\left(0.0821\frac{\text{L}\cdot\text{atm}}{\text{mol}\cdot\text{K}}\right)}$$

$$=\frac{(15)}{\left(\left(\frac{1}{100}+\frac{1}{200}\right)\right)\left(\frac{1}{\text{K}}\right)}$$

$$=\frac{(15)}{((0.010+0.0050))\left(\frac{1}{\text{K}}\right)}$$

$$=\frac{(15)\text{K}}{(0.015)}=1.0\times10^3\text{ K}$$

It is not necessary to write all these steps. Use shortcuts.

54. D—Stronger intermolecular forces lead to higher boiling points. Both molecules are nonpolar; therefore, the key intermolecular force is London dispersion. CBr_4 has more electrons, so it has stronger intermolecular forces.

55. B—The decrease in volume will induce an overall shift to the left, because there are more moles of gas on the right. The reverse rate cannot be zero (eliminating A). The reverse rate must be higher than the initial rate (eliminating C). A sudden change in volume will cause a sudden increase in rate (eliminating D), followed by a return to a steady rate.

56. B—N_2O_2 is an intermediate in the mechanism, detection of an intermediate supports a mechanism.

57. B—It is necessary to convert the temperature to kelvins and back again.

$$T_2 = (V_2T_1)/V_1$$
$$= [(15.00\text{ L}\times 500.\text{ K})/(7.50\text{ L})] - 273$$
$$= 727°\text{C}$$

Simplified by (15.00/7.50) = 2.00; therefore,

500. K × 2 = 1000 − 273 = 727°C

58. A—The original solution is a buffer, which will resist changes in pH until all the ammonia or ammonium ions are gone. The reaction of barium hydroxide with the ammonium ion is

$$Ba(OH)_2(aq) + 2\ NH_4^+(aq) \rightarrow Ba^{2+}(aq) + 2\ H_2O(l) + 2\ NH_4^+(aq).$$

There is 1.00 mole of ammonium ion present, which will require 0.500 mole of $Ba(OH)_2$ to completely react. After the ammonium ion completely reacts, the solution is no longer a buffer (to base) and the addition of more base will lead to a significant change in pH.

59. D—Subtract 54 × 2 mole N_2O_5 from the observed value (use Hess's law).

60. C—The decomposition of water is the reverse of the reaction shown; therefore, the enthalpy change is positive instead of negative. The amount of water decomposing is 1.00 mole, which is one-half the amount of water in the reaction. One-half the water will require one-half the energy.

AP Chemistry Practice Exam 2, Section II (Free Response)

Time—1 hour and 30 minutes

Answer the following questions in the time allowed. You may use a calculator and the resources at the back of the book. Write the answers on separate sheets of paper.

Question 1

$$C(s) + 2\ H_2(g) \rightleftharpoons CH_4(g)$$

The value of K_p for the above equilibrium is 0.262 at 1,270 K.

(a) Write the equilibrium constant expression for the above equation.
(b) What is the value of K_p for the following equilibrium at 1,270 K? Justify your answer.

$$CH_4(g) \rightleftharpoons C(s) + 2\ H_2(g)$$

(c) Methane gas is introduced to a rigid container. The container is heated to 1,270 K, and the initial partial pressure of the methane is 0.200 atm at this temperature. What is the partial pressure of H_2 once equilibrium has been established?
(d) In a separate experiment, a container with the initial equilibrium established is heated to 1,350 K. At the higher temperature, the amount of CH_4 present has decreased, and the amount of H_2 present has increased. Is the initial equilibrium endothermic or exothermic? Justify your answer.
(e) A small amount of carbon is introduced to a rigid container that is already at equilibrium. What changes, if any, would there be in the partial pressure of H_2? Justify your answer.

Question 2

The bromate ion will oxidize bromide ions to elemental bromine. It is necessary to add an acid to catalyze the reaction. Using the data in the following table, determine the rate law.

	INITIAL CONCENTRATIONS (MOLARITIES)			
EXPERIMENT	$[BrO_3^-]$ (M)	$[Br^-]$ (M)	$[H^+]$ (M)	RELATIVE RATE
1	0.10	0.50	0.30	1
2	0.10	0.50	0.60	4
3	0.20	0.50	0.60	8
4	0.10	0.50	0.60	4
5	0.10	1.00	0.30	2

(a) Determine the order of the reaction for BrO_3^-, Br^-, and H^+. Justify your answers.
(b) Write the rate law for the reaction.
(c) What are the units on the rate constant?
(d) Assuming the reaction is exothermic, draw and label the potential energy-level diagram for the reaction.

Question 3

$$5\ C_2O_4^{2-}(aq) + 2\ MnO_4^-(aq) + 16\ H^+(aq) \rightarrow Mn^{2+}(aq) + 10\ CO_2(g) + 8\ H_2O(l)$$

The above reaction is to be used in the analysis of a sample containing oxalate ion, $C_2O_4^{2-}$.

(a) Outline the general procedure for the standardization of a potassium permanganate, $KMnO_4$, solution beginning with primary standard sodium oxalate, NaC_2O_4.
(b) Outline the calculation of the concentration of the potassium permanganate solution.
(c) Show how to calculate the percent sodium oxalate in an unknown sample.
(d) List the problems arising if the primary sodium oxalate solution was not dried before being weighed.
(e) Normally, sulfuric acid, H_2SO_4, is used to supply the hydrogen ions for the reaction. What would be the problem with substituting hydrochloric acid, HCl, for sulfuric acid?

Question 4

$$4\ HCN(g) + 5\ O_2(g) \rightarrow 4\ CO_2(g) + 2\ H_2O(g) + 2\ N_2(g)$$

Thermodynamic values related to the above reaction are given in the following table.

SUBSTANCE	ΔH_f° (kJ/mol)	S° (J/mol K)
$CO_2(g)$	–393.5	213.7
$HCN(g)$	135	201.7
$H_2O(g)$	–241.83	188.72
$H_2O(l)$	–285.84	69.94
$O_2(g)$	0	205.0
$N_2(g)$	0	191.5

(a) Calculate the enthalpy change for this reaction.
(b) Calculate the entropy change for this reaction.
(c) Calculate the standard free energy for this reaction at 25°C.
(d) Calculate the standard free-energy change, at 25°C, if liquid water formed instead of water vapor.

Question 5

Answer each of the following with respect to chemical bonding and structure.

(a) The nitrite ion, NO_2^-, and the nitrate ion, NO_3^-, both play a role in nitrogen chemistry.
   i. Draw the Lewis (electron-dot) structure for the nitrite ion and the nitrate ion.
   ii. Predict which ion will have the shorter bond length, and justify your prediction.
(b) Using Lewis (electron-dot) structures, explain why the ClF_3 molecule is polar and the BF_3 molecule is not polar.

(c) Consider the following substances and their melting points:

SUBSTANCE	MELTING POINT (°C)
SrS	+2,000
KCl	770
H_2O	0.00
H_2S	–85.5
CH_4	–182

Explain the relative values of the melting points of these substances.

Question 6

Five beakers (A–E) are on a countertop. Each contains 200 mL of a 0.10 M solution. Beaker A contains $Ba(OH)_2$; beaker B contains $(NH_4)_2SO_4$; beaker C contains $(CH_3)_2CHOH$; beaker D contains K_3PO_4; and beaker E contains $FeSO_4$.

(a) Which beaker has the lowest pH? Explain.
(b) Which two beakers may be mixed to produce a gas with a characteristic odor? Write a chemical equation for the reaction.
(c) Which two solutions are basic?
(d) Which beaker will NOT give a precipitate containing the hydroxide ion when added to beaker A?

Question 7

A water-soluble sample of a solid containing some barium nitrate, $Ba(NO_3)_2$, is to be analyzed for barium. The barium is to be precipitated as barium sulfate, $BaSO_4$. Answer the following questions about this experiment.

(a) List the apparatus needed for this experiment.
(b) Outline the steps in this experiment.
(c) Set up the calculations needed to determine the percent barium in the sample.
(d) What changes would be required in the procedure if the original sample also contained lead, which also forms a precipitate with sulfate ion?

STOP. End of AP Chemistry Practice Exam 2.

› Answers and Explanations for Exam 2, Section II (Free Response)

Question 1

(a) $K_p = P_{CH_4}/P_{H2}^2 = 0.262$

You get 1 point for the correct equation. The "= 0.262" is optional.

(b) $K_{new} = 1/K_p = 1/0.262 = 3.82$

When reversing an equilibrium, you take the reciprocal of the equilibrium constant.

You get 1 point for the answer and 1 point for the explanation.

(c) $K_p = P_{CH_4}/P_{H2}^2 = 0.262$

	P_{H_2}	P_{CH_4}
Initial	0.000	0.200
Change	$+2x$	$-x$
Equilibrium	$2x$	$0.200 - x$

Substituting: $K_p = (0.200 - x)/(2x)^2 = 0.262$

$x = 0.170$

$P_{H2} = 2x = 0.340$ atm

You get 1 point for the correct answer, you get 1 point for the table, and you get 1 point for the setup. You could correctly use the equation and answer from part **c**.

(d) The reaction is exothermic. The change in the amounts of CH_4 and H_2 indicate the reaction has shifted to the left. Exothermic reactions shift to the left as the temperature increases.

You get 1 point for answering exothermic and 1 point for the explanation. You may get the explanation point even if you answered endothermic.

(e) There would be no change. Solids do not shift equilibria.

You get 1 point for "no change" and 1 point for the explanation.

Total your points for each part. There are 10 possible points. Subtract 1 point if you do not have the correct number of significant figures in all cases.

Question 2

(a) The orders for Br^- and BrO_3^- are 1. The order for H^+ is 2.

If you get all three of these correct, you get 1 point.

Comparing experiments 1 and 5: The concentration of Br^- is doubled, and the rate is doubled. This leads to Br^- having an order of 1.

You get 1 point for this reasoning.

Comparing experiments 2 and 3: The concentration of BrO_3^- is doubled, and the rate is doubled. This leads to BrO_3^- having an order of 1.

You get 1 point for this reasoning.

Comparing experiments 1 and 2: The concentration of H^+ is doubled, and the rate is quadrupled (22). This leads to H^+ having an order of 2.

You get 1 point for this reasoning.

(b) Rate = $k[Br^-][BrO_3^-][H^+]^2$

This answer is worth 1 point. You will still get 1 point if you use incorrect orders from part **a**.

(c) Rearrange the rate law to: $k = \dfrac{\text{Rate}}{[Br^-][BrO_3^-][H^+]^2}$

The rate has units of M s^{-1}, and everything else has units of M. Plugging the units in gives:

$$k = \frac{\text{Rate}}{[Br^-][BrO_3^-][H^+]^2} = \frac{\cancel{M}s^{-1}}{\cancel{M}\,M\,M^2} = M^{-3}s^{-1}$$

You get 1 point for the units.

(d)

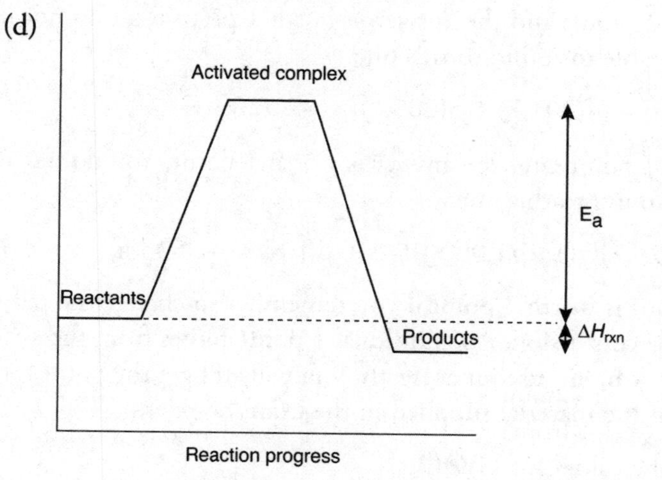

You get 1 point for showing the products being lower than the reactants. You get 1 point for labeling the activation energy. You get 1 point for correctly labeling ΔH.

Total your points for the different parts. There are 9 possible points. Subtract 1 point if you did not report the correct number of significant figures in part **c**.

Question 3

(a) 1. Samples of primary standard sodium oxalate are weighed into beakers, and deionized water is added to dissolve each of the samples.
2. A buret is rinsed with the potassium permanganate solution and then filled.
3. The potassium permanganate solution is titrated into the oxalate sample until a permanent pink color from the permanganate solution appears.

All three of these will get you 2 points. One or two of these will get you 1 point. Other items, such as heating the oxalate solution before the titration, are not necessary.

(b) (Grams sample) [(1 mol $Na_2C_2O_4$/134 g)][(2 mol MnO_4^-/5 mol $C_2O_4^{2-}$)][(1/L MnO_4^- added)]

This entire setup will get 2 points (the 134 is optional). You get 1 point if you miss a step.

(c) Percent sodium oxalate = (grams sodium oxalate/grams sample) × 100%

You get 1 point for this answer.

(d) If the primary standard were not dried, the sample would contain less sodium oxalate than indicated by the mass. The calculated concentration of the potassium permanganate solution would be too low.

This item is worth 1 point.

A low concentration for the potassium permanganate solution would yield a low percentage of sodium oxalate in the sample.

This item is worth 1 point.

(e) Hydrochloric acid might react with permanganate ion; therefore, there should be a test before using it.

This answer is worth 1 point.

Total your points for the various parts. There are 8 possible points.

Question 4

All necessary equations for this problem are in the exam booklet (and the tables at the end of this book).

(a) $\Delta H_{rxn}° = [4(-393.5) + 2(-241.83) + 2(0)] - [4(135) + 5(0)] = -2{,}598$ kJ

The setup (products – reactants) is worth 1 point, and the answer is worth 1 point. You do not need to get the exact answer, but you should be able to round to this one.

(b) $\Delta S_{rxn}° = [4(213.7) + 2(188.72) + 2(191.5)] - [4(201.7) + 5(205.0)] = -216.6$ J/K

The setup (products – reactants) is worth 1 point, and the answer is worth 1 point. You do not need to get the exact answer, but yours should round to this one.

(c) $\Delta G_{rxn}° = \Delta H_{rxn} - T\Delta S_{rxn}° = -2{,}598 \text{ kJ} - (298 \text{ K})(1 \text{ kJ}/1{,}000 \text{ J})(-216.6 \text{ J/K}) = -2{,}533$ kJ

The setup (putting values into the equation) is worth 1 point if you remember to change the temperature to kelvin and do the joule to kilojoule conversion. An additional 1 point comes from the answer. If you got the wrong value in either part **a** or **b**, but used it correctly, you will still get the point for the answer. The free-energy equation is part of the material supplied in the exam booklet.

(d) This requires a repeat of parts **a–c** using the values for H_2O(l).

$\Delta H_{rxn}° = [4(-393.5) + 2(-285.84) + 2(0)] - [4(135) + 5(0)] = -2{,}686$ kJ
$\Delta S_{rxn}° = [4(213.7) + 2(69.94) + 2(191.5)] - [4(201.7) + 5(205.0)] = -454.1$ J/K
$\Delta G_{rxn}° = \Delta H_{rxn} - T\Delta S_{rxn}° = -2{,}686 \text{ kJ} - (298 \text{ K})(1 \text{ kJ}/1{,}000 \text{ J})(-454.1 \text{ J/K}) = -2{,}551$ kJ

You get 1 point for the work and 1 point for the answer.

Total your points for each part. There are 8 possible points. Subtract 1 point if all reported answers did not have the correct number of significant figures.

Question 5

(a)

i. Nitrite ion: $[:\ddot{O}-\ddot{N}=\ddot{O}:]^-$ Nitrate ion: $[\ddot{O}=N(-:\ddot{O}:)-\ddot{O}:]^-$

Give yourself 1 point for each structure that is correct. The double bonds could be between the nitrogen and any of the oxygen atoms, not just the ones shown. Only one double bond per structure is acceptable.

ii. If you predicted the nitrite ion has the shorter bond length, you have earned 1 point. The explanation must invoke resonance. You do not need to show all the resonance structures. You need to mention that the double bond "moves" from one oxygen to another. In the nitrite ion, each N—O bond is a double bond half the time and a single bond the other half. This gives an average of 1.5 bonds between the nitrogen and each of the oxygen atoms. Similarly, for the nitrate ion, each N—O bond spends one-third of the time as a double bond, and two-thirds of the time as a single bond. The average N—O bond is 1.33. The larger the average number of bonds, the shorter the bond is.

This explanation will get you 1 point.

(b) Give yourself 1 point for each correct Lewis structure.

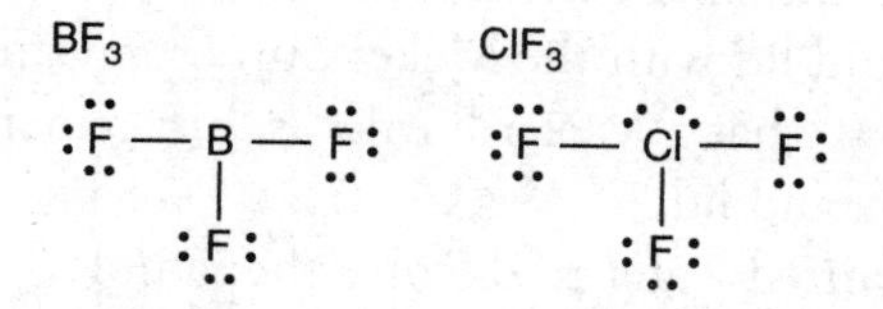

The BF_3, with three bonding pairs and no nonbonding pairs on the central atom, is not polar. The ClF_3, with five pairs about the central atom, is polar because of the two lone pairs.

Give yourself 1 point for this explanation.

(c) You get 1 point for saying that the two compounds with the highest melting points are ionic and the other compounds are molecular.

You get 1 point if you say that SrS is higher than KCl because the charges on the ions in SrS are higher.
You get 1 point if you say H_2O is higher than the lowest two because of hydrogen bonding.
You get 1 point if you say H_2S is higher than CH_4 because H_2S is polar and CH_4 is nonpolar.

Total your points for the different parts. There are a maximum of 11 points.

Question 6

(a) Beaker B has the lowest pH. The NH_4^+ ion is the conjugate acid of a weak base, and as an acid (weak), it will lower the pH. There are no other acids present.

You get 1 point for the correct beaker. You get 1 point for the explanation.

(b) Solutions A and B produce a gas with a characteristic odor when mixed. The reaction is:

$$NH_4^+(aq) + OH^-(aq) \rightarrow NH_3(g) + H_2O(l)$$

Or

$$(NH_4)_2SO_4(aq) + Ba(OH)_2(aq) \rightarrow 2\ NH_3(g) + 2\ H_2O(l) + BaSO_4(s)$$

You get 1 point for the correct two solutions, and you get 1 point for either of the balanced equations.

(c) Solutions A and D—solution A is a strong base and solution D contains the conjugate base, PO_4^{3-}, of a weak acid. The conjugate bases of weak acids undergo hydrolysis to produce basic solutions.

You get 1 point for each of these you get correct.

(d) Solution C will give no precipitate, as it is a nonelectrolyte that will not react. All other beakers, except E, will generate soluble hydroxide or release ammonia gas. Beaker E gives $Fe(OH)_2$, the only insoluble hydroxide that might form. Note that if you did not know about the solubility of barium hydroxide, the fact that the problem gave you a solution means that it must be soluble.

You get 1 point for choosing solution C.

Total your points for the different parts. There are 7 possible points.

Question 7

(a) analytical balance, beaker(s), crucible and cover, desiccator, drying oven, funnel, Meker burner, support stand, triangle, crucible support

You get 1 point for the following set: analytical balance, crucible and cover, and funnel. You get 1 point for any others you list. There is a maximum of 2 points. You can only get the second point if you have the entire first set.

(b)
1. *Weigh a sample of the solid (into a beaker).
2. Heat the crucible and lid with the Meker burner repeatedly until constant weight is achieved. (If you do not know what a Meker burner is, just remember, a burner is a burner is a burner.)
3. *Weigh the crucible and lid.
4. Add sufficient deionized water to dissolve the sample.
5. (Warm the solution.)
6. *Slowly add a solution containing the sulfate ion (usually sodium sulfate) to the solution.
7. Allow the precipitate to settle, and check for complete precipitation.
8. *Filter the precipitate and place it in the crucible.
9. Heat the crucible and lid with the sample to constant weight.
10. *Weigh the cooled crucible, lid, and sample.

You get 2 points for everything in order. Information in parentheses is optional. The first two items may be in any order. You get 1 point if the order is wrong or if any of the starred items is missing.

(c) Mass of precipitate (ppt) = (crucible + lid + sample) – (crucible + lid)

$$\text{Mass of barium in sample} = (\text{g ppt})\left(\frac{1\ \text{mol BaSO}_4}{233.34\ \text{g}}\right)\left(\frac{1\ \text{mol Ba}}{1\ \text{mol BaSO}_4}\right)\left(\frac{137.33\ \text{g Ba}}{1\ \text{mol Ba}}\right)$$

$$\text{Percentage of barium} = \left(\frac{\text{mass of barium in sample}}{\text{mass of original sample}}\right)\times 100\%$$

Give yourself 1 point for each correct equation. The values 137.33 and 233.34 are optional.

(d) Since lead also precipitates as the sulfate, it would be necessary to remove the lead before the precipitation of the barium with sulfate.

Give yourself 1 point for this answer.

There is a total of 8 points possible.

Scoring Sheet for Chemistry Practice Exam 2

Multiple-Choice Section

No. correct (60 max) ____ = ____ Subtotal A

Free-Response Section

Question 1 (10 points max): __ number correct = Subtotal Q1
Question 2 (9 points max): __ number correct = Subtotal Q2
Question 3 (8 points max): __ number correct = Subtotal Q3
Question 4 (8 points max): __ number correct = Subtotal Q4
Question 5 (11 points max): __ number correct = Subtotal Q5
Question 6 (7 points max): __ number correct = Subtotal Q6
Question 7 (8 points max): __ number correct = Subtotal Q7

Subtotal Q1 + Q2 + Q3 + Q4 + Q5 + Q6 + Q7 = ____ Subtotal B

Total Raw Score = Subtotal A + [1.11 × Subtotal B]

______ + ______ = ______

Approximate Conversion Scale:

Total Raw Score	**Approximate AP Grade**
68–120	5
52–67	4
40–51	3
24–39	2
0–23	1

Appendixes

SI UNITS

SI Prefixes

PREFIX	ABBREVIATION	MEANING
pico-	p	0.000000000001 or 10^{-12}
nano-	n	0.000000001 or 10^{-9}
micro-	μ	0.000001 or 10^{-6}
milli-	m	0.001 or 10^{-3}
centi-	c	0.01 or 10^{-2}
deci-	d	0.1 or 10^{-1}
deka-	da	10 or 10^{1}
hecto-	h	100 or 10^{2}
kilo	k	1,000 or 10^{3}
Mega-	M	1,000,000 or 10^{6}
Giga-	G	1,000,000,000 or 10^{9}
Tera-	T	1,000,000,000,000 or 10^{12}

SI Base Units and SI/English Conversions

Length

The base unit for length in the SI system is the *meter.*

1 kilometer (km) = 0.62 mile (mi)
1 mile (mi) = 1.61 kilometers (km)
1 yard (yd) = 0.914 meters (m)
1 inch (in) = 2.54 centimeters (cm)

Mass

The base unit for mass in the SI system is the *kilogram* (kg).

1 pound (lb) = 454 grams (g)
1 metric ton (t) = 10^{3} kg

Volume

The unit for volume in the SI system is the *cubic meter* (m^3).

1 dm^3 = 1 liter (L) = 1.057 quarts (qt)
1 milliliter (mL) = 1 cubic centimeter (cm^3)
1 quart (qt) = 0.946 liters (L)
1 fluid ounce (fl oz) = 29.6 milliliters (mL)
1 gallon (gal) = 3.78 liters (L)

Temperature

The base unit for temperature in the SI system is *kelvin* (K).

Celsius to Fahrenheit: °F = (9/5)°C + 32
Fahrenheit to Celsius: °C = (5/9)(°F − 32)
Celsius to kelvin: K = °C + 273.15

Pressure

The unit for pressure in the SI system is the *pascal* (Pa).

1 millimeter of mercury (mm Hg) = 1 torr
1 Pa = 1 N/m^2 = 1 $kg/m\ s^2$
1 atm = 1.01325×10^5 Pa = 760 torr
1 bar = 1×10^5 Pa

Energy

The unit for energy in the SI system is the *joule* (J).

1 J = 1 $kg\ m^2/s^2$ = 1 coulomb volt
1 calorie (cal) = 4.184 joules (J)
1 food Calorie (Cal) = 1 kilocalorie (kcal) = 4,184 joules (J)
1 British thermal unit (BTU) = 252 calories (cal) = 1,053 joules (J)

BALANCING REDOX EQUATIONS USING THE ION-ELECTRON METHOD

The following steps may be used to balance oxidation–reduction (redox) equations by the ion-electron (half-reaction) method. While other methods may be successful, none is as consistently successful as this particular method. The half-reactions used in this process will also be necessary when considering other electrochemical phenomena; thus the usefulness of half-reactions goes beyond balancing redox equations.

The basic idea of this method is to split a "complicated" equation into two parts called half-reactions. These simpler parts are then balanced separately, and recombined to produce a balanced overall equation. The splitting is done so that one of the half-reactions deals only with the oxidation portion of the redox process, whereas the other deals only with the reduction portion. What ties the two halves together is the fact that the total electrons lost by the oxidation process MUST equal the total gained by the reduction process (step 6).

It is very important that you follow each of the steps listed below completely, in order; do not try to take any shortcuts. There are many modifications of this method. For example, a modification allows you to balance all the reactions as if they were in acidic solution followed by a step, when necessary, to convert to a basic solution. Switching to a modification before you completely understand this method very often leads to confusion, and an incorrect result.

1. Assign Oxidation Numbers and Begin the Half-Reactions, One for Oxidation and One for Reduction

Beginning with the following example (phases are omitted for simplicity):

$$CH_3OH + Cr_2O_7^{2-} + H^+ \rightarrow HCOOH + Cr^{3+} + H_2O$$

(For many reactions, the substance oxidized, and the substance reduced will be obvious, so this step may be simplified. However, to be safe, at least do a partial check to confirm your predictions. Note: One substance may be both oxidized and reduced; do not let this situation surprise you—it is called disproportionation.)

Review the rules for assigning oxidation numbers if necessary, in the Basics chapter. These numbers are only used in this step. Do not force them into step 5.

Start the half-reactions with the entire molecules or ions from the net ionic form of the reaction. Do not go back to the molecular form of the reaction or just pull out atoms from their respective molecules or ions. Thus from the example above, the initial half-reactions should be:

$$CH_3OH \rightarrow HCOOH$$

$$Cr_2O_7^{2-} \rightarrow Cr^{3+}$$

The carbon is oxidized (C^{2-} to C^{2+}) and the chromium is reduced (Cr^{6+} to Cr^{3+}). Check to make sure you get the same oxidation numbers for the carbon and the chromium (hydrogen and oxygen are +1 and −2 respectively).

2. Balance All Atoms Except Oxygen and Hydrogen

(In many reactions this will have been done in step 1; because of this many people forget to check this step. This is a very common reason why people get the wrong result.)

In the above example, carbon (C) and chromium (Cr) are the elements to be considered. The carbon is balanced, so no change is required in the first half-reaction. The chromium needs to be balanced, and so the second half-reaction becomes:

$$Cr_2O_7^{2-} \rightarrow 2\ Cr^{3+}$$

Note: To carry out the next two steps correctly, it is necessary to know if the solution is acidic or basic. A basic solution is one that you are specifically told is basic, or one that contains a base of OH^- anywhere within the reaction. Assume that all other solutions are acidic (even if no acid is present).

3. Balance Oxygen Atoms

a. In Acidic Solutions Add 1 H_2O/O to the Side Needing Oxygen

b. In Basic Solutions Add 2 OH^- for Every Oxygen Needed on the Oxygen-Deficient Side, Plus 1 H_2O/O on the Opposite Side

Do not forget that two things (OH^- and H_2O) must be added in a basic solution. Also these must be added to opposite sides.

Examples:

acid: $$Cr_2O_7^{2-} \rightarrow 2\ Cr^{3+}$$

becomes:

$$Cr_2O_7^{2-} \rightarrow >2\ Cr^{3+} + 7\ H_2O$$

Now an example base:

base: $$Cr_2O_7^{2-} \rightarrow 2\ CrO_2^-$$

becomes:

$$3\ H_2O + Cr_2O_7^{2-} \rightarrow 2\ CrO_2^- + 6\ OH^-$$

4. Balance Hydrogen Atoms

a. In Acidic Solutions Add $H^+(aq)$

b. In Basic Solutions Add 1 H_2O/H Needed, Plus 1 OH^-/H on the Opposite Side

Again, do not forget that two things must be added in basic solutions (OH^- and H_2O). In this case they are still added to opposite sides, but with a different ratio.

Examples:

acid:

$$Cr_2O_7^{2-} \rightarrow 2\ Cr^{3+} + 7\ H_2O$$

becomes:

$$14\ H^+(aq) + Cr_2O_7^{2-} \rightarrow 2\ Cr^{3+} + 7\ H_2O$$

Now an example base:

base:

$$6\ OH^- + C_2H_5OH \rightarrow 2\ CO_2 + 3\ H_2O$$

becomes:

$$6\ OH^- + 6\ OH^- + C_2H_5OH \rightarrow 2\ CO_2 + 3\ H_2O + 6\ H_2O$$

If the basic step is done correctly, the oxygens should remain balanced. This may be used as a check at this point.

5. Balance Charges by Adding Electrons

The electrons must appear on opposite sides of the two half-reactions. They will appear on the left for the reduction, and on the right for the oxidation. Once added, make sure you check to verify that the total charge on each side is the same. Not being careful on this step is a major cause of incorrect answers. Do not forget to use both the coefficients and the overall charges on the ions (not the oxidation numbers from step 1).

Examples:

acid:

$$6\ e^- + 14\ H^+(aq) + Cr_2O_7^{2-} \rightarrow 2\ Cr^{3+} + 7\ H_2O$$

base:

$$6\ OH^- + 6\ OH^- + C_2H_5OH \rightarrow 2\ CO_2 + 3\ H_2O + 6\ H_2O + 12\ e^-$$

6. Adjust the Half-Reactions So That They Both Have the Same Number of Electrons

(Find the lowest common multiple, and multiply each of the half-reactions by the appropriate factor to achieve this value. This is the key step, as the number of electrons lost MUST equal the number gained.)

Example:

Lowest common multiple = 12

$$3 \times (H_2O + CH_3OH \rightarrow HCOOH + 4\ H^+(aq) + 4\ e^-)$$

$$2 \times (6\ e^- + 14\ H^+(aq) + Cr_2O_7^{2-} \rightarrow 2\ Cr^{3+} + 7\ H_2O)$$

giving:

$$3\ H_2O + 3\ CH_3OH \rightarrow 3\ HCOOH + 12\ H^+(aq) + 12\ e^-$$

$$12\ e^- + 28\ H^+(aq) + 2\ Cr_2O_7^{2-} \rightarrow 4\ Cr^{3+} + 14\ H_2O$$

7. Add the Half-Reactions and Cancel

(The electrons must cancel.)

Example (from step 6):

$$12\ e^- + 3\ H_2O + 3\ CH_3OH + 28\ H^+(aq) + 2\ Cr_2O_7^{2-}$$
$$\rightarrow 4\ Cr^{3+} + 14\ H_2O + 3\ HCOOH + 12\ H^+(aq) + 12\ e^-$$

becomes:

$$3\ CH_3OH + 16\ H^+(aq) + 2\ Cr_2O_7^{2-} \rightarrow 4\ Cr^{3+} + 11\ H_2O$$
$$+ 3\ HCOOH$$

8. Check to See If All Atoms Balance and That the Total Charge on Each Side Is the Same

This step will let you know whether you have done everything correctly.

If all the atoms and charges do not balance, you have made a mistake. Look over your work. If you have made an obvious mistake, then you should correct it. If the mistake is not obvious, it may take less time to start over from the beginning. The most common mistakes are made in steps 2 and 5, or step 3 in a basic solution.

Make sure you learn to apply each of the preceding steps. Look over the individual examples and make sure you understand them separately. Then make sure you learn the order of these steps. Finally, balance redox reactions; this will take a lot of practice. Make sure that you reach the point of being able to consistently balance equations without looking at the rules.

COMMON IONS

Ions Usually with One Oxidation State

Ion	Name	Ion	Name
Li^+	lithium ion	N^{3-}	nitride ion
Na^+	sodium ion	O^{2-}	oxide ion
K^+	potassium ion	S^{2-}	sulfide ion
Mg^{2+}	magnesium ion	F^-	fluoride ion
Ca^{2+}	calcium ion	Cl^-	chloride ion
Sr^{2+}	strontium ion	Br^-	bromide ion
Ba^{2+}	barium ion	I^-	iodide ion
Ag^+	silver ion		
Zn^{2+}	zinc ion		
Cd^{2+}	cadmium ion		
Al^{3+}	aluminum ion		

Cations with More than One Oxidation State

	+1		**+2**
Cu^{1+}	copper(I) ion or cuprous ion	Cu^{2+}	copper(II) ion or cupric ion
Hg_2^{2+}	mercury(I) ion or mercurous ion	Hg^{2+}	mercury(II) ion or mercuric ion
	+2		**+3**
Fe^{2+}	iron(II) ion or ferrous ion	Fe^{3+}	iron(III) ion or ferric ion
Cr^{2+}	chromium(II) ion or chromous ion	Cr^{3+}	chromium(III) ion or chromic ion
Mn^{2+}	manganese(II) ion or manganous ion	Mn^{3+}	manganese(III) ion or manganic ion
Co^{2+}	cobalt(II) ion or cobaltous ion	Co^{3+}	cobalt(III) ion or cobaltic ion
	+2		**+4**
Sn^{2+}	tin(II) ion or stannous ion	Sn^{4+}	tin(IV) ion or stannic ion
Pb^{2+}	lead(II) ion or plumbous ion	Pb^{4+}	lead(IV) ion or plumbic ion

Polyatomic Ions and Acids

Formula	**Name**	**Ion**	**Ion name**
H_2SO_4	sulfuric acid	SO_4^{2-}	sulfate ion
H_2SO_3	sulfurous acid	SO_3^{2-}	sulfite ion
HNO_3	nitric acid	NO_3^-	nitrate ion
HNO_2	nitrous acid	NO_2^-	nitrite ion

Formula	Name	Ion	Ion name
H_3PO_4	phosphoric acid	PO_4^{3-}	phosphate ion
H_2CO_3	carbonic acid	CO_3^{2-}	carbonate ion
$HMnO_4$	permanganic acid	MnO_4^-	permanganate ion
HCN	hydrocyanic acid	CN^-	cyanide ion
$HOCN$	cyanic acid	OCN^-	cyanate ion
$HSCN$	thiocyanic acid	SCN^-	thiocyanate ion
$HC_2H_3O_2$	acetic acid	$C_2H_3O_2^-$	acetate ion
$H_2C_2O_4$	oxalic acid	$C_2O_4^{2-}$	oxalate ion
H_2CrO_4	chromic acid	CrO_4^{2-}	chromate ion
$H_2Cr_2O_7$	dichromic acid	$Cr_2O_7^{2-}$	dichromate ion
$H_2S_2O_3$	thiosulfuric acid	$S_2O_3^{2-}$	thiosulfate ion
H_3AsO_4	arsenic acid	AsO_4^{3-}	arsenate ion
H_3AsO_3	arsenous acid	AsO_3^{3-}	arsenite ion

Oxyhalogen Acids

Formula	Oxy name	Ion	Ion name
$HClO$	hypochlorous acid	ClO^-	hypochlorite ion
$HClO_2$	chlorous acid	ClO_2^-	chlorite ion
$HClO_3$	chloric acid	ClO_3^-	chlorate ion
$HClO_4$	perchloric acid	ClO_4^-	perchlorate ion

Br or I can be substituted for chlorine Cl. F may form hypofluorous acid and the hypofluorite ion.

Other Ions

Ion	Ion name
O_2^{2-}	peroxide ion
OH^-	hydroxide ion
HSO_4^-	bisulfate ion; hydrogen sulfate ion
NH_4^+	ammonium ion
O_2^-	superoxide ion
HCO_3^-	bicarbonate ion; hydrogen carbonate ion
HPO_4^{2-}	hydrogen phosphate ion
$H_2PO_4^-$	dihydrogen phosphate ion

Ligands

Ligand	Formula(abbreviation)	Ligand name
Bromide ion	Br^-	bromo
Carbonate ion	CO_3^{2-}	carbonato
Chloride ion	Cl^-	chloro

Ligand	Formula(abbreviation)	Ligand name
Cyanide ion	CN^-	cyano
Fluoride ion	F^-	fluoro
Hydride ion	H^-	hydrido
Hydroxide ion	OH^-	hydroxo
Iodide ion	I^-	iodo
Nitrite ion	NO_2^-	nitrito
Oxalate ion	$C_2O_4^{2-}$	oxalato
Sulfide ion	S^{2-}	thio
Thiocyanate ion	SCN^-	thiocyanato
Ammonia	NH_3	ammine
Ethylenediamine	en	ethylenediamine
Water	H_2O	aqua

Colors of Common Ions in Aqueous Solution

Most common ions are colorless in solution; however, some have distinctive colors. These colors have appeared in questions on the AP exam.

Fe^{2+} and Fe^{3+}	various colors
Cu^{2+}	blue to green
Cr^{2+}	blue
Cr^{3+}	green or violet
Mn^{2+}	faint pink
Ni^{2+}	green
Co^{2+}	pink
MnO_4^-	dark purple
CrO_4^{2-}	yellow
$Cr_2O_7^{2-}$	orange

BIBLIOGRAPHY

Brown, Theodore L., H. Eugene Le May, Jr., and Bruce E Bursten. 2011. *Chemistry. The Central Science with Mastering Chemistry*, 12th ed. Upper Saddle Creek, NJ: Prentice Hall.

Cates, Charles R., Richard H. Langley, and John T. Moore. 2003. *Introductory Chemical Practice: A Quantitative Approach*, 6th ed. San Francisco, CA: Burgess Publishing.

Goldberg, David E. 2006. *Fundamentals of Chemistry*, 5th ed. New York: McGraw-Hill Education.

Hostage, David, and Martin Fossett. 2005. *Laboratory Investigations: AP Chemistry,* Saddle Brook, NJ: Peoples Publishing Group.

Moore, John T. 2011. *Chemistry for Dummies*. Hoboken, NJ: Wiley Publishing, Inc.

Moore, John T. 2010. *Chemistry Essentials for Dummies.* Hoboken, NJ: Wiley Publishing, Inc.

Moore, John T., and Richard Langley. 2007. *Chemistry for the Utterly Confused*. New York: McGraw-Hill Education.

Russo, Steve, and Mile Silver. 2010. *Introductory Chemistry*, 4th ed. Upper Saddle Creek, NJ: Prentice Hall.

Silberberg, Martin S. 2011. *Chemistry: The Molecular Nature of Matter and Change*, 6th ed. New York: McGraw-Hill Education.

Zumdahl, Steven S., and Susan A. Zumdahl. 2013. *Chemistry,* 9th ed. Belmont, CA: Brooks Cole.

WEBSITES

Here is a list of websites that contain information and links that you might find useful in your preparation for the AP Chemistry exam:

www.chemistry.about.com
www.webelements.com
www.collegeboard.com/student/testing/ap/sub_chem.html?chem
www.shs.nebo.edu/Faculty/Haderlie/apchem/apchem.html
www.chemistrygeek.com/ap.htm
www.chem.purdue.edu/apchem/MainPage/AP_References/KEYS_TO_PASS.html
www.rsc.org/

GLOSSARY

absolute zero Absolute zero is 0 K and is the point at which all molecular motion ceases.

acid dissociation constant (K_a) The acid dissociation constant is the equilibrium constant associated with a weak acid dissociation in water.

acidic A solution whose pH is *less* than 7.00 is said to be acidic.

acids Acids are proton (H^+) donors.

activation energy Activation energy is the minimum amount of energy that must be supplied to initiate a chemical reaction.

activity series for metals The activity series lists metals in order of decreasing ease of oxidation.

actual yield The actual yield is the amount of product that is actually formed in a chemical reaction.

alkali metals Alkali metals are in Group 1 on the periodic table.

alkaline earth metals Alkaline earth metals are in Group 2 on the periodic table.

alkanes Alkanes are hydrocarbons that contain only single covalent bonds within the molecule.

alkenes Alkenes are hydrocarbons that contain a carbon-to-carbon double bond.

alkynes Alkynes are hydrocarbons that contain a carbon-to-carbon triple bond.

alpha particle An alpha particle is essentially a helium nucleus with two protons and two neutrons.

amorphous solids Amorphous solids are solids that lack extensive ordering of the particles.

amphoteric Amphoteric substances will act as either an acid or a base, depending on whether the other species is a base on an acid.

amplitude Amplitude is the height of a wave and is related to the intensity (or brightness for visible light) of the wave.

amu An amu is $\frac{1}{12}$ the mass of a carbon atom that contains 6 protons and 6 neutrons (C-12).

angular momentum quantum number (l) The angular momentum quantum number is the quantum number that describes the shape of the orbital.

anions Anions are negatively charged ions.

anode The electrode at which oxidation is taking place is called the anode.

anode compartment The anode compartment is the electrolyte solution in which the anode is immersed.

aqueous solution An aqueous solution is a solution in which water is the solvent.

atomic number (Z) The atomic number of an element is the number of protons in the nucleus.

atomic orbital The atomic orbital is the region of space in which it is most likely to find a specific electron in an atom.

atomic solids In atomic solids, individual atoms are held in place by London forces.

Aufbau principle The Aufbau principle states that the electrons in an atom fill the lowest energy levels first.

Avogadro's law Avogadro's law states that there is a direct relationship between the volume and the number of moles of gas.

Avogadro's number Avogadro's number is the number of particles (atoms or molecules or ions) in a mole and is numerically equal to 6.022×10^{23} particles.

barometer A barometer is an instrument for measuring atmospheric pressure.

base dissociation constant, K_b The base dissociation constant is the equilibrium constant associated with the dissociation of a weak base in water.

bases Bases are defined as proton (H^+) acceptors.

basic A solution whose pH is *greater* than 7.00 is called basic.

beta particle A beta particle is an electron.

bimolecular reactions Bimolecular reactions are chemical reactions that involve the collision of two chemical species.

binary compounds Binary compounds are compounds that consist of only two elements.

body-centered unit cell A body-centered unit cell has particles located at the corners of a cube and in the center of the cube.

boiling The process of going from the liquid state to the gaseous state is called boiling.

boiling point The boiling point (b.p.) is the temperature at which a liquid boils.

bond order The bond order relates the bonding and antibonding electrons in the molecular orbital theory (# electrons in bonding MOs – # electrons in antibonding MOs)/2.

Boyle's law Boyle's law states that there is an inverse relationship between the volume and pressure of a gas, if the temperature and amount are kept constant.

buffer capacity The buffer capacity is the ability of the buffer to resist a change in pH.

buffers Buffers are solutions that resist a change in pH when an acid or base is added to them.

calorie The calorie is the amount of energy needed to raise the temperature of 1 gram of water 1°C.

calorimetry Calorimetry is the laboratory technique used to measure the heat released or absorbed during a chemical or physical change.

capillary action Capillary action is the spontaneous rising of a liquid through a narrow tube against the force of gravity.

catalyst A catalyst is a substance that speeds up the reaction rate and is (at least theoretically) recoverable at the end of the reaction in an unchanged form.

cathode The cathode is the electrode in an electrochemical cell at which reduction takes place.

cathode compartment The cathode compartment is the electrolyte solution in which the cathode is immersed.

cations Cations are positively charged ions.

cell notation Cell notation is a shorthand notation for representing a galvanic cell.

Charles's law Charles's law states that there is a direct relationship between the volume and temperature of a gas, if the pressure and amount are kept constant.

chemical equilibrium A chemical equilibrium has been reached when two exactly opposite reactions are occurring at the same place, at the same time, and with the same rates of reaction.

colligative properties Colligative properties are solution properties that are simply dependent upon the *number* of solute particles, and not the type of solute.

colloids Colloids are homogeneous mixtures in which solute diameters fall in between solutions and suspensions.

combination reactions Combination reactions are reactions in which two or more reactants (elements or compounds) combine to form one product.

combined gas equation The combined gas equation relates the pressure, temperature, and volume of a gas, assuming the amount is held constant.

combustion reactions Combustion reactions are redox reactions in which the chemical species rapidly combines with oxygen and usually emits heat and light.

common-ion effect The common-ion effect is an application of Le Châtelier's principle to equilibrium systems of slightly soluble salts.

complex A complex is composed of a central atom, normally a metal, surrounded by atoms or groups of atoms called ligands.

compounds Compounds are pure substances that have a fixed proportion of elements.

concentrated Concentrated is a qualitative way of describing a solution that has a relatively large amount of solute in comparison to the solvent.

concentration Concentration is a measure of the amount of solute dissolved in the solvent.

concentration cell A concentration cell is an electrochemical cell in which the same chemical species are used in both cell compartments, but differ in concentration.

conjugate acid–base pair This is an acid–base pair that differs by only a single H^+.

continuous spectrum A continuous spectrum is a spectrum of light much like the rainbow.

coordinate covalent bonds Coordinate covalent bonds are covalent bonds in which one of the atoms furnishes both of the electrons for the bond.

coordination compounds Coordination compounds are a type of complex in which a metal atom is surrounded by ligands.

coordination number Coordination number is the number of ligands that can covalently bond to the metal ion in the complex ion.

covalent bonding In covalent bonding, one or more electron pairs are shared between two atoms.

crisscross rule The crisscross rule can be used to help determine the formula of an ionic compound.

critical point The critical point of a substance is the point on the phase diagram beyond which the gas and liquid phases are indistinguishable from each other.

crystal lattice The crystal lattice is a three-dimensional structure that crystalline solids occupy.

crystalline solids Crystalline solids display a very regular ordering of the particles (atoms, molecules, or ions) in a three-dimensional structure called the crystal lattice.

Dalton's law Dalton's law states that in a mixture of gases (A + B + C . . .) the total pressure is simply the sum of the partial pressures (the pressures associated with each individual gas).

decomposition reactions Decomposition reactions are reactions in which a compound breaks down into two or more simpler substances.

diamagnetism Diamagnetism is the repulsion of a molecule from a magnetic field due to the presence of all electrons in pairs.

dilute Dilute is a qualitative term that refers to a solution that has a relatively small amount of solute in comparison to the amount of solvent.

dimensional analysis Dimensional analysis, sometimes called the factor label method, is a method for generating a correct setup for a mathematical problem.

dipole–dipole intermolecular force Dipole–dipole intermolecular forces occur between polar molecules.

double displacement (replacement) or metathesis reaction A double displacement (replacement) or metathesis reaction is a chemical reaction where at least one insoluble product is formed from the mixing of two solutions.

effective nuclear charge The overall attraction that an electron experiences is called the effective nuclear charge. This is less than the actual nuclear charge, because other electrons interfere with the attraction of the protons for the electron being considered.

electrochemical cells Electrochemical cells use indirect electron transfer to produce electricity by a redox reaction, or they use electricity to produce a desired redox reaction.

electrochemistry Electrochemistry is the study of chemical reactions that produce electricity and chemical reactions that take place because electricity is supplied.

electrode The electrode is that solid part of the electrochemical cell that conducts the electrons that are involved in the redox reaction.

electrode compartment The solutions in which the electrodes are immersed are called the electrode compartments.

electrolysis Electrolysis is a reaction in which electricity is used to decompose a compound.

electrolyte An electrolyte is a substance which, when dissolved in solution or melted, conducts an electrical current.

electrolytic cells Electrolytic cells use electricity from an external source to produce a desired redox reaction.

electromagnetic spectrum The electromagnetic spectrum is radiant energy, composed of gamma rays, X-rays, ultraviolet light, visible light, etc.

electron affinity The electron affinity is the energy change that results from adding an electron to an atom or ion.

electron capture Electron capture is a radioactive decay mode that involves the capturing of an electron from the energy level closest to the nucleus (1s) by a proton in the nucleus.

electron cloud The electron cloud is a volume of space in which the probability of finding the electron is high.

electronegativity The electronegativity (EN) is a measure of the attractive force that an atom exerts on a bonding pair of electrons.

electronic configuration The electronic configuration is a condensed way of representing the pattern of electrons in an atom.

elementary step Elementary steps are the individual reactions in the reaction mechanism or pathway.

empirical formula The empirical formula is a chemical formula that tells us which elements are present in the compound and the simplest whole-number ratio of elements.

endothermic Endothermic reactions absorb energy from their surroundings.

endpoint The endpoint of a titration is the point signaled by the indication that an equivalent amount of base has been added to the acid sample, or vice versa.

enthalpy The enthalpy change, ΔH, is the heat gained or lost by the system during constant pressure conditions.

entropy Entropy (S) is a measure of the disorder of a system.

equilibrium constant The quantity calculated when the equilibrium concentrations of the chemical species are substituted into the reaction quotient.

equivalence point The equivalence point is that point in the titration where the moles of H^+ in the acid solution have been exactly neutralized with the same number of moles of OH^-.

excited state An excited state of an atom is an energy state of higher energy.

exothermic An exothermic reaction releases energy (heat) to its surroundings.

face-centered unit cell The face-centered unit cell has particles at the corners and one in the center of each face of the cube, but not in the center of the cube.

First Law of Thermodynamics The First Law of Thermodynamics states that the total energy of the universe is constant.

formation constant The formation constant is the equilibrium constant for the formation of a complex ion from a metal ion and ligands.

frequency The frequency, ν, is defined as the number of waves that pass a point per second.

functional group Functional groups are reactive groups on a compound that react in a characteristic way no matter what the rest of the molecule consists of.

galvanic (voltaic) cells Galvanic (voltaic) cells are electrochemical cells that produce electricity by a redox reaction.

gamma emission Gamma emission is a radioactive decay process in which high-energy, short-wavelength photons that are similar to X-rays are given off.

gas A gas is a state of matter that has neither definite shape nor volume.

Gay-Lussac's law Gay-Lussac's law describes the direct relationship between the pressure of a gas and its Kelvin temperature, if the volume and amount are held constant.

Gibbs free energy The Gibbs free energy (G) is a thermodynamic function that combines the enthalpy, entropy, and temperature. ΔG is the best indicator of whether or not a reaction will be spontaneous.

Graham's law Graham's law says that the speed of gas diffusion (mixing of gases due to their kinetic energy) or effusion (movement of a gas through a tiny opening) is inversely proportional to the square root of the gases' molecular mass.

ground state The ground state of an atom is the lowest energy state that the electron can occupy.

groups Groups (families) are the vertical columns on the periodic table.

half-life The half-life, $t_{1/2}$, is the amount of time that it takes for a reactant concentration to decrease to one-half its initial concentration.

halogens Halogens are in Group 17 on the periodic table.

heat capacity Heat capacity is the quantity of heat needed to change the temperature by 1 K.

heat of vaporization The heat of vaporization is the heat needed to transform a liquid into a gas.

Henderson–Hasselbalch equation The Henderson–Hasselbalch equation can be used to calculate the pH of a buffer.

Henry's law The solubility of a gas will increase with increasing partial pressure of the gas.

Hess's law Hess's law states that if a reaction occurs in a series of steps, then the enthalpy change for the overall reaction is simply the sum of the enthalpy changes of the individual steps.

heterogeneous catalyst A heterogeneous catalyst is a catalyst that is in a different phase or state of matter from the reactants.

homogeneous catalyst A homogeneous catalyst is a catalyst that is in the same phase or state of matter as the reactants.

Hund's rule Hund's rule states that electrons are added to the orbitals, half filling them all before any pairing occurs.

hybrid orbitals Hybrid orbitals are atomic orbitals formed as a result of the mixing of the atomic orbitals of the atoms involved in a covalent bond.

hydrocarbons Hydrocarbons are organic compounds containing only carbon and hydrogen.

hydrogen bonding Hydrogen bonding is a specific type of dipole–dipole attraction in which a hydrogen atom is polar-covalently bonded to one of the following extremely electronegative elements: O, N, or F.

ideal gas An ideal gas is a gas that obeys the five postulates of the Kinetic Molecular Theory of Gases.

ideal gas equation The ideal gas equation relates the temperature, volume, pressure, and amount of a gas, and has the mathematical form of $PV = nRT$.

indicators Indicators are substances that change their color during a titration to indicate the endpoint.

inert (inactive) electrode An inert (inactive) electrode is a solid conducting electrode in an electrochemical cell that does not take part in the redox reaction.

inner transition elements The inner transition elements are the two horizontal groups that have been pulled out of the main body of the periodic table.

integrated rate law The integrated rate law relates the change in the concentration of reactants or products over time.

intermediates Intermediates are chemical species that are produced and consumed during the reaction, but that do not appear in the overall reaction.

intermolecular forces Intermolecular forces are attractive or repulsive forces between molecules caused by partial charges.

ion–dipole intermolecular force Ion–dipole intermolecular forces are attractive forces that occur between ions and polar molecules.

ion-induced dipole intermolecular forces Ion-induced dipole intermolecular forces are attractive forces that occur between an ion and a nonpolar molecule.

ion-product The ion-product has the same form as the solubility product constant, but represents a system that is not at equilibrium.

ionic bond Ionic bonds result from some metal losing electrons to form cations and some nonmetal gaining those electrons to form an anion.

ionic equation The ionic equation shows the soluble reactants and products in the form of ions.

ionic solids Ionic solids have their lattices composed of ions held together by the attraction of opposite charges of the ions.

ionization energy The ionization energy (IE) is the energy needed to completely remove an electron from an atom in the vapor state.

isoelectronic Isoelectronic means having the same electronic configuration.

isomers Isomers are compounds that have the same molecular formulas but different structural formulas.

isotopes Isotopes are atoms of the same element (same number of protons) that have differing numbers of neutrons.

joule (J) The joule is the SI unit of energy.

kinetic energy Kinetic energy is energy of motion.

Kinetic Molecular Theory The Kinetic Molecular Theory attempts to represent the properties of gases by modeling the gas particles themselves at the microscopic level.

kinetics Kinetics is the study of the speed of reactions.

Law of Conservation of Matter The Law of Conservation of Matter says that, in ordinary chemical reactions, matter is neither created nor destroyed.

Le Châtelier's principle Le Châtelier's principle states that if a chemical system at equilibrium is stressed (disturbed), it will reestablish equilibrium by shifting of the reactions involved.

Lewis electron-dot structure The Lewis electron-dot structure is a structural formula that represents the element and its valence electrons.

limiting reactant The limiting reactant is the reactant that is used up first in a chemical reaction.

line spectrum A line spectrum is a series of fine lines of colors representing wavelengths of photons that are characteristic of a particular element.

liquid A liquid is a state of matter that has a definite volume but no definite shape.

macromolecules Macromolecules are extremely large molecules.

magnetic quantum number (m_l) The magnetic quantum number describes the orientation of the orbital around the nucleus.

main-group elements Main-group elements are the groups on the periodic table that are labeled 1–2 and 13–18.

manometer A manometer is an instrument used to measure the gas pressure inside a container.

mass number The mass number is the sum of the protons and neutrons in an atom.

mass percent The mass percentage of a solution is the mass of the solute divided by the mass of the solution and then multiplied by 100% to get percentage.

mass-volume percent The mass-volume percent of a solution is the mass of the solute divided by the volume of the solution and then multiplied by 100% to yield percentage.

mechanism The mechanism is the sequence of steps that a reaction undergoes in going from reactants to products.

melting point The temperature at which a solid converts into the liquid state is called the melting point (m.p.) of the solid.

metallic bonding In metallic bonding the electrons of the atoms are delocalized and are free to move throughout the entire solid.

metallic solids Metallic solids have metal atoms occupying the crystal lattice and held together by metallic bonding.

metalloids Metalloids are a group of elements that have properties of both metals and nonmetals.

metals Metals are normally solids (mercury being an exception), shiny, and good conductors of heat and electricity. They can be hammered into thin sheets (malleable) and extruded into wires (ductile). Chemically, metals tend to lose electrons in reactions.

metathesis reaction In a metathesis reaction two substances exchange bonding partners.

molality (*m*) Molality is defined as the moles of solute per kilogram of solvent.

molar heat capacity The molar heat capacity (C) is the amount of heat needed to change the temperature of 1 mole of a substance by 1 K.

molar mass The mass in grams of 1 mole of a substance.

molarity (M) or sometimes [ ] Molarity is a concentration term that represents the moles of solute per liters of solution.

mole The mole (mol) is defined as the amount of a substance that contains the same number of particles as atoms in exactly 12 g of carbon-12.

molecular equation The molecular equation is an equation in which both the reactants and products are shown in the undissociated form.

molecular formula The molecular (actual) formula shows which elements are in the compound and the actual number of atoms of each element.

molecular orbital theory The molecular orbital (MO) theory of covalent bonding proposes that atomic orbitals combine to form molecular orbitals that encompass the entire molecule.

molecular solids Molecular solids have their lattices composed of molecules held in place by London forces, dipole–dipole forces, and hydrogen bonding.

molecule A molecule is a covalently bonded compound.

monomers Macromolecules are composed of repeating units, called monomers.

Nernst equation The Nernst equation allows the calculation of the cell potential of a galvanic cell that is not at standard conditions.

net ionic equation The net ionic equation is written by dropping out the spectator ions and showing only those chemical species that are involved in the chemical reaction.

network covalent solids Network covalent solids have covalent bonds joining the atoms together in an extremely large crystal lattice.

neutral Neutral is 7.00 on the pH scale.

neutralization reactions Neutralization reactions are acid–base reactions in which an acid reacts with a base to give a salt and usually water.

noble gases Noble gases are in Group 18 on the periodic table. They are very unreactive owing to their filled valence shell.

nonelectrolytes Nonelectrolytes are substances that do not conduct electricity when dissolved in water or melted.

nonmetals Nonmetals have properties that are generally the opposite of metals. Some are gases, are poor conductors of heat and electricity, are neither malleable nor ductile, and tend to gain or share electrons in their chemical reactions.

nonpolar covalent bond In a nonpolar covalent bond the electrons are shared equally by the two atoms involved in the bond.

nuclear belt of stability The nuclear belt of stability is a plot of the # neutrons versus the # protons for the known stable isotopes.

nucleus The nucleus is a dense core of positive charge at the center of the atom that contains most of the mass of the atom.

octet rule The octet rule states that during chemical reactions, atoms lose, gain, or share electrons in order to achieve a filled valence shell, to complete their octet.

orbital An orbital or wave function is a quantum mechanical mathematical description of the location of electrons. The electrons in a particular subshell are distributed among these volumes of space of equal energies.

order of reaction The order of reaction is the exponent in the rate equation that indicates what effect a change in concentration of that particular reactant species will have on the reaction rate.

organic chemistry Organic chemistry is the study of the chemistry of carbon.

osmosis Osmosis is the passing of solvent molecules through a semipermeable membrane.

osmotic pressure The osmotic pressure is the amount of pressure that must be exerted on a solution in order to prevent osmosis of solvent molecules through a semipermeable membrane.

oxidation Oxidation is the loss of electrons.

oxidation numbers Oxidation numbers are bookkeeping numbers that allow chemists to do things like balance redox equations.

oxidizing agent The oxidizing agent is the reactant being reduced.

paramagnetism Paramagnetism is the attraction of a molecule to a magnetic field and is due to unpaired electrons.

pascal The pascal is the SI unit of pressure.

percent yield The percent yield (% yield) is the actual yield divided by the theoretical yield, with the result multiplied by 100.

periods Periods are the horizontal rows on the periodic table that have consecutive atomic numbers.

phase changes Phase changes are changes of state.

phase diagram A phase diagram is a graph representing the relationship of the states of matter of a substance to temperature and pressure.

pi (π) bonds Pi bonds result from the overlap of atomic orbitals on both sides of a line connecting two atomic nuclei.

polar covalent bonds Polar covalent bonds are covalent bonds in which there is an unequal sharing of the bonding pair of electrons.

polyprotic acids Polyprotic acids are acids that can donate more than one proton.

potential energy Potential energy is stored energy.

positron A positron is essentially an electron that has a positive charge instead of a negative one.

precipitate A precipitate is an insoluble product that forms in a solution; the formation of a solid from ions in solution.

precipitation reactions Precipitation reactions are reactions that involve the formation of an insoluble compound, a precipitate, from the mixing of two soluble compounds.

pressure Pressure is the force exerted per unit of surface area.

principal quantum number (*n*) The principal quantum number describes the size of the orbital and relative distance from the nucleus.

proof The proof of an aqueous ethyl alcohol solution is twice the volume percent.

quantized Quantized means that there could be only certain distinct energies associated with a state of the atom.

quantum numbers Quantum numbers are used to describe each electron within an atom corresponding to the orbital size, shape, and orientation in space.

radioactivity Radioactivity is the spontaneous decay of an unstable isotope to a more stable one.

rate constant (*k*) The rate constant is a proportionality constant that appears in the rate law and relates the concentration of reactants to the speed of reaction.

rate-determining step The rate-determining step is the slowest one of the reaction steps and controls the rate of the overall reaction.

rate equation The rate equation relates the speed of reaction to the concentration of reactants and has the form: Rate $= k[A]^m[B]^n$. . . . where k is the rate constant and m and n are the orders of reaction with respect to that specific reactant.

reactants The starting materials in a chemical reaction, which get converted into different substances called products.

reaction intermediate A reaction intermediate is a substance that is formed but then consumed during the reaction mechanism.

reaction mechanism The reaction mechanism is the sequence of individual reactions that occur in an overall reaction in going from reactants to products.

reaction quotient The reaction quotient, Q, is the numerical value that results when non-equilibrium concentrations are inserted into the equilibrium expression. When the system reaches equilibrium, the reaction quotient becomes the equilibrium constant.

reactive site The reactive site of a molecule is the place at which the reaction takes place.

redox reactions Redox reactions are chemical reactions in which electrons are lost and gained.

reducing agent The reactant undergoing oxidation in a redox reaction is called the reducing agent.

reduction Reduction is the gain of electrons in a redox reaction.

resonance Resonance is a way of describing a molecular structure that cannot be represented by a single Lewis structure. Several different Lewis structures are used, each differing only by the position of electron pairs.

reverse osmosis Reverse osmosis takes place when the pressure on the solution side exceeds the osmotic pressure and solvent molecules are forced back through the semipermeable membrane into the solvent side.

root mean square speed The average velocity of the gas particles is called the root mean square speed.

salt bridge A salt bridge is often an inverted U-tube that contains a gel containing a concentrated electrolyte solution, used in an electrochemical cell to maintain electrical neutrality in the cell compartments.

saturated hydrocarbons Saturated hydrocarbons are hydrocarbons that are single bonded to the maximum number of other atoms.

saturated solution A solution in which one has dissolved the maximum amount of solute per given amount of solvent at a given temperature is called a saturated solution.

Second Law of Thermodynamics The Second Law of Thermodynamics states that all processes that occur spontaneously move in the direction of an increase in entropy of the universe (system + surroundings).

semipermeable membrane A semipermeable membrane is a thin porous film that allows the passage of solvent molecules but not solute particles.

shells The electrons in an atom are located in various energy levels or shells that are located at different distances from the nucleus.

SI system The system of units used in science is the SI system (Système International), which is related to the metric system.

sigma (σ) bonds Sigma bonds have the orbital overlap on a line drawn between the two nuclei.

simple cubic unit cell The simple cubic unit cell has particles located at the corners of a simple cube.

single displacement (replacement) reactions Single displacement reactions are reactions in which atoms of an element replace the atoms of another element in a compound.

solid A solid is a state of matter that has both a definite shape and a definite volume.

solubility product constant (K_{sp}) The solubility product constant is the equilibrium constant associated with sparingly soluble salts and is the product of the ionic concentrations, each one raised to the power of the coefficient in the balanced chemical equation.

solute The solute is the component of the solution that is present in smallest amount.

solution A solution is defined as a homogeneous mixture composed of solvent and one or more solutes.

solvation Solvation is the forming of a layer of bound solvent molecules around a solute.

solvent The solvent is that component of a solution that is present in largest amount.

specific heat capacity (or specific heat) (c) The specific heat capacity is the quantity of heat needed to raise the temperature of 1 g of the substance by 1 K.

spectator ions Spectator ions are ions that are not actually involved in the chemical reaction taking place, but simply maintain electrical neutrality.

speed of light (c) The speed of light is the speed at which all electromagnetic radiation travels in a vacuum, 3.0×10^8 m/s.

spin quantum number (m_s) The spin quantum number indicates the direction the electron is spinning.

standard cell potential ($E°$) The standard cell potential is the potential (voltage) associated with an electrochemical cell at standard conditions.

standard enthalpy of formation The standard enthalpy of formation of a compound ($\Delta H_f°$) is the change in enthalpy when 1 mol of the compound is formed from its elements and when all substances are in their standard states.

standard molar entropies ($S°$) Standard molar enthalpies of elements and compounds are the entropies associated with 1 mole of a substance in its standard state.

standard reduction potentials The standard reduction potential is the voltage associated with a half-reaction shown in the form of reduction.

state function A state function is a function that doesn't depend on the pathway, only the initial and final states.

stoichiometry Stoichiometry is the calculation of the amount (mass, moles, particles) of one substance in the chemical reaction through the use of another.

strong acid A strong acid is an acid that ionizes completely in solution.

strong base A strong base is a base that ionizes completely in solution.

strong electrolytes Strong electrolytes completely ionize or dissociate in solution.

structural isomers Structural isomers are compounds that have the same molecular formula but differ in how the atoms are attached to each other.

sublimation Sublimation is going directly from the solid state to the gaseous state without ever having become a liquid.

subshells Within the shells, the electrons are grouped in subshells of slightly different energies.

supersaturated solution A supersaturated solution has more than the maximum amount of solute dissolved in the solvent at a given temperature.

surface tension Surface tension is the amount of force that is required to break through the molecular layer at the surface of a liquid.

surroundings The surroundings is a thermodynamic term meaning the part of the universe that is not the system that is being studied.

suspension A heterogeneous mixture in which the particles are large (in excess of 1,000 nm).

system The system is a thermodynamics term meaning the part of the universe that we are studying.

ternary compounds Ternary compounds are those containing three (or more) elements.

theoretical yield The theoretical yield is the maximum amount of product that can be formed.

thermochemistry Thermochemistry is the part of thermodynamics dealing with the changes in heat that take place during chemical processes, for example, ΔH_{fusion}.

thermodynamics Thermodynamics is the study of energy and its transformations.

titrant The titrant is that solution in a titration that has a known concentration.

titration A titration is a laboratory procedure in which a solution of known concentration is used to determine the concentration of an unknown solution.

transition elements Groups 3–12 on the periodic table are called the transition elements.

transmutation Transmutation is a nuclear reaction that results in the creation of one element from another one.

triple point The triple point of a substance is the combination of temperature and pressure on a phase diagram at which all three states of matter can exist in equilibrium.

Tyndall effect The Tyndall effect is exhibited when a light is shone through a colloid and is visible, owing to the reflection of the light off the larger colloid particles.

unimolecular reactions Unimolecular reactions are reactions in which a single chemical species decomposes or rearranges.

unit cells Unit cells are the repeating units in a crystal lattice.

unsaturated Unsaturated organic compounds have carbons that do not have the maximum number of bonds to other atoms; there is at least one carbon-to-carbon double or triple bond present.

unsaturated solution An unsaturated solution has less than the maximum amount of solute dissolved in a given amount of solvent.

valence bond theory The valence bond theory describes covalent bonding as the overlap of atomic orbitals to form a new kind of orbital, a hybrid orbital.

valence electrons Valence electrons are the electrons in the outermost energy level (outermost shell). Valence electrons are normally considered to be only the s and p electrons in the outermost energy level.

van der Waals equation van der Waals equation is an equation that is a modification of the ideal gas equation to compensate for the behavior of real gases.

van't Hoff factor (i) The van't Hoff factor is the ratio of moles of solute particles formed to moles of solute dissolved in solution.

vapor pressure The pressure exerted by the gaseous molecules that are at equilibrium with a liquid in a closed container.

viscosity Viscosity is the resistance to flow of a liquid.

volume percent The volume percent of the solution is the volume of the solute divided by the volume of the solution and then multiplied by 100% to generate the percentage.

VSEPR theory The VSEPR (valence-shell electron-pair repulsion) theory says that the electron pairs around a central atom will try to get as far as possible from each other in order to minimize the repulsive forces. This theory is used to predict molecular geometry.

water dissociation constant (K_w) The water dissociation constant is the equilibrium constant associated with the ionization of pure water.

wave function The wave function is a mathematical description of the electron's motion.

wavelength (λ) Wavelength is the distance between two identical points on a wave.

weak acid A weak acid is an acid that only partially ionizes in solution.

weak base A weak base is a base that only partially ionizes in solution.

weak electrolytes Weak electrolytes only partially ionize or dissociate in solution.

AVOIDING "STUPID MISTAKES" ON THE FREE-RESPONSE SECTION

We (the authors) have been grading the free-response part of the AP Chemistry exam for quite a while. Between the two of us, we have nearly 20 years of grading experience—that's more than 100,000 exams! Over the years, we have seen quite a number of careless mistakes made by students. These mistakes resulted from not being careful rather than not being prepared for the exam. Here are some practical tips to avoid the most common careless errors.

- **Don't forget to state the units of measurement.** Many students would have gotten more credit if they had shown the units, both in the calculations and in the final answer. The units help you stay on the right track and help the grader determine if (or where) you went wrong.
- **Use the formula given.** If the exam gives you a chemical formula, don't use a different formula in your answer. In general, do not alter anything given to you on the exam. For example, we have seen $Ba(NO_3)_2$ become $Ba(NO_2)_2$.
- **Be careful with the math.** We have seen many errors involving the simplest math such as 12 mL + 3 mL = 0.042 L (rather than 0.015 L).
- **Don't confuse molarity and moles.** The units M and [ ] are identical (molarity) and are completely different from moles.
- **Show your work for conversions.** For example, if you are changing grams to moles and make a simple mistake, showing your work (labeled) may get you partial credit.
- **Don't argue with the test.** This is an argument you cannot win. For example, if the question asks for calculations, you are unlikely to get full credit without any calculations even if you have the right answer. It won't help to write that you feel the calculations are unnecessary.
- **Be careful in applying gas laws.** Gas laws can be very useful. However, they should *never* be used when there is not a gas in the problem. Having a volume included in the question information doesn't necessarily mean you are dealing with a gas.
- **Be careful making comparisons.** We have seen many students incorrectly say that 10^{-8} is smaller than 10^{-12} and actually write $10^{-8} < 10^{-12}$. We have even seen students write the relationship correctly ($10^{-8} > 10^{-12}$) but still state that 10^{-8} is smaller.
- **Be careful using 22.4 L/mol.** You will probably not need to use this on the exam. But if you do want to use this value, you *must* have a gas and this gas *must* be at 0°C (273 K) and 1 atm (STP). If you forget the values for STP, they can be found on the exam. We have seen quite a few students incorrectly use this value at 298 K.
- **There are no trick questions on the exam.** If you think you have found a trick question, you need to reevaluate your thinking and reread the question.
- **Don't confuse solutions and precipitates in solution.** They are different phases and are not interchangeable. The color of one is not necessarily the color of the other.
- **Be careful describing reactions.** If the problem gives you, for example, a sodium nitrate solution, part of your answer describing a reaction cannot be "the sodium nitrate dissolves." You already have a solution, so the process of dissolving happened before you got to the problem. Furthermore, dissolving should not be treated as a reaction.
- **Be careful using positive and negative charges.** In the following equation, each reactant and product is wrong: $NH_4 + NO_3 \rightarrow NH_4^+NO_3^-$, and will not substitute for the

correct $NH_4^+ + NO_3^- \rightarrow NH_4NO_3$. Remember, ionic equations, of any type, have ions (with charges) on one or both sides of the reaction arrow.

- **Don't do a calculator dump** (write down every number displayed by your calculator). For example, your final answer will not be 3.27584827 g.
- **Keep in mind the meaning of "observe."** If the problem asks about observation, tell what you would actually observe (see, hear, or smell). You will *not* see a compound separating into ions; usually you will *not* see the excess reagent, and you will *not* see the atoms forming bonds. In contrast, you might observe a compound dissolving.
- **Remember a solvent is usually not a reactant.** Therefore changing the grams of solvent to moles is probably wrong. (However, you will need to know the moles of solvent if you are looking for a mole fraction.)
- **Think before creating mole ratios.** Since the solvent is not a reactant, a mole ratio relating the solvent to anything else in the problem is most likely wrong. We have seen many students change the grams of water to moles and then use these moles in a mole ratio to relate to some other substance in the problem.
- **Don't go off on a tangent.** Stay focused on answering the original question.
- **Double-check the numbers you use.** We have seen many cases where the problem gave a number like 2.75×10^{-18}, and the student worked the problem with 2.75×10^{-8}. If you show your work, it will be obvious to the grader that you miscopied the value and you might pick up some points; otherwise, you just have a wrong answer.
- **Remember that sometimes not all of the information given is needed to solve the problem.** For example, in the equilibrium problem, many times the temperature is given but it is not actually part of the calculations.
- **Only round your final answer.** Don't round off the results of intermediate calculations; only use rounding after you've gotten your final answer.
- **Be careful in reading graphs.** Especially take care in reading the scales. We have seen students write down that 0.5 is between 1.0 and 2.0.
- **Don't confuse intermolecular and intramolecular forces.** These are two different concepts and are not interchangeable.

In addition to avoiding the careless mistakes mentioned above, here are some easy ways to help improve your score on the free-response questions:

- **Show your work.** In most cases, no work, no credit.
- **Use the space provided for answers.** It helps you and the grader if you answer the question in the space provided instead of crowding the answers between the questions. You will have more than enough room on the following page(s). It also helps to label the parts (a, b, etc.) and to answer the parts in order.
- **Make sure your answer can be easily read.** It will really help the grader—and your score—if you write legibly, in a normal size (not too small, please), and use a pencil or pen that writes dark enough to be easily read.
- **Don't use periodic trends and general rules as explanations.** General rules such as "like dissolves like" are never explanations. They may help you in answering the multiple-choice part of the exam, but will be of little benefit by themselves in the free-response section.
- **Don't confuse "define" and "describe."** They are two different processes. If you are asked to describe or explain, simply giving a definition will earn you very few points.
- **Use only standard abbreviations.** Your instructor may understand your abbreviations, but the grader may not. If you want to use abbreviations in a response, be sure to define them.
- **Don't ramble.** Normally an explanation or justification can be done in five sentences or less. Your answers should be clear, concise, and to the point.

EXAM RESOURCES

Keywords and Equations: For Use with Free-Response Questions Only

Basics

T = temperature
V = volume
KE = kinetic energy
n = moles
D = density
t = time
m = mass
v = velocity
P = pressure
M = molar mass

Boltzmann's constant, $k = 1.38 \times 10^{-23}$ J K^{-1}
electron charge = -1.602×10^{-19} coulombs
1 electron volt per atom = 96.5 kJ mol^{-1}
Avogadro's number = 6.022×10^{23} mol^{-1}
K = °C + 273 $\quad D = m/V$

Gases

u_{rms} = root mean square speed
r = rate of effusion
STP = 0.000° C and 1.000 atm
$PV = nRT$
$(P + n^2a/V^2)(V - nb) = nRT$
$P_A = P_{total} \times X_A$, where X_A = moles A/total moles
$P_{total} = P_A + P_B + P_C + \ldots$
$P_1V_1/T_1 = P_2V_2/T_2$

$u_{rms} = \sqrt{3kT/m} = \sqrt{3RT/M}$

KE per molecule = $\frac{1}{2}mv^2$

KE per mol = $\frac{3}{2}RT$

$r_1/r_2 = \sqrt{M_2/M_1}$

1 atm = 760 mm Hg
= 760 torr
Gas constant, $R = 8.31$ J mol^{-1} k^{-1}
= 0.0821 L atm mol^{-1} K^{-1}
= 8.31 volt coulomb mol^{-1} K^{-1}
= 62.36 L torr mol^{-1} K^{-1}

Thermodynamics

$S°$ = standard entropy
$G°$ = standard free energy
c = specific heat capacity
$H°$ = standard enthalpy
q = heat
C_p = molar heat capacity at constant pressure

$\Delta S° = \Sigma\, S°$ products $- \Sigma\, S°$ reactants
$\Delta H° = \Sigma\, \Delta H_f°$ products $- \Sigma\, \Delta H_f°$ reactants
$\Delta G° = \Sigma\, \Delta G_f°$ products $- \Sigma\, \Delta G_f°$ reactants

$\Delta G° = \Delta H° - T\Delta S°$
$= -RT \text{ in } K = -2.303\ RT \log K$
$= -nFE°$
$\Delta G = \Delta G° + RT \ln Q = \Delta G° + 2.303\ RT \log Q$
$q = mc\Delta T$
$C_p = \Delta H / \Delta T$

Light and Electrons

E = energy
p = momentum
m = mass
ν = frequency
v = velocity
$E = h\nu$
λ = wavelength
n = principal quantum number
$c = \lambda\nu$

$E_n = (-2.178 \times 10^{-18}/n^2)$J
Speed of light, $c = 3.0 \times 10^8\ \text{ms}^{-1}$
Planck's constant, $h = 6.63 \times 10^{-34}$ Js

Solutions

Molarity, M = moles solute per liter solution

Kinetics

$\ln[A]_t - \ln[A]_o = -kt$

$$\frac{1}{[A]_t} - \frac{1}{[A]_o} = kt$$

$$\ln k = \frac{-E_a}{R}\left(\frac{1}{T}\right) + \ln A$$

E_a = activation energy
k = rate constant
A = frequency factor

Electrochemistry

I = current (amperes)
$E°$ = standard reduction potential
q = charge (coulombs)
K = equilibrium constant
Faraday's constant, F = 96,500 coulombs per mole of electrons

$$I = q/t \quad \log k = \frac{nE°}{0.0592}$$

Equilibrium

Q = reaction quotient

$$Q = \frac{[C]^c[D]^d}{[A]^a[B]^d}, \text{where } aA + bB \rightleftharpoons cC + dD$$

equilibrium constants:
K_a (weak acid)
K_b (weak base)
K_w (water)
K_p (gas pressure)
K_c (molar concentrations)

$$K_a = \frac{[H^+][A^-]}{[HA]} \quad K_b = \frac{[OH^-][HB^+]}{[B]} \quad K_p = \frac{(P_c)^c(P_d)^d}{(P_a)^a(P_b)^b}$$

$K_w = [OH^-][H^+] = 1.0 \times 10^{-14} = K_a \times K_b$ at 25°C

$pH = -\log [H^+]$, $pOH = -\log [OH^-]$
$14 = pH + pOH$

$$pH = pK_a + \log \frac{[A^-]}{[HA]}$$

$$pOH = pK_b + \log \frac{[HB^+]}{[B]}$$

$pK_a = -\log K_a$, $pK_b = -\log K_b$
$K_p = K_c(RT)^{\Delta n}$, where Δn = moles product gas − moles reactant gas

Experimental

Beer's Law: $A = abc$ (A = absorbance; a = molar absorbtivity; b = path length; c = concentration)

Periodic Table of the Elements

May be used with all questions.

The periodic table

Key: Element Name / Atomic number / Symbol / Atomic weight (mean relative mass)

1	2		3	4	5	6	7	8	9	10	11	12	13	14	15	16	17	18
Hydrogen 1 **H** 1.008																		Helium 2 **He** 4.002602(2)
Lithium 3 **Li** 6.94	Beryllium 4 **Be** 9.012182(3)												Boron 5 **B** 10.81	Carbon 6 **C** 12.011	Nitrogen 7 **N** 14.007	Oxygen 8 **O** 15.999	Fluorine 9 **F** 18.9984032(5)	Neon 10 **Ne** 20.1797(6)
Sodium 11 **Na** 22.98976928(2)	Magnesium 12 **Mg** 24.3050(6)												Aluminium 13 **Al** 26.9815386(2)	Silicon 14 **Si** 28.085	Phosphorus 15 **P** 30.973762(2)	Sulfur 16 **S** 32.06	Chlorine 17 **Cl** 35.45	Argon 18 **Ar** 39.948(1)
Potassium 19 **K** 39.0983(1)	Calcium 20 **Ca** 40.078(4)		Scandium 21 **Sc** 44.955912(6)	Titanium 22 **Ti** 47.867(1)	Vanadium 23 **V** 50.9415(1)	Chromium 24 **Cr** 51.9961(6)	Manganese 25 **Mn** 54.938045(5)	Iron 26 **Fe** 55.845(2)	Cobalt 27 **Co** 58.933195(5)	Nickel 28 **Ni** 58.6934(4)	Copper 29 **Cu** 63.546(3)	Zinc 30 **Zn** 65.38(2)	Gallium 31 **Ga** 69.723(1)	Germanium 32 **Ge** 72.63(1)	Arsenic 33 **As** 74.92160(2)	Selenium 34 **Se** 78.96(3)	Bromine 35 **Br** 79.904(1)	Krypton 36 **Kr** 83.798(2)
Rubidium 37 **Rb** 85.4678(3)	Strontium 38 **Sr** 87.62(1)		Yttrium 39 **Y** 88.90585(2)	Zirconium 40 **Zr** 91.224(2)	Niobium 41 **Nb** 92.90638(2)	Molybdenum 42 **Mo** 95.96(2)	Technetium 43 **Tc** [97.91]	Ruthenium 44 **Ru** 101.07(2)	Rhodium 45 **Rh** 102.90550(2)	Palladium 46 **Pd** 106.42(1)	Silver 47 **Ag** 107.8682(2)	Cadmium 48 **Cd** 112.411(8)	Indium 49 **In** 114.818(3)	Tin 50 **Sn** 118.710(7)	Antimony 51 **Sb** 121.760(1)	Tellurium 52 **Te** 127.60(3)	Iodine 53 **I** 126.90447(3)	Xenon 54 **Xe** 131.293(6)
Caesium 55 **Cs** 132.9054519(2)	Barium 56 **Ba** 137.327(7)	57–70 *	Lutetium 71 **Lu** 174.9668(1)	Hafnium 72 **Hf** 178.49(2)	Tantalum 73 **Ta** 180.94788(2)	Tungsten 74 **W** 183.84(1)	Rhenium 75 **Re** 186.207(1)	Osmium 76 **Os** 190.23(3)	Iridium 77 **Ir** 192.217(3)	Platinum 78 **Pt** 195.084(9)	Gold 79 **Au** 196.966569(4)	Mercury 80 **Hg** 200.59(2)	Thallium 81 **Tl** 204.38	Lead 82 **Pb** 207.2(1)	Bismuth 83 **Bi** 208.98040(1)	Polonium 84 **Po** [209]	Astatine 85 **At** [210]	Radon 86 **Rn** [222]
Francium 87 **Fr** [223.02]	Radium 88 **Ra** [226.03]	89–102 **	Lawrencium 103 **Lr** [262.11]	Rutherfordium 104 **Rf** [265.12]	Dubnium 105 **Db** [268.13]	Seaborgium 106 **Sg** [271.13]	Bohrium 107 **Bh** [270]	Hassium 108 **Hs** [277.15]	Meitnerium 109 **Mt** [276.15]	Darmstadtium 110 **Ds** [281.16]	Roentgenium 111 **Rg** [280.16]	Copernicium 112 **Cn** [285.17]	Ununtrium 113 **Uut** [284.18]	Ununquadium 114 **Uuq** [289.19]	Ununpentium 115 **Uup** [288.19]	Ununhexium 116 **Uuh** [293]	Ununseptium 117 **Uus** [294]	Ununoctium 118 **Uuo** [294]

*lanthanoids	Lanthanum 57 **La** 138.90547(7)	Cerium 58 **Ce** 140.116(1)	Praseodymium 59 **Pr** 140.90765(2)	Neodymium 60 **Nd** 144.242(3)	Promethium 61 **Pm** [144.91]	Samarium 62 **Sm** 150.36(2)	Europium 63 **Eu** 151.964(1)	Gadolinium 64 **Gd** 157.25(3)	Terbium 65 **Tb** 158.92535(2)	Dysprosium 66 **Dy** 162.500(1)	Holmium 67 **Ho** 164.93032(2)	Erbium 68 **Er** 167.259(3)	Thulium 69 **Tm** 168.93421(2)	Ytterbium 70 **Yb** 173.054(5)
**actinoids	Actinium 89 **Ac** [227.03]	Thorium 90 **Th** 232.03806(2)	Protactinium 91 **Pa** 231.03588(2)	Uranium 92 **U** 238.02891(3)	Neptunium 93 **Np** [237.05]	Plutonium 94 **Pu** [244.06]	Americium 95 **Am** [243.06]	Curium 96 **Cm** [247.07]	Berkelium 97 **Bk** [247.07]	Californium 98 **Cf** [251.08]	Einsteinium 99 **Es** [252.08]	Fermium 100 **Fm** [257.10]	Mendelevium 101 **Md** [258.10]	Nobelium 102 **No** [259.10]